Broadening Cultural Horizons in Social Marketing

Rachel Hay · Lynne Eagle · Abhishek Bhati
Editors

Broadening Cultural Horizons in Social Marketing

Comparing Case Studies from Asia-Pacific

Springer

Editors
Rachel Hay
James Cook University Australia
Douglas, Townsville, QLD, Australia

Lynne Eagle
James Cook University Australia
Douglas, Townsville, QLD, Australia

Abhishek Bhati
James Cook University
Singapore, Singapore

ISBN 978-981-15-8516-6 ISBN 978-981-15-8517-3 (eBook)
https://doi.org/10.1007/978-981-15-8517-3

This Springer imprint is published by the registered company Springer Nature Singapore Pte Ltd.
The registered company address is: 152 Beach Road, #21-01/04 Gateway East, Singapore 189721, Singapore

Introduction

Broadening Cultural Horizons in Social Marketing: Social Marketing Case Studies in the Asia-Pacific region brings together a unique series of health, social and environmental case studies and empirical papers from community change agents from across Asia Pacific. In a collaboration to share the latest tools and approaches to changing behaviours for societal well-being, the book provides insight from a cross-cultural perspective into the application of social marketing to address gaps in changing lifestyles among its targeted community. It allows the reader to learn new insights from academic, practitioner and industry experts. The book uses case studies to challenge people's beliefs about societal well-being through cross-cultural perspectives. The case studies propose to use theory in new ways, providing the reader with an insight into theory application to the area of social marketing.

The content's focus is to bring ideas on social change from around the Asia-Pacific into the social marketing community. It presents case studies on topics relevant to social marketing and social change. In particular, it highlights unique challenges around behaviour and social change across the Asia-Pacific, showcasing successful social marketing interventions across the region. This book is intended for academics, graduates, policymakers, researchers, practitioners and industry specialists.

Broadening Cultural Horizons in Social Marketing: Social Marketing Case Studies contains a variety of case studies and empirical papers on topics such as health and well-being, policy, legislation and upstream social marketing. It includes innovative new methods in social marketing and social change research in undeveloped, underdeveloped, and developing countries. The case studies and empirical papers also cover topics such as the environment and sustainability, cross-cultural issues in social marketing, individual, group, community and/or organizational well-being, and transformative services and midstream social marketing.

Part I of the series is dedicated to environmental issues in renewable energy, reef protection and food waste. Part II contains a range of healthcare-related social marketing issues including co-creation in social media health promotion, men's counselling towards the prevention of violence against women, organ donation and coordinated mental health crisis communications. The final category in this edited volume is devoted to behavioural change in the societal context. Several case studies show innovative campaigns, especially run by the government, resulting in desired behavioural change in targeted communities.

In this edited volume, internationally recognized expert contributors write thirteen case studies, covering specific environmental, social and health issues related to social marketing with a specific emphasis on cross-cultural perspectives in Asia Pacific. We hope that these interesting cases will encourage further discourse in the ever-interesting research field of behavioural change. Above all, we are most grateful to our contributors' insights and for their time and effort in developing these useful cases. We hope you will enjoy the cases and become inspired to contribute to behavioural change through social marketing.

Contents

Editors and Contributors

About the Editors

Dr. Rachel Hay is a Lecturer in Marketing with the College of Business Law and Governance, James Cook University. Dr. Hay recently completed her Ph.D. 'The engagement of women and technology in agriculture'. Her research interests centre on technology adoption and agriculture, social marketing and readability. Within this area, she is active in research relating to technology adoption by women in agriculture, and marketing communication effects and effectiveness, including readability in populations sectors that face literacy and numeracy challenges. Her recent projects focus on trans-disciplinary approaches to sustained behaviour change in social marketing and environmental protection interventions.

Prof. Lynne Eagle is a Professor in Marketing at James Cook University, Townsville, Australia, and an Adjunct Professor at Canterbury University, Christchurch, New Zealand. She holds a Ph.D. from the University of Auckland, in her native New Zealand. Her research interests centre on marketing communication effects and effectiveness, including the impact of persuasive communication on children; the impact of new, emerging and hybrid media forms and preferences for the use of formal and informal communications channels, along with trans-disciplinary approaches to sustained behaviour change in social marketing/health promotion/environmental protection campaigns. She has published in a wide range of academic journals, including the Journal of Advertising and European Journal of Marketing, led the development of both Marketing Communications and Social Marketing texts and contributed several book chapters for other texts, as well as writing commissioned expert papers and presenting numerous research papers at international conferences.

Prof. Abhishek Bhati is the Campus Dean of James Cook University Singapore campus. As Campus Dean JCU Singapore, Abhishek contributes to JCU's initiatives in Tropical Asia. Prof. Abhishek Bhati's research investigates resilience planning, sustainable development of cities and scholarship of learning and teaching. In particular, he is interested in technology and the role it has as a catalyst for industry changes; 'Smart City' as a mechanism for future sustainable development; and the need to deliver United Nations Sustainable Development Goals (SDGs).

Contributors

Ali Ahani Griffith Business School, Griffith, QLD, Australia

Abhishek Bhati James Cook University, Townsville, Australia

Shailey Chawla James Cook University, Singapore, Singapore

Shinyi Chin School of Media & Communication, RMIT University, Melbourne, Australia

Laura Dunstan NQ Dry Tropics, Townsville, Australia

Lynne Eagle James Cook University, Townsville, Australia

Rene Erhardt Deakin University, Melbourne, Australia

Rachel Hay James Cook University, Townsville, Australia

Jeawon Kim Griffith University, Brisbane, Australia

Kathy Knox Social Marketing@Griffith, Griffith Business School, Nathan, QLD, Australia;
Griffith University, Brisbane, Australia

Nirmala Ika Kusumaningrum Clinical Psychologist University of Indonesia, Depok, Indonesia

Brennan Linda School of Media & Communication, RMIT University, Melbourne, Australia

Carolyn Loton Juntos Marketing, Sydney, Australia

Irma S. Martam Clinical Psychologist University of Indonesia, Depok, Indonesia

Tracy McCaffrey Department of Nutrition, Dietetics & Food, Monash University, Notting Hill, Australia

Breda McCarthy James Cook University, Townsville, Australia

Annika Molenaar Department of Nutrition, Dietetics & Food, Monash University, Notting Hill, Australia

Melissa Neal Murrumbidgee Primary Health Network, Wagga Wagga, Australia

Filippa Neilsen Griffith Business School, Griffith, QLD, Australia

Lucy Nyundo National Institute of Public Administration, Lusaka, Zambia

Joy Parkinson Social Marketing@Griffith, Griffith Business School, Nathan, QLD, Australia

Lisa Pulman NQ Dry Tropics, Townsville, Australia

Mike Reid Finance & Marketing, RMIT University School of Economics, Melbourne, Australia

Sharyn Rundle-Thiele Griffith University, Brisbane, Australia

Jagdeep Kaur Sabharwal James Cook University, Singapore, Singapore

Sue Sargent Sue Sargent Consulting Services, Bundaberg, Australia

Anouk Sherman Department of Nutrition, Dietetics & Food, Monash University, Notting Hill, Australia

Phill Sherring Ministry for Primary Industries, Wellington, New Zealand

Helen Truby Department of Nutrition, Dietetics & Food, Monash University, Notting Hill, Australia;
School of Human Movement and Nutrition Sciences, The University of Queensland, Brisbane, Australia

Yoko Uryuhara Faculty of Commerce, Doshisha University, Kyoto, Japan

Nathan Waltham Centre for Tropical Water and Aquatic Ecosystem Research, James Cook University, Townsville, Australia

Maxine Whittaker James Cook University, Townsville, Australia

List of Figures

List of Tables

Part I
Environmental

Winds of Change: A Study of the Acceptability of Renewable Energy in Regional Australia chapter discusses stakeholder's resistance to the generation of renewable energy in Australia. Opposition appears to centre on dislike of the visual impact of windmills, turbine noise, potential dangers to wildlife and claimed human health impacts. Several studies have concluded that there is no medical evidence of a direct link between turbine operation and human health.

Finding the Sweet Spot chapter seeks to establish whether social marketing can encourage farmers to change practices for water quality outcomes in the Great Barrier Reef catchment. It applies a community-based social marketing approach to increase adoption of practice changes to reduce excess irrigation and nutrients in the runoff flowing into nearby wetlands to respond to increasing pressure from the cumulative effects of climate change, land-based runoff, coastal land-use change and direct use impacts.

The final chapter in the category, *Waste Not Want Not: A co-created food waste pilot*, reports on the application of co-design, which centres programme design on the target to identify potential strategies, one of which was used to design and implement a social marketing programme. The outcome of the programme aims to understand if co-designed social marketing can reduce food waste, focusing research attention on one food waste behaviour.

Chapter 1
Winds of Change: Engaging with Conflicting Perspectives in Renewable Energy

Breda McCarthy and Lynne Eagle

Abstract This study examines public attitudes towards climate change and policies to limit greenhouse gases such as through the expansion of renewable energy sources. The study includes a critical examination of the apparent contradiction between belief in climate change and generalized support for renewable energy, as revealed in this survey, and of opposition to the placement of large-scale renewable energy generators such as wind farms. The literature identifies significant psychogenic influences leading to opposition to specific wind farm sitting as well as strong policy support for coal mining in Australia, but these appear to be ignored or dismissed by environmentalists, leading to ineffective communication on the issues and, ultimately, to entrenched opposition to an energy transition. We use social marketing benchmarks to identify engagement strategies that may increase understanding of the need to reduce fossil fuel consumption, alleviate anxiety about wind farm impacts and improve acceptance of renewable energy generation strategies overall.

Keywords Renewable energy · Acceptance of wind farms · Fossil fuels · Social marketing

1.1 Introduction

Climate change, and the need to reduce carbon dioxide (CO_2) and other greenhouse gas emissions, is one of the most urgent issues facing the global community. Climate change is often called a 'wicked problem' since policymaking around climate change is characterized by complexity, uncertainly and divergence of values (Head 2008). The United Nations (UN), guided by reports from the *Intergovernmental Panel of Climate Change* (IPCC), has set a target of limiting average warming to 2 °C above pre-industrial temperatures (IPCC 2014). Recent reports highlight that even a warming of 1.5 °C above pre-industrial levels could result in climate impacts that may be long lasting or irreversible, such as the loss of some ecosystems (IPCC

B. McCarthy (✉) · L. Eagle
James Cook University, Townsville, Australia
e-mail: breda.mccarthy@jcu.edu.au

R. Hay et al. (eds.), *Broadening Cultural Horizons in Social Marketing*,
https://doi.org/10.1007/978-981-15-8517-3_1

2018). There is concern that irreversible 'tipping points' could be activated on Earth, such as the melting of the Greenland and Antartic ice sheets and dieback of the Amazon rainforest (Lenton 2011). Experiences of extreme weather and climate events, such as heat waves, hurricanes, cyclones and floods over recent years in Australia, the USA and elsewhere, highlight that delaying action on climate change will come at a great cost to society. The socioeconomic impacts of natural disasters, caused by storms, floods and bushfires are extensive, with billions in financial costs arising from damage in Australia alone (UNISDR 2015). The energy sector is the largest contributor to global greenhouse gas (GHG) emissions (IPCC 2014). Globally around half of global emissions were the result of electricity and heat production in 2014, with transport, buildings, industry, agriculture, residential and commercial sectors also contributing to emissions (Ritchie and Roser 2019). Global energy-related CO_2 emissions grew by 1.7Gt CO_2 in 2018 to reach a historic high of 33.1Gt CO_2, due to a robust global economy as well as from weather conditions in some parts of the world that led to increased energy demand for heating and cooling (IEA 2018). Hence, there is a strong focus on reforming the world's energy system and implementing measures such as replacing fossil fuels with renewables, as well as reducing energy consumption, increasing energy efficiency, deploying nuclear power and exploiting carbon dioxide capture and storage (Edenhofer and Flachsland 2013; Kallies 2016; IPCC 2014).

Renewable Energy (RE), as outlined in the Intergovernmental Panel of Climate Change (IPCC) report (2014), refers to energy sources such as bioenergy, direct solar energy, geothermal energy, hydropower and ocean and wind energies. Sustainable energy transitions, which typically means moving away from fossil fuels towards renewable resources (Dowling et al. 2018), have been the subject of intense debate in academic and political spheres (Geels 2014; Hall and Taplin 2008). In Australia, a country rich in both fossil fuels and renewable energy, electricity is mainly generated from fossil fuels (Djerf-Pierre et al. 2015). Within this context, this chapter examines acceptance of renewable energy in regional Australia, along with the narratives and tensions that underlie the phase-out or closure of coal-fired stations along with the building of large-scale renewable energy infrastructures. Examining these tensions will provide policymakers in countries that are slow to embrace an energy transition with a greater understanding of barriers to, and potential enablers of, attitudinal and behavioural change. The social sciences are therefore now seen as a key contributor towards understanding the sociocultural complexities (as opposed to the technical barriers) of a renewable energy transition (Minsch et al. 2012; Sovacool 2014). We draw on this field, particularly social marketing, to analyse the reasons for opposition and to identify new approaches to engaging with opponents in order to increase understanding and acceptance around the need to transition away from coal, alleviate anxiety about wind farm impacts and, ultimately, to support a renewable energy transition.

1.2 Explaining Support for Coal: Politics, Ideology and Economic Rationalism

Coal-fired electricity generation is the single largest source of the global temperature increase, accounting for 30% of global CO_2 emissions (IEA 2018). Coal is particularly rich in carbon, and the burning of black coal generates carbon dioxide that is more than twice its weight (Hong and Slatick 1994). In a growing number of countries, the elimination of coal-fired generation is a key climate goal, while in others coal is abundant and affordable and remains the key source of electricity (IEA 2018). With the commitment of the Australian government to the Paris Agreement (UNFCCC 2015), scholars and policy advisors suggest that carbon pricing, such as an emission trading scheme or a national clean energy target, is essential (Edenhofer and Flachsland 2013; Finkel 2017; Queensland Renewable Energy Expert Panel 2016; Meadowcroft 2011). However, the federal government has shown inconsistent support for climate change (Nelson 2016), and according to the Clean Energy Council (2019), this uncertainty surrounding energy and climate policy is a deterrent to investors. In 2018, the Federal government did not implement its own National Energy Guarantee (NEG), hence a bipartisan solution on energy policy remains elusive (Clean Energy Council 2019). During the Federal election in 2019, the National-Liberal Party pledged its support for the Paris Agreement and to the carbon emission reduction target, which is 26% by the year 2030. However, it did not rule out the building of new coal-fired power stations (Chang 2019). Unlike the opposition party, labour, it did not raise the emission reduction target and the focus was on measures to reign in rising electricity prices (Clean Energy Council 2019). According to some commentators, the election of Scott Morrison, a Prime Minister who famously brought a lump of coal into Parliament, is a signal of a divided society and suggests that progressive climate policy is not on the political agenda (Lucas 2019).

Barriers such as the strength of the coal lobby, the emphasis placed on coal's contribution to the Australian economy and the way in which politics impedes an energy transition in a fossil-fuel dependent nation are well documented in the literature (Dowling et al. 2018; Cheung and Davies 2017; Muenstermann 2012). Cheung and Davies (2017), after analysing four Australian Prime Ministers, suggest that personal ideology drives energy politics. They suggest that there is an inherent contradiction with respect to Australia's energy policy, because despite actions towards renewables, there remains an on-going political disposition to subside the fossil fuel industry. Likewise, Warren et al. (2016) conclude that the failure to integrate climate and energy goals has hampered a renewable energy transition, evident in the contests between the neoliberalism and sustainable development discourses. Hence, the twin 'pillars' of energy policy, affordability and security of supply, have been given overwhelming priority over climate interests. As noticed by Sovacool (2016), political leadership and adequate incentives are essential elements for an energy transition. Traditionally, management has been strongly influenced by the philosophy of neoconservatism or economic rationalism (Dunphy 2003), and arguably, the coal industry is no exception. CEOs may accept the need for change, but ultimately the logic of profit, shareholder return and market competition drive their acceptance, rather than a clean energy agenda.

Globally, renewable energy sources increased by 4% in 2018, now accounting for over 25% of global power output (International Energy Agency 2018). There are significant differences across countries in terms of per capita installed renewable energy generation. Australia's reliance on non-renewables for electricity generation is shown in Table 1.1. It shows that Australia relies much more heavily on coal than all other member countries of the OECD. Natural gas has a 17.9% share of the energy mix. Gas is seen as a transition fuel, meaning that it allows for a reduction in emissions from power generation through a gradual substitution of coal (Guidolin and Alpcan 2019). The share of renewables in the energy mix is dominated by hydro, wind and solar. Australia has a slightly lower share of renewables in the energy mix, compared with other OECD countries, even though Australia has the highest solar radiation per square meter of any continent, and some of the best solar and wind resources in the world (Geoscience Australia 2010). In recent years, performance in renewables has improved significantly. An increase in 21% of total power generation in 2018 now puts renewables at its ever highest level (Clean Energy Council 2019).

Notes: Combustible renewables, according to the International Energy Agency (2019), refer to solid biofuels, biogases, liquid biofuels and municipal renewable

Table 1.1 Energy resources (2018)

Australia	Gigawatt hours (GWh)	Share—Australia (%)	Share—OECD total (%)
Conventional thermal	194,509	80.4	58.2
Coal	146,439	60.6	25.3
Oil	1,204	0.5	1.9
Natural gas	43,341	17.9	27.7
Combustible Renewables	3,525	1.5	2.7
Other combustibles	–	–	0.6
Nuclear	–		17.4
Hydro	17,642	7.3	13.8
Wind	17,414	7.2	6.9
Solar	12,275	5.1	3.0
Geothermal	0	0.0	0.5
Other renewables	–	–	0.0
Non-specified	–	–	0.3
Renewables	50,856	21.0	26.9
Non-renewables	190,985	79.0	73.1

Source International Energy Agency (2019)

Table 1.2 Coal power generation: gigawatts per capita ranking (selected countries) 2018

Country	GWh	GWh per capita	Ranking
Australia	146,439	5953	1
USA	1,183,559	3617	2
Germany	224,160	2708	3
Canada	55,571	1499	4
Greece	15,679	1456	5
Portugal	11,576	1123	6
Denmark	6103	1062	7
Spain	37,430	801	8
Ireland	3328	696	9
New Zealand	1519	317	10
Sweden	1964	196	11

Source International Energy Agency (2019). *Note* per capita figures are author-derived

wastes. Other combustibles refer to production from all other combustible fuels (such as industrial and non-renewable municipal solid wastes). Other renewables refer to electricity generated from tide, wave, ocean and other non-combustible sources.

Table 1.2 shows that Australia ranks number 1 out of a list of OECD countries in terms of coal power generation. The high level of dependence of Australia on 'extraction and production of non-renewable resources' (Djerf-Pierre et al. 2015, p. 635), such as coal for export, is put forward as a reason for its historically low level of renewables in the energy mix.

1.3 Social Acceptance and Wind Energy in Australia

Widespread public support is needed when developing large-scale energy infrastructures, such as wind farms (Batel and Devine-Wright 2015). Social acceptance is a concept that significantly shapes the implementation of renewable energy technologies and achievement of renewable energy targets (Moula et al. 2013; Wolsink 2012, 2013). Scholars (Batel and Devine-Wright 2015; Batel et al. 2013) have written extensively about public responses to large-scale energy infrastructures. Despite increased academic attention, no clear definition of social acceptance of renewable energy technologies exists (Wüstenhagen et al. 2007). According to Wolsink (2010, p. 303), "*Social acceptance is not simply a set of static attitudes of individuals; instead it refers more broadly to social relationships and organizations, and it is dynamic as it is shaped in learning processes*". Scholars note that the drivers of social acceptance have not received adequate attention in the literature (Friedl and Reichl 2016).

There is a considerable body of literature noting that, while support for renewable energy is, in general, high, substantial opposition becomes evident regarding the sitting of generation facilities such as specific wind farms (see, for example, Bell et al. 2013; Batel and Devine-Wright 2015). The 'social gap' or 'attitude-behaviour gap' between general support for the concept versus opposition to specific citing's of facilities is receiving increasing focus (Lennon and Scott 2015; Larson and Krannich 2016), with the recognition that "*the social dimension of the implementation of wind farms has emerged as a factor of at least equal importance to technology*" (Fournis and Fortin 2017).

High levels of localized resistance–particularly to wind power–were predicted to significantly hamper renewable energy targets set by the Australian federal, state and territory governments (Hall et al. 2015). Community opposition to sitting decisions has been somewhat dismissively classified as NIMBYISM ('Not In My Back Yard'), and the term has been used uncritically in both policy and academic material (Beben 2015; Petrova 2016). Opposition appears to centre on four factors: dislike of their visual impact (Knopper and Ollson 2011), turbine noise (Botelho et al. 2015; McCunney et al. 2014), potential dangers to wildlife (see, for example, Marques et al. 2014) and claimed human health impacts. Government inquiries and reviews have been conducted in several countries, including Australia, and they have concluded that there is no medical evidence of a direct link between turbine operation and human health, although the poor quality of the data has been noted (Chapman et al. 2014; Knopper and Ollson 2011).

Wind energy is seen as a key component of sustainable power development, decreasing reliance on fossil fuels and thus reducing greenhouse gas emissions and helping address climate change challenges (Crichton and Petrie 2015a, b; Duić 2015). Wind energy is the fastest-growing electricity generation form in the world (Blanes-Vidal and Schwartz 2016) and is claimed to be the most profitable (Read et al. 2013). It is also claimed to be able to produce over five times the current global energy demand (Sahu 2018). A study in Australia showed that the cost of new wind and solar is now lower than the cost of new coal generation (CSIRO 2018).

There is a wide variation in the amount of electricity generated by wind: in 2010, it was estimated that Denmark generated 25.5% of its electricity from wind energy sources, compared with only 2% in Australia, although there were substantial differences by state, with South Australia generating 20% of electricity from wind (Aparicio et al. 2012). By 2018, Australia's wind energy production stood at 7.2% of total electricity generation, which is slightly above the OECD average at 6.9% (IEA 2019). Table 1.3 shows the gigawatts per capita ranking of OECD countries and Australia is ranked ninth out of 11 countries. While this is an improvement on previous years, Australia still lags far behind Denmark, a country that has demonstrated rapid increases in wind power and has made a commitment to increase wind power share to 50% of electricity consumption by 2020 (Hvelplund et al. 2017).

Table 1.4 shows that wind generation in Australia in 2018 was 16,172 gigawatt hours and a breakdown by state is given. Wind generation now accounts for 33.5% of renewable energy generation and another 24 wind farms, with a combined capacity of 5.69 GW, were financially committed or under construction at the end of 2018 (Clean Energy Council 2019).

Table 1.3 Wind power generation: gigawatts per capita ranking (selected countries) 2018

Country	GWh	GWh per capita	Ranking
Denmark	13,899	2418	1
Ireland	8391	1754	2
Sweden	16,623	1663	3
Germany	110,891	1339	4
Portugal	12,,513	1214	5
Spain	49,571	1061	6
USA	275,064	841	7
Canada	29,357	792	8
Australia	17,414	708	9
Greece	6300	585	10
New Zealand	2047	427	11

Source International Energy Agency (2019). *Note* per capita figures are author derived

Table 1.4 Current wind generation (Clean Energy Council 2019)

	Current (end of 2018)	
	Share (%)	Capacity (GWh)
Breakdown by state		
South Australia	35.1	5692
Western Australia	9.9	1594
New South Wales	19.3	3124
Victoria	28.0	4528
Tasmania	6.8	1093
Queensland	0.9	141
Total National	100	16,172

Although high levels of localized resistance were predicted to hamper renewable energy investment (Hall et al. 2015), the Clean Energy Council found that incentives provided by the Federal Renewable Energy Target (RET) and other state policies propelled investment in renewable energy in 2018. Queensland, the second largest state by area in Australia, is increasingly hosting large-scale projects. The government also delivered its *Solar Farm Guidelines* in 2018, designed to ensure that large-scale projects maintain a strong social license to operate (Clean Energy Council 2019). The next section examines the factors that affect social acceptance in more detail.

1.4 Explaining Opposition to Large-Scale Energy Infrastructures: Health and Environmental Impacts, NIMBYISM and Information Deficit Assumptions

There is an increasing body of work showing that developing renewable energy as a climate mitigation option leads to a range of cobenefits for society, e.g. reduced health impacts due to lower air pollution (Xue et al. 2015). Renewable energy supply options have many advantages, but they also have drawbacks and differ with regards to their overall health and environmental impacts (Ellabban et al. 2014). Debates around granting a social license to (the now already established) coal mines were intense, with academics drawing attention to externalities, unpaid social and environmental costs, such as pollution, coal mining accidents and occupational health hazards (Byrnes et al. 2013; Lockie et al. 2008). Coal is a source of mercury and other toxic metals, harming ecosystems and potentially human life; the surface mining of coal causes a substantial change in land use and leads to mining waste (IPCC 2014). There are similar debates surrounding the impacts of wind energy today. Opposition towards wind energy appears to be driven more by emotions and psychological issues, rather than rationality (discussed in the next section). Given the intended expansion of wind energy, it is important to understand the nature and impact of any organized opposition to its development. These factors are discussed in more detail below.

Some early analyses gave localized opposition to the development of the pejorative term NIMBYISM ('Not In My Back Yard'), representing selfishness, ignorance and irrationality, but this concept is now acknowledged as not providing an effective framework for understanding the actual reasons for the opposition (Petrova 2016). Additional criticisms are that the term is simplistic and inaccurate (Burningham et al. 2015), and that the actual causes of opposition are obscured rather than explained (Wolsink 2012). In spite of this, the term continues to be used in an uncritical way in both policy documents and academic material (see, for example, Beben 2015). Often neglected in the debates is an "*inverse NIMBY syndrome*" where those living close to wind farms are the strongest supporters, often after initial concerns based on perceptions of potential negative impacts have not been realized (Warren and Birnie 2009; Enevoldsen and Sovacool 2016).

Even without invoking the NIMBY perspective, there are assumptions that wind power opposition is 'deviant', founded on ignorance and misinformation (Aitken 2010).

The assumption that opposition to a concept, issue or development is based partly on an information deficit (i.e. ignorance) and that it can be overcome by simply providing information is widespread, encompassing issues as diverse as climate change and vaccine hesitancy (Brulle et al. 2012; Yaqub et al. 2014) as well as for renewable energy (Bidwell 2016). Reliance on this assumption may explain why communication among stakeholders involved in specific scenarios is acknowledged as frequently inefficient and ineffective (Chen et al. 2015). Wind farm opponents may be a minority of any specific population but can have significant influence on sitting decision processes and approvals (Bidwell 2016), resulting in cost escalations, project delays or rejection (Enevoldsen and Sovacool 2016). Social pressure from others has been shown to predict intentions to oppose wind farm development (Read et al. 2013), therefore information provision alone is unlikely to overcome these pressures. Like the NIMBY concept, the information deficit assumption continues to feature in policy documents across topics as diverse as sustainability, climate change and vaccination hesitancy (Eagle et al. 2015; Lorenzoni et al. 2007; Marteau, Sowden, & Armstrong (2002).

In addition to the NIMBYISM and information deficit perspectives presented above, opposition to wind farm centres on four main factors:

(1) dislike of their visual impact (Knopper and Ollson 2011)
(2) noise from the turbines (Botelho et al. 2015; McCunney et al. 2014)
(3) potential dangers to wildlife (see, for example, Marques et al. 2014) and
(4) a range of human health impacts.

The first three factors are now reviewed briefly before a more detailed examination of the claims regarding health effects.

Visual Impact: Visual impact is a complex area, with much debate based on subjective views such as whether they are ugly or attractive in appearance (Wilson and Dyke 2016), with former Australian Prime Minister Tony Abbott publicly declaring in 2013 that wind farms were '*ugly and noisy*' (Green 2016). Others see wind turbines as graceful and visually 'magnificent' (Wilson and Dyke 2016). Some concern relates to the sitting of wind farms in areas that are '*ecologically sensitive and valued for their scenic qualities*' (Larson and Krannich 2016). There is a growing body of literature investigating the impacts of wind farms on tourism (see, for example, Riddington et al. 2010; Broekel and Alfken 2015). A detailed discussion of this sector is beyond the scope of this paper. Concern about visual impact may also reflect 'place attachment' (i.e. emotional bonds that form between people and their physical surroundings), and this is increasingly seen as a more significant explanation for resistance to local development than NIMBYISM (Devine-Wright 2009);

Noise: Noise appears to be addressed in multiple ways, including modern designs but, more importantly, on regulations regarding turbine size and distance ("set back") from homes in order to reduce annoyance from turbine noise (Hall et al. 2013; Onakpoya et al. 2015).

Wildlife impacts: Claims of mass deaths of wildlife due to turbines appear to be overstated. A review of Canadian avian mortality found that turbine-related deaths were much lower than cat predation, collisions with windows, vehicles and transmission lines (Calvert et al. 2013). This does not mean that efforts should not be made to minimize turbine-related fatalities: harm minimization strategies appear to be part of modern planning processes (Marques et al. 2014; May et al. 2015) and a detailed discussion of this area is beyond the scope of this paper.

The focus on the next section is on the adverse health impacts of wind due to the significance of claims made in this area.

1.5 Claimed Health Impacts of Wind Energy

The claim regarding adverse health effects from wind farms is an area in which contested, and indeed polarized views regarding the nature and quality of evidence are evident (Blanes-Vidal and Schwartz 2016). This extends even to government-sponsored reviews. For example, a 2015 Australian Senate inquiry revealed deep divisions between the Senate panel members. The first part of the report criticized previous reviews that did not find evidence of a direct link between wind farms and health issues, yet in another part of the report, it was concluded that there was no evidence of any health-related impacts from wind farms. It was also suggested that some studies (that did not identify adverse health impacts) were of limited value, as their authors did not have medical qualifications. This contesting and devaluing of expertise from those whose views diverge from a preferred stance are unfortunately not uncommon (Lennon and Scott 2015).

There have been frequent claims, supported in part by some academic studies, that wind turbines are a threat to human health, with more than 240 claimed health problems including '*sleep disturbance, headaches, nausea, tinnitus, ear pressure, vertigo or dizziness, visual blurring, irritability, memory and concentration problems, panic episodes, tachycardia, and body vibration*' (Tonin et al. 2016). Additional problems identified by other authors include *fatigue, inability to concentrate, depression, irritability, aggressiveness, chest pains, vomiting* and *annoyance*, although the link between this latter problem and specific medical conditions is unclear (Blanes-Vidal and Schwartz 2016; Chapman et al. 2013; Havas and Colling 2011). However, the studies on which the academic-based negative health claims are criticized for not controlling for potential confounding effects may offer alternative explanations of adverse health effects as reported by those living near wind farms (Blanes-Vidal and Schwartz 2016).

An analysis of complaints about noise or health issues relating to wind turbines living near 51 Australian wind farms found that the pattern of complaints did not match either the establishment or location of the wind farms (Chapman et al. 2013). These authors, in common with others who have conducted systematic reviews in this area conclude that wind turbine syndrome is a '*communicated disease*' (Knopper and Ollson 2011; Schmidt and Klokker 2014), with its foundations in psychological rather than pathological factors. Thus, anxiety and expectations of negative effects appear to be a factor in reported symptoms (Crichton et al. 2014). The reported effects are thus frequently referred to as reflecting a 'nocebo effect', whereby adverse health effects are produced by the expectation that these effects will occur (Colloca and Miller 2011; Faasse and Petrie 2013). This phenomenon has been recognized in the health academic literature for two decades (Benson 1997; Spiegel 1997). In the specific context of wind farms, the effect is clearly stated as follows:

'The nocebo effect is a negative reaction from exposure to an innocuous substance due to expectations of harm. It is the converse to a placebo, which is an inert substance that creates either a beneficial response or no response in a patient. The nocebo effect is psychogenic in nature and is a reaction to a patient's expectations and perceptions of how an exposure to a substance will affect them' (Tonin et al. 2016).

News and social media coverage of issues can result in large-scale nocebo effects, increasing expectations that reported adverse effects will in fact occur (Faasse and Petrie 2013).

A range of government inquiries and reviews has been conducted in several countries including the UK, Canada, Belgium and Australia. There is general agreement that there is no medical evidence of a direct link between turbine operation and human health but rather an association with a range of psychosocial factors including, among others, annoyance, although the poor quality of data available has been noted (Chapman et al. 2014; Knopper and Ollson 2011). For example, in a review commissioned by the Australian National Health and Medical Research Council (NHMRC), the following comments were made:

'It is a significant limitation of the available evidence that it was not known whether any of the observed health effects in residents were present or occurring at a different intensity prior to wind turbine exposure (i.e. demonstrating appropriate temporal proximity)' (Merlin et al. 2013).

In March 2016, the NHMRC awarded AU$3.3 million over a five year period '*to enrich the evidence-based understanding of the effects of wind farms on human health*' with particular emphasis on infrasound (National Health and Medical Research Council 2016). Infrasound (low-frequency, sub-audible sound) has been claimed to have adverse health effects (Pierpont 2009). However double-blind tests exposing people to infrasound and sham infrasound, i.e. silence, have found a correlation between reporting of symptoms and expectation of negative effects (Crichton et al. 2014), with internet-sourced information linking infrasound to health risks creating negative expectations then symptoms (Crichton and Petrie 2015a, b), suggesting high nocebo effects.

The next section explores the impact of the mass media on people's attitudes towards energy sources.

1.6 Information Sources and Procedural Measures to Resolve Conflict

Mass media are important sources of information for the general population across a wide range of topic areas including health, science and environmental issues (Dahlstrom et al. 2012) as well as energy issues (Deignan and Hoffman-Goetz 2015) and can have an impact on attitude and beliefs. In relation to energy sources, the media can unintentionally amplify the anxieties people have in relation to their own personal health or the health of their families:

'be used as potential cases for appropriate illness behaviour responses and can initially alarm those at risk…Too often it is the media-created event to which people respond rather than the objective situation itself' (Chapman et al. 2013).

The news media can intentionally, and unintentionally, spread misinformation, as can internet sites including social media and once misinformation has been accepted, new information that is not compatible with previous information may struggle for acceptance even if the new information is correct (Lewandowsky et al. 2012). It is therefore concerning that a Canadian study of newspaper coverage of health impacts of wind turbine-sensationalized negative claims (Deignan and Hoffman-Goetz 2015). It has been found that media coverage and interaction with lobby groups can increase symptoms (Rubin et al. 2014; McCunney et al. 2014). This has the potential not only to add to personal distress, but also to have a detrimental impact on policy decisions:

'The public's lack of understanding of many issues negatively affects the ability of the government to represent the will of the people. Yet, evidence suggests that decision-makers will still side with the public over scientists and experts, even when it is probable the public does not understand the issue' (Stoutenborough and Vedlitz 2016).

The media claim that they are providing balance in presenting all views 'objectively' and that this stance is a fundamental principle of journalism (Clarke 2008). This strategy may lead to intentional or unintentional bias especially if one view is clearly in the minority but still receives equal coverage, as has been noted in other areas such as climate change coverage (Boykoff and Mansfield 2008) and human vaccine controversies (Picard and Yeo 2011). Further, the news media maintains this balance irrespective of evidence that may support or refute claims (Boykoff and Boykoff 2004; Gross 2009; Lewandowsky et al. 2012).

In developing communication strategies to address wind farm anxiety, there is a need for engagement with those affected, recognizing that these people are experiencing genuine, distressing symptoms. There is also the need to recognize that strong emotions are involved (Cass and Walker 2009; Walker et al. 2010). As noted by Crichton and Petrie (2015a, b), it is necessary to acknowledge the health issues experienced by people who claim to be affected by wind farms:

'Understanding what might be causing symptom reports is critical to inform successful interventions to alleviate distress and symptom reporting in communities in which wind farms are proposed and operating. It is noteworthy that experiencing symptoms is a common phenomenon, and is not in and of itself indicative of illness' (Crichton and Petrie 2015a, b).

Those who dismiss complaints on the grounds of lack of direct medical evidence, while not wrong, are unhelpful in developing the type of genuine public engagement recommended by others (see, for example, Devine-Wright 2011; Shaw et al. 2015). There is a need to determine people's values as these underpin personal norms and shape attitudes towards wind farm developments (Bidwell 2013; Steg et al. 2014).

Community compensation (Terwel et al. 2014) is seen as a means to help avoid or reduce local public opposition to energy projects, although the use of financial incentives is viewed cynically in some quarters. In the words of one commentator, *"It would seem 'wind turbine syndrome' can be prevented by the wonder drug called money"* (Chapman 2013) and another assessment is:

'those who benefit economically from wind turbines (e.g. those who have leased their property to wind farm developers) report significantly lower levels of annoyance than those who received no economic benefits, despite increased proximity to the turbines and exposure to similar (or louder) sound levels' (Knopper et al. 2015).

Social acceptance of large-scale wind energy projects is not a given, but the literature shows that procedural issues show promise in resolving conflict. These measures include the following: identifying the expectations and interests of different stakeholders; providing balanced and objective information; gaining trust at local level; adopting benefit-sharing mechanisms and developing mechanisms for articulating conflict and engaging in negotiation (Hall 2014; Hall et al. 2013; Howard 2015; Terwel et al. 2014). While public concerns about large-scale energy infrastructures cannot be entirely eliminated, addressing these concerns is a key part of many jurisdictions' planning, sitting, and permitting processes (IPCC 2014).

1.7 Research Objectives and Methodology

This study aims to evaluate consumers' support for renewable energy sources, including wind farms, and examine priorities placed on factors such as human health, the economy and climate change mitigation. A quantitative methodology was chosen and a questionnaire was developed, which is the norm in renewable energy studies (Stoutenborough et al. 2015; Dockerty et al. 2012). Data were collected through field sampling and online distribution. A total of 325 usable responses were obtained. Respondents came from a regional city in Queensland, Australia, with an even spread of males and females and a variety of ages and income groups. Questions were informed by the literature (Poortinga et al. 2006; Eagle et al. 2016), noting that the various energy technologies have different environmental, economic and social impacts. 5-point Likert-type scales were used (with 1 = strongly disagree and 5 = strongly agree), with one question specifically measuring support for wind energy and a second question providing benchmark data on attitudes. A rank order question was used (1 = most important, 6 = least important) to examine perceived priorities for generation decisions. Ethics approval was secured from the Human Ethics Committee at the authors' university (H6601).

1.8 Results

The section presents the results from the survey on acceptance of energy supply sources, energy-related attitudes and public policy preferences.

Acceptance of energy supply sources by Australian consumers

Table 1.5 shows the level of support for each energy source. It indicates a high level of support for renewable energy, with solar and wind energy receiving the highest mean scores. Nuclear energy was by far the least favoured technology. Battery storage enjoyed strong support from the sample, along with hydroelectric and marine power.

Table 1.5 Acceptance of energy technologies (5 = strongly agree)

Energy	Mean	n	SD
Solar (producing energy from the sun)	4.70	322	0.53
Wind (producing energy from the wind)	4.62	323	0.71
Marine power (generated from the movement of tides, waves or ocean)	4.37	319	0.87
Hydroelectric power (energy generated from flowing water)	4.27	319	0.76
Battery storage (a storage device connected to any source of energy, including solar and hydro)	4.07	324	0.89
Biomass (energy produced from sugar cane, landfill gas, wood or sorghum crops specifically grown for energy)	4.01	322	0.81
Geothermal (generated from energy stored in the Earth, such as hot, dry rock)	3.81	323	0.99
Fuel cell technology (generated from hydrogen)	3.61	329	0.92
Natural gas (drilling wells into the ground to reach the gas, including coal seam and shale gas)	2.98	322	1.21
Oil (producing energy from oil reserves)	2.72	314	1.19
Coal (i.e. coal-fired power stations)	2.63	323	1.07
Nuclear (i.e. generated from nuclear fission)	2.55	320	1.32

The public clearly preferred renewable energy to fossil fuels (which received lower scores).

Figure 1.1 shows the support for wind in more detail, with 230 respondents who strongly support wind, 74 who support wind and 12 who are neutral about wind. These groups are labelled 'wind gusters' (n = 230), 'wind breezers' (n = 74) and

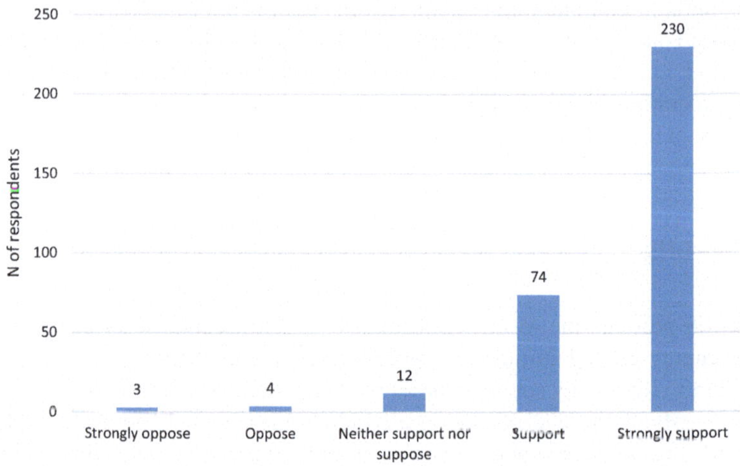

Fig. 1.1 Support for wind energy

'wind draggers' (n = 12) based on level of support for wind. Further analysis of this data is presented below.

1.9 Consumers' Attitudes Towards Energy Supply Sources

Responses in relation to the social, ecological and economic indicators of acceptance are shown in Table 1.6. Respondents agreed with the statement that 'Queensland's renewable energy sources (solar, wind) should be fully exploited', and that Queensland 'is rich in renewable resources'. Respondents also agreed with statements such as 'it is our responsibility to develop renewable energy for future generations' and 'high levels of energy use will impact future generations'. The majority of respondents believed in human-induced climate change ($\mu = 4.19$). Respondents disagreed with the statement 'there is no link between electricity used in the home and climate change' ($\mu = 2.43$). In relation to fossil fuels, they were neutral with regards to their environmental impacts and support for the economy, and interestingly they agreed with the statement that 'we are using up supplies of fossil fuels (coal, oil, gas) too fast'.

Table 1.6 Attitudes towards energy supply sources (5 = strongly agree)

Attitudinal item	Mean	n	SD
Queensland's renewable energy sources (solar, wind) should be fully exploited	4.33	323	0.95
Queensland is rich in renewable energy sources (e.g. solar, wind)	4.10	320	0.98
It is our responsibility to develop renewable energy for future generations	4.45	323	0.74
High levels of energy use will impact future generations' standard of living	4.27	322	0.85
Human-induced climate change is occurring at some level	4.19	323	0.85
There is no link between electricity used in the home and climate change	2.43	321	1.19
Investment in renewable energy is a means of stimulating economic growth	3.95	318	0.92
We are using up supplies of fossil fuels (i.e. coal, oil, gas) too fast	3.89	321	1.19
Fossil fuels (i.e. coal, gas, oil) should not be avoided because they support the economy	2.74	322	1.14
The environmental impacts associated with coal-fired power stations are often overstated	2.69	323	1.22

Table 1.7 Consumers' attitudes towards government investment priorities

	1	2	3	4	5	6
Investment areas	% (n)	% (n)	% (n)	% (n)	% (n)	% (n)
Natural environment	29.0 (61)	26.7 (56)	17.6 (37)	13.8 (29)	9.5 (20)	3.3 (7)
Climate change	27.6 (58)	13.3 (28)	15.7 (33)	14.3 (30)	12.4 (26)	16.7 (35)
Human health and safety	19.5 (41)	20.0 (42)	23.8 (50)	21.9 (46)	9.5 (20)	5.2 (11)
Cost of electricity	11.4 (24)	5.2 (11)	12.3 (26)	7.1 (15)	25.7 (54)	38.1 (80)
Pollution	9.5 (20)	27.6 (58)	22.8 (48)	24.3 (51)	9.0 (19)	7.6 (16)
Economy	2.8 (6)	7.1 (15)	8.6 (18)	18.6 (39)	33.8 (71)	29 (61)

Note 1 = ranked as first preference to 6 = ranked as last preference for government investment

1.10 Preferences for Government Investment

The survey harnessed insights into what consumers believe are investment priorities for the government in terms of energy supply. Respondents were asked to rank six different factors (in order of importance) in determining which methods of energy generation should be prioritized. Table 1.7 displays the proportion of ranked priority for the six investment factors.

Respondents' preferences for investment priorities varied greatly. Effects on the natural environment were perceived as more important than other investment areas with 117 respondents ranking this in their top two preferences. Then, helping to prevent climate change and effects on human health and safety were of second most important; these factors had relatively equal proportions of respondents across the top four preferences. Level of pollution was of medium importance, whereas the cost of electricity to consumers and effects on the economy were the lowest ranked investment priority areas.

Finally, analysis was performed on respondents who displayed varying levels of support for wind energy, and one group in particular, people who were neutral about wind, was of interest. Those who were neutral about wind ('wind dragger') had different priorities from those who strongly supported wind ('wind guster'). The 'wind dragger' ranked human health and climate change first, whereas the strong supporters of wind ranked the natural environment and climate change first. Looking at the second ranked factor, human health was ranked alongside the natural environment and pollution as priorities for the 'wind dragger'. Pollution also featured highly as the third ranked factor for this group (Table 1.8).

1.11 Discussion and Conclusion

Our study demonstrates a high level of social acceptance for wind energy and renewable energy overall. These results confirm previous literature which reports strong support for renewable energy (Devine-Wright 2007; Dockerty et al. 2012; Stoutenborough et al. 2015; Sütterlin and Siegrist 2017; Truelove 2012; Warren et al. 2005).

Table 1.8 Support for wind and attitudes towards government investment priorities

Rank	Support for wind	Natural environment	Human health	Climate change	Level of pollution
First	Wind 'dragger'	10.0% (1)	30.0% (3)	30.0% (3)	0.0% (0)
	Wind 'breezer'	19.5% (8)	22.0% (9)	22.0% (9)	22.0% (9)
	Wind 'guster'	31.3% (51)	19.0% (31)	29.4% (48)	7.4% (12)
Second	Wind 'dragger'	22.2% (2)	22.2% (2)	0.0% (0)	22.2% (2)
	Wind 'breezer'	34.1% (14)	19.5% (8)	12.2% (5)	22.0% (9)
	Wind 'guster'	25.3% (41)	19.8% (32)	13.6% (22)	29.6% (48)
Third	Wind 'dragger'	0.0% (0)	0.0% (0)	12.5% (1)	50.0% (4)
	Wind 'breezer'	17.1% (7)	19.5% (8)	14.6% (6)	12.2% (5)
	Wind 'guster'	18.9% (31)	23.8% (39)	15.9% (26)	23.8% (39)

The survey revealed positive attitudes towards renewable energy and recognition of benefits such as supporting the economy, exploiting abundant natural resources, climate change mitigation and providing for the needs of future generations. Other studies also highlight common indicators of acceptance such as climate change concern, perceptions of cost and economic impact (Devine-Wright 2008; Carr-Cornish et al. 2011; Moula et al. 2013). Contrary to expectations from the academic literature (for example, Marques et al. 2014; McCunney et al. 2014), we find no evidence of the NIMBYISM ('Not in My Back Yard') concept (Pidgeon and Demski 2012). The overwhelming support for wind energy is not too surprising. Queensland is the largest state in Australia and wind farms, which are few in number, are located in sparsely populated rural areas; hence, proximity to wind farms is not an issue for this sample of predominantly urban respondents. Although the perceived risks associated with the exploitation of renewable energy are relatively low (Bronfman et al. 2012), there has been considerable discussion in the literature on sources of community opposition to wind farms, such as health risks (Hall et al. 2013). This survey reveals some degree of concern with human health and safety. Our study revealed three groups of wind supporters, the 'wind gusters', 'wind breezers' and 'wind draggers', and analysis suggests that concern about human health and safety distinguishes the strong supporters from those who are neutral about wind. The following section offers recommendations as to how low support for wind farms (albeit in a minority of respondents) and potential health concerns could be addressed.

Social marketing has a role to play in facilitating an energy transition. Given the contradiction between public support for RE, and the reluctance of the Federal government to support renewable energy development at the expense of coal generation, Social marketing would appear to have an important role to play in strengthening social acceptance. Hence, investors and other proponents of wind energy need to do more in terms of recognizing emotional barriers to wind energy acceptance and communicating effectively with policymakers, the media and the general public. Muenstermann (2012) argues that it is critical to counter misinformation from the coal lobby, and this could occur through inoculation strategies such as advocated by Cook et al. (2017) at downstream, midstream and upstream levels of society (Gordon 2013; Luca et al. 2016). An explanation of these terms is given below:

- *Upstream*: influencing the environment in which behaviour occurs including policymakers, the media, lobby groups and influential organizations
- *Midstream*: working with partners, communities and institutions (for example, schools and other educational agencies)
- *Downstream*: working with specific individuals and groups of individuals (including families, peers and immediate neighbourhoods or communities) (Dibb 2014; Kamin and Anker 2014; Russell-Bennett et al. 2013).

Traditionally, social marketing activity has encompassed a wide range of health-related behaviour, such as smoking cessation, sun safety, genetic screening and so on. While the focus is on encouraging sustained, positive behavioural change among individuals and groups, social marketing also encompasses environmental factors and major shifts in society, such as encouraging an energy transition. It is now recognized that the role of social marketing goes beyond individual behavioural change (Saunders et al. 2015). Social marketing offers a framework for designing proenvironmental behavioural change programs and is flexible enough to be applied to a range of policy issues (Corner and Randall 2011). However, it is not a panacea, and the role of legislation and incentives in conjunction with both education and social marketing must be recognized (Rothschild 1999; Sheavly and Register 2007). As noted by Šćepanović et al. (2017), policymakers can help communities achieve positive social outcomes through information-based or reward-based interventions. Hence, it is argued that the benefits of wind farms and other forms of renewable energy (i.e., climate change mitigation, low pollution, security of supply, abundance of the natural energy resource, local employment benefits) need to be presented to Australian society to reinforce acceptance. In addition, it is argued that the negative impacts of wind farms (i.e., health and safety, impacts on natural environment) be acknowledged rather than ignored. Social marketing has featured in some debates as a means of 'overcoming' opposition and reducing conflict (see, for example, Beben 2015; Chen et al. 2015; Menegaki 2012), although some social marketing campaigns could be counterproductive and even strengthen opposition (Corner and Randall 2011). It is recognized that reliance on the mere provision of information, based on the assumption that a lack of knowledge (i.e. an 'information deficit') is the reason for any lack of support for wind energy, will not be effective; information provision alone will not change attitudes (Costello et al. 2009; Owens and Driffill 2008; Semenza et al. 2008). Similar to other researchers in the area of wind energy (Hall 2014; Hall et al. 2013; Howard 2015; Terwel et al. 2014), we recognize that a focus on procedural issues is crucial to social acceptance. These may include community consultation, securing the trust of local residents and using incentives, such as jobs for local residents during the construction and maintenance phase of wind farm development or even some form of local ownership of wind projects (Hvelplund et al. 2017).

In relation to health, the literature highlights the significance of psychogenic influences on the opposition to wind farms (Cass and Walker 2009; Walker et al. 2010). Whether or not wind turbines pose a risk to human health is a matter of heated debate and still poses a barrier to acceptance of wind farms; thus, investors, developers and local government need to understand more comprehensively how large-scale wind energy infrastructures could potentially harm constituents' well-being, if not on physical health grounds, but emotionally. As in clinical practice, there is a need for 'reassuring, empathetic, and supportive communication' (Colloca and Miller 2011, p. 602). There is a growing acknowledgement that adopting the principles or benchmarks of social marketing (Andreasen 1995; National Social Marketing Centre n.d; French and Russell-Bennett 2015), particularly when underpinned by theory-driven approaches, can lead to more persuasive messages rather than information-only messages. For instance, adopting a customer orientation, undertaking qualitative and quantitative research, fully understanding people's anxieties about wind energy, maximizing benefits and minimizing costs of a renewable energy transition, segmenting audiences and tailoring messages based on values, knowledge and attitudes and pre-testing communications.

In conclusion, social acceptance of wind energy, and recognition of the need to reduce reliance on coal-fired electricity, is crucial if Australia is to make a sustainable energy transition.

References

Aitken M (2010) Wind power and community benefits: challenges and opportunities. Energy Policy 38(10):6066–6075

Andreasen AR (1995) Marketing social change: changing behavior to promote health, social development, and the environment. Jossey-Bass, San Francisco, p 101

Aparicio N, MacGill I, Abbad JR, Beltran H (2012) Comparison of wind energy support policy and electricity market design in Europe, the United States, and Australia. IEEE Trans Sustain Energy 3(4):809–818

Batel S, Devine-Wright P (2015) A critical and empirical analysis of the national-local 'gap' in public responses to large-scale energy infrastructures. J Environ Plan Manag 58(6):1076–1095

Batel S, Devine-Wright P, Tangeland T (2013) Social acceptance of low carbon energy and associated infrastructures: a critical discussion. Energy Policy 58:1–5

Beben R (2015) The role of social marketing in overcoming the NIMBY syndrome. J Positive Manag 6(1):3

Bell D, Gray T, Haggett C, Swaffield J (2013) Re-visiting the 'social gap': public opinion and relations of power in the local politics of wind energy. Environ Polit 22(1):115–135

Benson H (1997) The nocebo effect: history and physiology. Prev Med 26(5):612–615

Bidwell D (2013) The role of values in public beliefs and attitudes towards commercial wind energy. Energy Policy 58:189–199

Bidwell D (2016) The effects of information on public attitudes toward renewable energy. Environ Behav 48(6):743–768

Blanes-Vidal V, Schwartz J (2016) Adverse health effects attributed to wind turbines: have we been missing important co-exposures? In: ISEE conference abstracts

Botelho A, Arezes P, Bernardo C, Dias H, Pinto LMC (2015) Effects of turbine noise from wind farms on local residents' decision to adopt mitigation measures. Paper presented at the proceedings 8th international conference on environmental and geological science and engineering (EG'15)

Boykoff MT, Boykoff JM (2004) Balance as bias: Global warming and the US prestige press. Global Environ change 14(2):125–136

Boykoff MT, Mansfield M (2008) 'Ye Olde Hot Aire'*: reporting on human contributions to climate change in the UK tabloid press. Environ Res Lett 3(2):024002

Broekel T, Alfken C (2015) Gone with the wind? The impact of wind turbines on tourism demand. Energy Policy 86:506–519

Bronfman NC, Jiménez RB, Arévalo PC, Cifuentes LA (2012) Understanding social acceptance of electricity generation sources. Energy Policy 46:246–252

Brulle RJ, Carmichael J, Jenkins JC (2012) Shifting public opinion on climate change: an empirical assessment of factors influencing concern over climate change in the US, 2002–2010. Clim Change 114(2):169–188

Burningham K, Barnett J, Walker G (2015) An array of deficits: unpacking NIMBY discourses in wind energy developers' conceptualizations of their local opponents. Soc Nat Resour 28(3):246–260

Byrnes L, Brown C, Foster J, Wagner LD (2013) Australian renewable energy policy: barriers and challenges. Renewable Energy 60:711–721

Calvert A, Bishop C, Elliot R, Krebs E, Kydd T, Machtans C, Robertson G (2013) A synthesis of human-related avian mortality in Canada. Avian Conserv Ecol 8(2)

Carr-Cornish S, Ashworth P, Gardner J, Fraser SJ (2011) Exploring the orientations which characterise the likely public acceptance of low emission energy technologies. Climatic Change 107(3-4):549–565

Cass N, Walker G (2009) Emotion and rationality: the characterisation and evaluation of opposition to renewable energy projects. Emot Space Soc 2(1):62–69

Chapman S, Joshi K, Fry L (2014) Fomenting sickness: nocebo priming of residents about expected wind turbine health harms. Front Pub Health 2:279

Chang C (2019) Federal election 2019: the energy policies of labour and the coalition. https://www.news.com.au/finance/money/costs/federal-election-2019-the-energy-policies-of-labor-and-the-coalition/news-story/89f6a8bc58e181f70158f3c71b6e89e5

Chapman S (2013) Turbine hosts 'not bothered. Farmonline, sourced from the conversation. http://www.farmonline.com.au/news/agriculture/general/opinion/turbine-hosts-not-bothered/2671889.aspx. Accessed 1 Dec 2017

Chapman S, George AS, Waller K, Cakic V (2013) The pattern of complaints about Australian wind farms does not match the establishment and distribution of turbines: support for the psychogenic, 'communicated disease' hypothesis. PLoS ONE 8(10).e76584

Chen JL, Liu HH, Chuang CT (2015) Strategic planning to reduce conflicts for offshore wind development in Taiwan: a social marketing perspective. Mar Pollut Bull 99(1–2):195–206

Cheung G, Davies PJ (2017) In the transformation of energy systems: what is holding Australia back? Energy Policy 109:96–108

Clarke CE (2008) A question of balance: the autism-vaccine controversy in the British and American Elite Press. Sci Commun 30(1):77–107

Clean Energy Council (2019) Clean energy Australia report 2019. https://assets.cleanenergycouncil.org.au/documents/resources/reports/clean-energy-australia/clean-energy-australia-report-2019.pdf

Colloca L, Miller FG (2011) Role of expectations in health. Curr Opin Psychiatry 24:149–155

Corner A, Randall A (2011) Selling climate change? The limitations of social marketing as a strategy for climate change public engagement. Glob Environ Change 21(3):1005–1014

Cook J, Lewandowsky S, Ecker UK (2017) Neutralizing misinformation through inoculation: exposing misleading argumentation techniques reduces their influence. PLoS ONE 12(5):e0175799

Costello A, Abbas M, Allen A, Ball S, Bell S, Bellamy R, Patterson C (2009) Managing the health effects of climate change. The Lancet 373(9676):1693–1733

Crichton F, Petrie KJ (2015a) Health complaints and wind turbines: the efficacy of explaining the nocebo response to reduce symptom reporting. Environ Res 140:449–455

Crichton F, Dodd G, Schmid G, Gamble G, Petrie KJ (2014) Can expectations produce symptoms from infrasound associated with wind turbines? Health Psychol 33(4):360

Crichton F, Petrie KJ (2015b) Accentuate the positive: counteracting psychogenic responses to media health messages in the age of the Internet. J Psychosom Res 79(3):185–189

CSIRO (2018) Annual update finds renewables are cheapest new build power. https://www.csiro.au/en/News/News-releases/2018/Annual-update-finds-renewables-are-cheapest-new-build-power. Accessed 21 Dec 2018

Dahlstrom MF, Dudo A, Brossard D (2012) Precision of information, sensational information, and self-efficacy information as message-level variables affecting risk perceptions. Risk Anal Int J 32(1):155–166

Deignan B, Hoffman-Goetz L (2015) Emotional tone of Ontario newspaper articles on the health effects of industrial wind turbines before and after policy change. J Health Commun 20(5):531–538

Devine-Wright P (2007) Reconsidering public attitudes and public acceptance of renewable energy technologies: a critical review. Beyond Nimbyism: a multidisciplinary investigation of public engagement with renewable energy technologies, p 15

Devine-Wright P (2008) Reconsidering public acceptance of renewable energy technologies: a critical review. Delivering a low carbon electricity system: technologies, economics and policy, pp 1–15

Devine-Wright P (2009) Rethinking NIMBYism: the role of place attachment and place identity in explaining place-protective action. J Community Appl Soc Psychol 19(6):426–441

Devine-Wright P (2011) Public engagement with large-scale renewable energy technologies: breaking the cycle of NIMBYism. Wiley Interdis Rev Clim Change 2(1):19–26

Dibb S (2014) Up, up and away: social marketing breaks free. J Mark Manag 30(11–12):1159–1185

Djerf-Pierre M, Cokley J, Kuchel LJ (2015) Framing renewable energy: a comparative study of newspapers in Australia and Sweden. Environ Commun 1–22

Dockerty T, Appleton K, Lovett A (2012) Public opinion on energy crops in the landscape: considerations for the expansion of renewable energy from biomass. J Environ Plan Manag 55(9):1134–1158

Dowling R, McGuirk P, Maalsen S (2018) Multiscalar governance of urban energy transitions in Australia: the cases of Sydney and Melbourne. Energy Res Soc Sci 44:260–267

Duić N, Urbaniec K, Huisingh D (2015) Components and structures of the pillars of sustainability. J Cleaner Prod 88:1–12

Dunphy D (2003) Corporate sustainability: challenge to managerial orthodoxies. J Manag Org 9(1):2

Eagle L, Hamann M, Low D (2016) The role of social marketing, marine turtles and sustainable tourism in reducing plastic pollution. Mar Pollut Bull 107(1):324–332

Eagle L, Low D, Case P, Vandommele L (2015) Attitudes of undergraduate business students toward sustainability issues. Int J Sustain High Educ 16(5):650–668

Edenhofer O, Flachsland C (2013) Transforming the global energy system: pathways towards a sustainable energy supply. Global trends

Ellabban O, Abu-Rub H, Blaabjerg F (2014) Renewable energy resources: current status, future prospects and their enabling technology. Renew Sustain Energy Rev 39:748–764

Enevoldsen P, Sovacool BK (2016) Examining the social acceptance of wind energy: practical guidelines for onshore wind project development in France. Renew Sustain Energy Rev 53:178–184

Faasse K, Petrie KJ (2013) The nocebo effect: Patient expectations and medication side effects. Postgrad Med J 89:540–546
Finkel A (2017) Independent review into the future security of the national electricity market—blueprint for the future—a snapshot. Department of the Environment and Energy, Australian Government, Canberra. https://www.energy.gov.au/government-priorities/energy-markets/independent-review-future-security-national-electricity-market
Fournis Y, Fortin MJ (2017) From social 'acceptance' to social 'acceptability' of wind energy projects: towards a territorial perspective. J Environ Plan Manag 60(1):1–21
French J, Russell-Bennett R (2015) A hierarchical model of social marketing. Journal of Social Marketing 5(2):139–159
Friedl C, Reichl J (2016) Realizing energy infrastructure projects—a qualitative empirical analysis of local practices to address social acceptance. Energy Policy 89:184–193
Geels FW (2014) Regime resistance against low-carbon transitions: introducing politics and power into the multi-level perspective. Theory Cult Soc 31:21–40
Geoscience Australia (2010) Australia energy resource assessment. Department of Resources, Energy and Tourism, Australian Government, Canberra. https://d28rz98at9flks.cloudfront.net/70142/70142_complete.pdf
Gordon R (2013) Unlocking the potential of upstream social marketing. Eur J Mark 47(9):1525–1547
Green M (2016) Community power. Nat Energy 1:16014
Gross L (2009) A broken trust: lessons from the vaccine–autism wars. PLoS Biol 7(5):1–7
Guidolin M, Alpcan T (2019) Transition to sustainable energy generation in Australia: interplay between coal, gas and renewables. Renew Energy 139:359–367
Hall NL (2014) Can the "social licence to operate" concept enhance engagement and increase acceptance of renewable energy? A case study of wind farms in Australia. Soc Epistemol 28(3–4):219–238
Hall N, Ashworth P, Devine-Wright P (2013) Societal acceptance of wind farms: analysis of four common themes across Australian case studies. Energy Policy 58:200–208
Hall NL, Taplin R (2008) Room for climate advocates in a coal-focused economy? NGO influence on Australian climate policy. Aust J Soc Iss 43(3):359–379
Hall N, Lacey J, Carr-Cornish S, Dowd A-M (2015) Social licence to operate: understanding how a concept has been translated into practice in energy industries. J Clean Prod 86:301–310
Havas M, Colling D (2011) Wind turbines make waves: why some residents near wind turbines become ill. Bull Sci Technol Soc 31(5):414–426
Head BW (2008) Wicked problems in public policy. Public Policy 3(2):101
Hvelplund F, Østergaard PA, Meyer NI (2017) Incentives and barriers for wind power expansion and system integration in Denmark. Energy Policy 107:573–584
Hong BD, Slatick ER (1994) Carbon dioxide emissions factors for coal. US Energy Information Agency (EIA), Washington DC. http://www.eia.gov/coal/production/quarterly/co2_article/co2.html
Howard T (2015) Olivebranches and idiot's guides: frameworks for community engagement in Australian wind farm development. Energy Policy 78:137–147
International Energy Agency (IEA) (2018) Global energy & CO_2 status report. https://www.iea.org/geco/renewables/
International Energy Agency (IEA) (2019) Monthly electricity statistics 2019. https://www.iea.org/media/statistics/surveys/electricity/mes.pdf
IPCC (2014) Climate change 2014: synthesis report. Contribution of working groups I, II and III to the fifth assessment report of the intergovernmental panel on climate change. Core writing team: Pachauri RK, Meyer LA (eds) IPCC, Geneva, Switzerland, 151 pp. https://www.ipcc.ch/report/ar5/syr/
IPCC (2018) Summary for policymakers. In: Global warming of 1.5 °C. An IPCC special report on the impacts of global warming of 1.5 °C above pre-industrial levels and related global greenhouse gas emission pathways, in the context of strengthening the global response to the threat of climate

change, sustainable development, and efforts to eradicate poverty. In: Masson-Delmotte V, Zhai P, Pörtner H-O, Roberts D, Skea J, Shukla PR, Pirani A, Moufouma-Okia W, Péan C, Pidcock R, Connors S, Matthews JBR, Chen Y, Zhou X, Gomis MI, Lonnoy E, Maycock, Tignor M, Waterfield T (eds) World Meteorological Organization, Geneva, Switzerland, 32 pp. https://www.ipcc.ch/sr15/

Kallies A (2016) A barrier for Australia's climate commitments? Law, the electricity market and transitioning the stationary electricity sector. UNSW Law J 39(4):1547–1582

Kamin T, Anker T (2014) Cultural capital and strategic social marketing orientations. J Soc Mark 4(2):94–110

Knopper LD, Ollson CA, McCallum LC, Aslund MLW, Berger RG, Souweine K, McDaniel M (2015) Wind turbines and human health. Front Pub Health 63:1–20

Knopper LD, Ollson CA (2011) Health effects and wind turbines: a review of the literature. Environ Health 10(1):1

Larson EC, Krannich RS (2016) "A great idea, just not near me!" understanding public attitudes about renewable energy facilities. Soc Nat Resour 29(12):1436–1451

Lennon M, Scott M (2015) Contending expertise: an interpretive approach to (Re) conceiving wind power's 'planning problem'. J Environ Plan Policy Manag 17(5):593–616

Lenton TM (2011) Early warning of climate tipping points. Nat Clim Change 1(4):201

Lewandowsky S, Ecker UH, Seifert CM, Schwarz N, Cook J (2012) Misinformation and its correction: continued influence and successful debiasing. Psychol Sci Public Interest 13(3):106–131

Lockie S, Franetovich M, Sharma S, Rolfe J (2008) Democratisation versus engagement? Social and economic impact assessment and community participation in the coal mining industry of the Bowen Basin, Australia. Impact Assess Proj Apprais 26(3):177–187

Lorenzoni I, Nicholson-Cole S, Whitmarsh L (2007) Barriers perceived to engaging with climate change among the UK public and their policy implications. Glob Environ Change 17(3–4):445–459

Luca NR, Hibbert S, McDonald R (2016) Midstream value creation in social marketing. J Mark Manag 32(11–12):1145–1173

Lucas C (2019). Not everyone cares about climate change, but reproach won't change their minds. https://theconversation.com/not-everyone-cares-about-climate-change-but-reproach-wont-change-their-minds-118255

Marques AT, Batalha H, Rodrigues S, Costa H, Pereira MJR, Fonseca C, Bernardino J (2014) Understanding bird collisions at wind farms: an updated review on the causes and possible mitigation strategies. Biol Cons 179:40–52

Marteau TM, Sowden AJ, Armstrong D (2002) Implementing research findings into practice: beyond the information deficit model. Getting research findings into practice. London: BMJ Books, 68–76

McCunney RJ, Mundt KA, Colby WD, Dobie R, Kaliski K, Blais M (2014) Wind turbines and health: a critical review of the scientific literature. J Occup Environ Med 56(11):e108–e130

May R, Reitan O, Bevanger K, Lorentsen SH, Nygård T (2015) Mitigating wind-turbine induced avian mortality: sensory, aerodynamic and cognitive constraints and options. Renew Sustain Energy Rev 42:170–181

Merlin T, Newton S, Ellery B, Milverton J, Farah C (2013) Systematic review of the human health effects of wind farms. National Health & Medical Research Council

Meadowcroft J (2011) Engaging with the politics of sustainability transitions. Environ Innov Soc Trans 1(1):70–75

Menegaki AN (2012) A social marketing mix for renewable energy in Europe based on consumer stated preference surveys. Renew Energy 39(1):30–39

Minsch J, Goldblatt DL, Flüeler T, Spreng D (2012) The indispensable role of social science in energy research. In: Tackling long-term global energy problems. Springer, Dordrecht, pp 23–43

Muenstermann I (2012) Australia's climate change, wind farming, coal industry and the 'big carbon plan': mine coal, sell coal, repeat until rich. Rural Soc 21(3):231–249

Moula M, Maula J, Hamdy M, Fang T, Jung N, Lahdelma R (2013) Researching social acceptability of renewable energy technologies in Finland. Int Jo Sustain Built Environ 2:89–98

National Health and Medical Research Council (2016) NHMRC awards funding into wind farms and human health. Canberra. Press release 22 March 2016. https://www.nhmrc.gov.au/media/releases/2016/nhmrc-awards-funding-wind-farms-and-human-health. Accessed 1 Dec 2017
National Social Marketing Centre (n.d). NSMC benchmark criteria. NSMC, London. https://www.thensmc.com/content/nsmc-benchmark-criteria-1
Nelson T (2016) Redesigning a 20th century regulatory framework to deliver 21st century energy technology. J Bioecon 1–18
Onakpoya IJ, O'Sullivan J, Thompson MJ, Heneghan CJ (2015) The effect of wind turbine noise on sleep and quality of life: a systematic review and meta-analysis of observational studies. Environ Int 82:1–9
Owens S, Driffill L (2008) How to change attitudes and behaviours in the context of energy. Energy Policy 36(12):4412–4418
Picard RG, Yeo M (2011) Medical and health news and information in the UK media: the current state of knowledge. A report of the Reuters Institute for the study of Journalism
Petrova MA (2016) From NIMBY to acceptance: toward a novel framework—VESPA—for organizing and interpreting community concerns. Renew Energy 86:1280–1294
Pierpont N (2009) Wind turbine syndrome. K-selected books
Pidgeon N, Demski CC (2012) From nuclear to renewable: energy system transformation and public attitudes. Bull Atom Sci 68(4):41–51
Poortinga W, Pidgeon NF, Lorenzoni I (2006) Public perceptions of nuclear power, climate change and energy options in Britain: summary findings of a survey conducted during October and November 2005. Technical report (Understanding Risk Working Paper 06-02). Centre for Environmental Risk, Norwich
Queensland Government (n.d) Powering Queensland plan. https://www.dnrme.qld.gov.au/energy/initiatives/powering-queensland
Queensland Renewable Energy Expert Panel (2016) Final report—credible pathways to a 50% renewable energy target for Queensland. https://www.dews.qld.gov.au/electricity/solar/solar-future/expert-panel
Read DL, Brown RF, Thorsteinsson EB, Morgan M, Price I (2013) The theory of planned behaviour as a model for predicting public opposition to wind farm developments. J Environ Psychol 36:70–76
Riddington G, McArthur D, Harrison T, Gibson H (2010) Assessing the economic impact of wind farms on tourism in Scotland: GIS, surveys and policy outcomes. Int J Tourism Res 12(3):237–252
Ritchie H, Roser M (2019) CO_2 and other greenhouse gas emissions. Published online at OurWorldInData.org. https://ourworldindata.org/co2-and-other-greenhouse-gas-emissions
Rothschild ML (1999) Carrots, sticks, and promises: a conceptual framework for the management of public health and social issue behaviors. J Mark 63(4):24–37
Rubin GJ, Burns M, Wessely S (2014) Possible psychological mechanisms for "wind turbine syndrome". On the windmills of your mind. Noise Health 16:116–22. http://www.noiseandhealth.org/article.asp?issn=1463-1741;year=2014;volume=16;issue=69;spage=116;epage=122;aulast=Rubin
Russell-Bennett R, Wood M, Previte J (2013) Fresh ideas: services thinking for social marketing. J Soc Mark 3(3):223–238
Sahu BK (2018) Wind energy developments and policies in China: a short review. Renew Sustain Energy Rev 81:1393–1405
Saunders SG, Barrington DJ, Sridharan S (2015) Redefining social marketing: beyond behavioural change. J Soc Mark 5(2):160–168
Šćepanović S, Warnier M, Nurminen JK (2017) The role of context in residential energy interventions: a meta review. Renew Sustain Energy Rev 77:1146–1168
Semenza JC, Hall DE, Wilson DJ, Bontempo BD, Sailor DJ, George LA (2008) Public perception of climate change: voluntary mitigation and barriers to behavior change. Am J Prev Med 35(5):479–487

Shaw K, Hill SD, Boyd AD, Monk L, Reid J, Einsiedel EF (2015) Conflicted or constructive? Exploring community responses to new energy developments in Canada. Energy Res Soc Sci 8:41–51

Sheavly SB, Register KM (2007) Marine debris & plastics: environmental concerns, sources, impacts and solutions. J Polym Environ 15(4):301–305

Schmidt JH, Klokker M (2014) Health effects related to wind turbine noise exposure: a systematic review. PLoS ONE 9(12):e114183

Steg L, Bolderdijk JW, Keizer K, Perlaviciute G (2014) An integrated framework for encouraging pro-environmental behaviour: The role of values, situational factors and goals. J Environ Psychol 38:104–115

Stoutenborough JW, Vedlitz A (2016) The role of scientific knowledge in the public's perceptions of energy technology risks. Energy Policy 96:206–216

Stoutenborough JW, Shi L, Vedlitz A (2015) Probing public perceptions on energy: support for a comparative, deep-probing survey design for complex issue domains. Energy 81:406–415

Sovacool BK (2014) What are we doing here? Analyzing fifteen years of energy scholarship and proposing a social science research agenda. Energy Res Soc Sci 1:1–29

Sovacool BK (2016) How long will it take? Conceptualizing the temporal dynamics of energy transitions. Energy Res Soc Sci 13:202–215

Spiegel D (1997) Trauma, dissociation, and memory

Sütterlin B, Siegrist M (2017) Public acceptance of renewable energy technologies from an abstract versus concrete perspective and the positive imagery of solar power. Energy Policy 106:356–366

Terwel BW, Koudenburg FA, ter Mors E (2014) Public responses to community compensation: the importance of prior consultations with local residents. J Commun Appl Soc Psychol 24(6):479–490

Tonin R, Brett J, Colagiuri B (2016) The effect of infrasound and negative expectations to adverse pathological symptoms from wind farms. J Low Freq Noise Vib Active Control 35(1):77–90

Truelove HB (2012) Energy source perceptions and policy support: image associations, emotional evaluations, and cognitive beliefs. Energy Policy 45:478–489

United Nations Office for Disaster Risk Reduction (UNISDR) (2015) Australia—disaster and risk profile. UNISDR, Geneva. http://www.preventionweb.net/english/countries/statistics/?cid=9

United Nations Framework Convention on Climate Change (UNFCCC) (2015) Adoption of the Paris agreement. UNFCCC, Geneva. https://unfccc.int/resource/docs/2015/cop21/eng/l09r01.pdf

Walker G, Devine-Wright P, Hunter S, High H, Evans B (2010) Trust and community: exploring the meanings, contexts and dynamics of community renewable energy. Energy Policy 38(6):2655–2663

Warren CR, Birnie RV (2009) Re-powering Scotland: wind farms and the 'energy or environment?' Debate. Scott Geogr J 125(2):97–126

Warren B, Christoff P, Green D (2016) Australia's sustainable energy transition: the disjointed politics of decarbonisation. Environ Innov Soc Trans 21:1–12

Warren CR, Lumsden C, O'Dowd S, Birnie RV (2005) 'Green on green': public perceptions of wind power in Scotland and Ireland. J Environ Plan Manag 48(6):853–875

Wilson GA, Dyke SL (2016) Pre-and post-installation community perceptions of wind farm projects: the case of Roskrow Barton (Cornwall, UK). Land use Policy 52:287–296

Wolsink M (2010) Contested environmental policy infrastructure: Socio-political acceptance of renewable energy, water, and waste facilities. Environ Impact Assess Rev 30(5):302–311

Wolsink M (2012) The research agenda on social acceptance of distributed generation in smart grids: renewable as common pool resources. Renew Sustain Energy Rev 16(1):822–835

Wolsink M (2013) The next phase in social acceptance of renewable innovation. EDI Q 5(1):10–13

Wüstenhagen R, Wolsink M, Bürer MJ (2007) Social acceptance of renewable energy innovation: an introduction to the concept. Energy Policy 35(5):2683–2691

Xue B, Ma Z, Geng Y, Heck P, Ren W, Tobias M, Fujita T et al (2015) A life cycle co-benefits assessment of wind power in China. Renew Sustain Energy Rev 41:338–346

Yaqub O, Castle-Clarke S, Sevdalis N, Chataway J (2014) Attitudes to vaccination: a critical review. Soc Sci Med 112:1–11

Dr. McCarthy is a Senior Lecturer of Marketing at James Cook University (JCU). She has over 80 publications, including articles spanning A, B and C rated journals that are ranked by the Australian Business Dean's Council (ABDC). Her research interests include consumer behaviour, sustainable food consumption, food waste, and networks. She has published papers on consumer decision-making around food purchasing, shopping and food waste behaviours, in journals such as the *Australasian Marketing Journal, Rural Society and the British Food Journal.*

Prof. Lynne Eagle is a Professor in Marketing at James Cook University, Townsville, Australia and an Adjunct Professor at Canterbury University, Christchurch, New Zealand. She holds a Ph.D. from the University of Auckland in her native New Zealand. Her research interests centre on marketing communication effects and effectiveness, including the impact of persuasive communication on children; the impact of new, emerging and hybrid media forms and preferences for the use of formal and informal communications channels, along with trans-disciplinary approaches to sustained behaviour change in social marketing / health promotion / environmental protection campaigns. She has published in a wide range of academic journals, including the Journal of Advertising and European Journal of Marketing, led the development of both Marketing Communications and Social Marketing texts and contributed several book chapters for other texts as well as writing commissioned expert papers and presenting numerous research papers at international conferences.

Chapter 2
Finding the Sweet Spot: Can Social Marketing Encourage Cane Farmers to Change Practices for Water Quality Outcomes in the Great Barrier Reef Catchment?

Laura Dunstan, Lisa Pulman, Nathan Waltham, and Sue Sargent

Abstract The Great Barrier Reef is under increasing pressure from the cumulative effects of climate change, land-based runoff, coastal land-use change and direct use impacts. Many water quality behaviour change programs fail to achieve sustained change using traditional methods. NQ Dry Tropics received funding from the Queensland Government Reef Water Quality Program to trial methods with 14 farmers at two wetland sites, to achieve water quality outcomes in the sugarcane farming sector of the Lower Burdekin. These methods sought to increase the adoption of practice changes to reduce excess irrigation and nutrients in the runoff flowing into nearby wetlands using an adapted community-based social marketing methodology to understand an individual's perceived benefits and barriers to practice change, and create tailored extension, engagement, and communication tools. This paper will evaluate the effectiveness of using social marketing techniques to change the attitudes and behaviour of the sugarcane farmers.

Keywords Great Barrier Reef · Water quality · Sugarcane farmers · Practice change · Wetlands · Community-based social marketing · Behaviour change · Wetland restoration

---

L. Dunstan (✉) · L. Pulman
NQ Dry Tropics, Townsville, Australia
e-mail: Laura.c.dunstan@gmail.com

N. Waltham
Centre for Tropical Water and Aquatic Ecosystem Research, James Cook University, Townsville, Australia

S. Sargent
Sue Sargent Consulting Services, Bundaberg, Australia

R. Hay et al. (eds.), *Broadening Cultural Horizons in Social Marketing*,
https://doi.org/10.1007/978-981-15-8517-3_2

2.1 Introduction

Worldwide, changes to coastal land use continue to threaten ecosystem conditions, either through reduced water quality conditions (hypoxia, eutrophication, sedimentation), lost sensitive wetland habitats (e.g. mangroves, seagrass, and freshwater wetlands) and altered hydrology (reducing frequency, duration and timing of connectivity) (Beesley et al. 2012; Gopal 2013; Herbert et al. 2015; Nagelkerken et al. 2015; Waltham et al. 2019). Managers are implementing land use practice changes, and, in combination with restoration efforts, it is hoped that these efforts together will reverse or halt further ecosystem decline (Weinstein and Litvin 2016; Zedler 2016). While these efforts are important and increasingly necessary, data that measure the successful return on the investment are lacking (Sheaves et al. 2014; Creighton et al. 2015; Creighton et al. 2016; Litvin et al. 2018).

In Australia, coastal land use change following the expansion of agricultural production, urban sprawl and industrial development have resulted in the loss of sensitive and important coastal wetland ecosystems (Creighton et al. 2016). An example of this coastal ecosystem change has been along the Great Barrier Reef floodplains (Waltham et al. 2019). Water quality continues to be a major threat to the health and resilience of the Great Barrier Reef (GBR) (Waterhouse et al. 2016a).

North Queensland's Burdekin River delivers the second-highest contribution of nutrient loads (nitrogen and phosphorus) to the GBR (25%) and the Burdekin region is one of the highest priorities for investment to reach targets established in the Reef 2050 Water Quality Improvement Plan (State of Queensland 2017; Waters et al. 2014; Waterhouse et al. 2016a). The region is, therefore, important for research investment, especially in the area of social and ecological change for water quality outcomes. Wetlands in the Lower Burdekin play an important function in the landscape and for the health of the GBR (Eberhard et al. 2017; State of Queensland 2016b; Waltham and Fixler 2017). They assist in the removal of excess nutrients, chemicals, pollutants, and sediment before they reach sensitive seagrass communities and near-shore coral reefs (Eberhard et al. 2017; Schaffelke et al. 2017). Regional development, notably the sugarcane industry, has resulted in altered wetland hydrology, loss of seasonality, loss and disturbance of native species and habitat, water quality decline, increases in pest species and excessive weed growth (Waltham et al. 2017; Waterhouse et al. 2017).

Most of the region's wetlands are permanently inundated and choked with weeds associated with the transport of irrigation water and excess runoff from sugarcane farms. This affects the wetland's health, viability, and water quality treatment function (Waltham et al. 2019). Research and modelling show that water quality outcomes for the GBR can be improved by restoring the health and function of coastal wetlands (Adame et al. 2010; Adame et al. 2012; State of Queensland, 2016a; Haynes et al. 2007; Johnston et al. 1990; Johnson 1991). An important part of this restoration is reducing the impact of runoff by improving sugarcane farming practices. Significant reductions in nitrogen and excess water runoff can be achieved through improving irrigation and nitrogen efficiency (NQ Dry Tropics 2016).

Sugarcane farmers in the Burdekin are faced with increasing pressure and public scrutiny regarding the effect their industry has on the water quality of the GBR. Traditional behaviour change techniques including information campaigns, financial incentives, and nudging are not always enough to achieve long-term improved behaviour change (Eagle et al. 2016; Gneezy et al. 2011; Greiner and Gregg 2011; Steg 2008). Strategies that facilitate voluntary behaviour change by focussing on the target groups attitudes, beliefs and perspectives have been shown to increase the potential for sustained positive behaviour change (Bryant et al. 2009; Haldeman and Turner 2009; Neiger et al. 2003).

NQ Dry Tropics is the natural resource management (NRM) body for the Burdekin Dry Tropics region. Australia has 56 regional NRM organisations that are a mixture of government agencies and non-government organisations that manage Australia's natural resources. NQ Dry Tropics were funded by the Queensland Government Reef Water Quality Program to trial methods to increase adoption of practice change that would also, by connections in the landscape, improve the ecological function of coastal wetland ecosystems and contribute to improving Reef water quality outcomes. McKenzie-Mohr's community-based social marketing (CBSM) was used as a framework to test the hypothesis that social marketing methodology can encourage practice change by removing barriers to a behaviour change while simultaneously enhancing the benefits. This paper will focus on the effectiveness of using CBSM to change attitudes and behaviour of the 14 sugarcane farmers that participated in the project.

2.2 Methods

A project team was formed with representatives from the state government, universities, science organisations, industry, and community groups who guided, advised and collaboratively delivered the project with NQ Dry Tropics.

The project was based at two wetland sites, the Horseshoe Lagoon (HSL) site located in the Burdekin River Irrigation Area (six farmers with 325 hectares under cane production) and the Lilliesmere Lagoon (LL) site in the Delta Area (eight farmers with 566 hectares under cane production) (see Fig. 2.1). Wetland selection was informed by the Burdekin Water Quality Improvement Plan 2016 and learnings from previous NQ Dry Tropics projects. A water quality evaluation program for each wetland site was established as part of the National Environmental Science Program Tropical Water Quality Hub Project 3.2.1 'Science evaluation of management solutions for coastal wetland systems repair'.

Farmer selection was based on proximity to the wetlands and farm runoff entering the wetlands. Cane farmers were initially invited to participate by the Burdekin Bowen Integrated Floodplain Management Advisory Committee (BBIFMAC) field officer who is a well known, respected and trusted member of the cane farming community. The engagement focused on the incentives (such as receiving free water quality monitoring data while maintaining farmer data and identity anonymity) and the project's motivations (increasing farmer profitability and productivity while

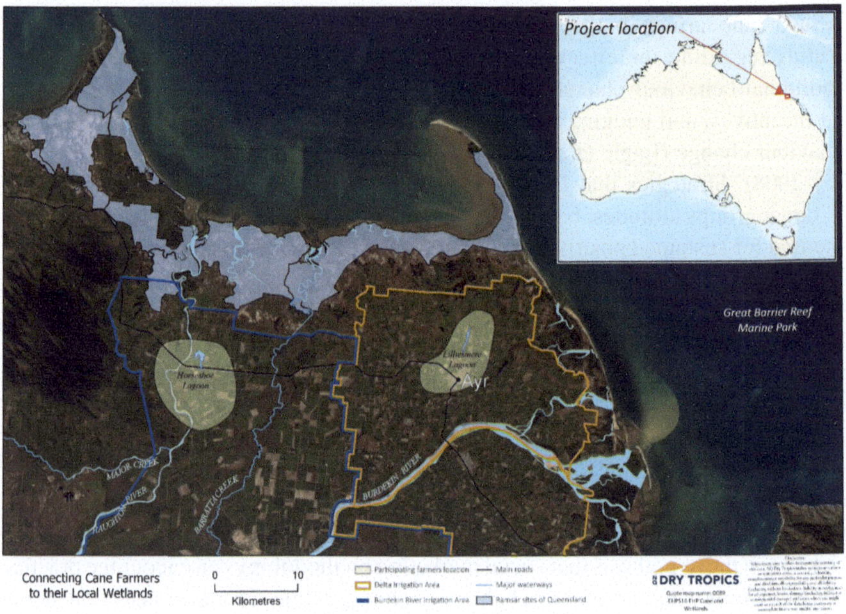

Fig. 2.1 Project sites, horseshoe lagoon and lilliesmere lagoon, QLD

improving the health and function of wetlands). Fourteen farmers were selected, each with different environmental understanding and knowledge, farm management practices, soil types, and, most importantly, different levels of interest and willingness to change their farming practices. Some had participated in previous practice change projects, and for others, this was their first time. This selection method differed from other projects, which invite farmers based on their interest and ability to participate (often using cash incentives) rather than location (with incentives based on data and information).

During face-to-face engagements between the NQ Dry Tropics and the 14 farmers, a social Knowledge, Attitudes, Skills, and Aspirations (KASA) survey was conducted in March 2017 and June 2019. An industry-standard, Paddock to Reef Water Quality Risk Framework Questions (P2RQ) were asked in July 2017, July 2018 and June 2019, a communication and engagement survey was conducted in November 2017 and November 2018, and evaluation surveys were conducted after individual events. The social survey had seven quantitative and four qualitative questions. The results were analysed by members of the project team. An external social scientist conducted a statistical analysis of the quantitative questions.

The initial social and P2RQ surveys were used to establish baseline farming practices, current understandings, attitudes, awareness, and beliefs regarding cane farming, wetlands, the GBR and the connections between them. To create a social study of barriers and benefits of practice change, this information was collated and interpreted to form the baseline information for each farmer.

Information from the surveys, social study, and a research and literature review were used to create an engagement strategy. It was initially proposed that three to four practice changes per farmer based on farmer interest, appropriateness for the farm, and what would achieve the best water quality outcome would be selected. These changes would be aligned with the relevant P2RQ, and then individual barriers and benefits to undertaking the changes would be determined. The CBSM impact, probability, and penetration research (a research method to select behaviours) would then be undertaken to determine the best practice change per farmer. However, during the initial one-on-one engagements, it was noted that many farmers did not know what they wanted to change, felt they had their practices as good as they could get them (with the information they had available) and needed monitoring data to help them identify where efficiencies could be made. As a result, the practice change was determined after the first 6 months of monitoring data was provided to the farmers. This ensured that the changes were appropriate for the individual farmer's needs and the farm's characteristics.

The engagement strategy for each farmer was analysed to identify the relevant CBSM strategies and the farmers' information was deidentified to ensure confidentiality. The strategy stated the practice change to be trialled on each property, the behaviour change tools for extension, education and communication, and the CBSM tool they aligned with (commitment, social norms, social diffusion, prompts, communication, incentives, and convenience).

A social monitoring, evaluation, and project adaptation framework was developed to guide the design of communication and engagement tools and events, evaluate the effectiveness of these activities, and adapt the engagement strategy every six months.

2.3 Results

2.3.1 Knowledge Attitudes, Skills and Aspirations (KASA) Surveys

The results were analysed in multiple ways, comparing the second set of answers (2019) to an individual's first survey responses (2017) to evaluate the effectiveness of the project's CBSM strategy on participants (see Table 2.1).

Results were also analysed to compare the efficacy of the CBSM strategy based on whether a landholder had previously participated in a practice change project or not (see Table 2.2), and the effects of participant age, farm size, the location of the property and attendance at communication events.

Q1: Do cane farmers play a role in improving the health and function of local wetlands and downstream areas?

The responses changed as a result of being part of the project. The post-scores were significantly higher than the pre-scores for this question, indicating that the farmers were more willing to agree that they play a role in the health and function of

Table 2.1 Paired t-tests from Knowledge, Attitudes, Skills and Aspirations (KASA) social surveys 2017 and 2019, identifying responses to questions that changed as a result of being a part of the project (N = 14)

KASA social survey question	Mean* ± Standard deviation		$t(13)$	p
	Pre-project	Post-project		
Q1: Do cane farmers play a role in improving the health and function of local wetlands and downstream areas?	5.08 ± 1.00	6.25 ± 0.42	3.79	0.002
Q2: On a scale of 1–7, how would you describe your level of contribution to improving the health of local wetlands?	5.1 ± 1.00	5.83 ± 0.77	2.38	0.033
Q3: On a scale of 1–7, how would you describe your level of knowledge about the effects of nutrients and irrigation runoff on the wetlands?	5.0 ± 1.52	5.62 ± 0.90	2.40	0.032
Q4: On a scale of 1–7, how much of a negative impact do excess nutrient and irrigation runoff have on wetlands?	4.71 ± 1.07	4.50 ± 0.73	-0.68	0.506
Q5: On a scale of 1–7, how important are the wetlands to the health of the Great Barrier Reef?	5.93 ± 0.91	6.25 ± 0.69	1.91	0.079
Q7: On a scale of 1–7, what do you think the quality of water in the Lagoon (either HSL or LL) is currently like?	4.86 ± 1.17	5.63 ± 0.56	2.18	0.048
Q8: Are cane farmers sufficiently acknowledged for their role in protecting local wetlands?	2.36 ± 1.22	2.33 ± 1.26	-0.07	0.945

*For repeated measures, paired t-tests were used. Responses for the two participants that did not complete the post-survey were replaced with the mean of all responses. For independent groups, independent t-tests were used

local wetlands and downstream areas as a result of being part of this project. At the end of the project, 100% of the farmers involved agreed that they play a role. The farmers' responses to the question also varied according to participation in previous projects. Prior to starting the project, respondents that had previously participated were significantly less likely to acknowledge the role that they play in improving the health and function of local wetlands compared with those that have not. However, there were no differences in the post-scores.

Q2: On a scale of 1–7, how would you describe your level of contribution to improving the health of local wetlands?

The responses changed as a result of being part of the project. The post-scores were significantly higher than the pre-scores. This indicates that over the life of the project, participating farmers believed that they had increased their contribution to improving local wetland health. At the end of the project, they felt that, on average, they have a medium to high level of contribution to improving the local wetland health. The extent to which respondents felt that they contributed to the health of local wetlands did not vary as a function of their previous participation in projects

Table 2.2 Independent t-tests from Knowledge, Attitudes, Skills and Aspirations (KASA) social surveys 2017 and 2019, to evaluate the effect of previous participation in a practice change project on participant survey responses (N = 14)

KASA social survey question	Response	Mean* ± Standard deviation		*t*(12)	p
		Previous participant	Not previously participated		
Q1: Do cane farmers play a role in improving the health and function of local wetlands and downstream areas?	Pre-survey	4.57 ± 0.95	5.86 ± 0.14	3.25	0.006
	Post-survey	6.18 ± 0.14	6.32 ± 0.22	0.63	0.542
Q2: On a scale of 1–7, how would you describe your level of contribution to improving the health of local wetlands?	Pre-survey	5.00 ± 2.00	5.14 ± 0.48	0.24	0.814
	Post-survey	6.40 ± 0.31	5.26 ± 0.20	−4.21	0.001
Q3: On a scale of 1–7, how would you describe your level of knowledge about the effects of nutrients and irrigation runoff on the wetlands?	Pre-survey	5.86 ± 0.48	4.14 ± 2.81	−2.50	0.028
	Post-survey	6.06 ± 0.18	5.16 ± 1.08	−2.19	0.049
Q4: On a scale of 1–7, how much of a negative impact do excess nutrient and irrigation runoff have on wetlands?	Pre-survey	5.00 ± 0.86	4.43 ± 1.29	−1.07	0.337
	Post-survey	4.50 ± 0.92	4.50 ± 0.25	0.00	1.00
Q5: On a scale of 1–7, how important are the wetlands to the health of the Great Barrier Reef?	Pre-survey	6.29 ± 0.57	5.57 ± 0.95	−1.53	0.076
	Post-survey	6.46 ± 0.26	6.04 ± 0.68	−1.17	0.264
Q7: On a scale of 1–7, what do you think the quality of water in the Lagoon (either HSL or LL) is currently like?	Pre-survey	4.29 ± 1.57	5.43 ± 0.62	2.04	0.064
	Post-survey	5.73 ± 0.15	5.52 ± 0.50	−0.70	0.495
Q8: Are cane farmers sufficiently acknowledged for their role in protecting local wetlands?	Pre-survey	2.14 ± 1.81	2.57 ± 1.29	0.64	0.531
	Post-survey	2.19 ± 1.81	5.48 ± 1.59	0.41	0.689

*For repeated measures, paired t-tests were used. Responses for the two participants that did not complete the post-survey were replaced with the mean of all responses. For independent groups, independent t-tests were used in the pre-survey, however, it did change in the post-survey; such that those that had previously participated in projects scored higher in the post-survey compared with those that have not.

Q3: On a scale of 1–7, how would you describe your level of knowledge about the effects of nutrients and irrigation runoff on the wetlands?

The responses changed as a result of being part of the project. The post-scores were significantly higher than the pre-scores. This indicates that the farmers increased their knowledge of the effects of nutrients and irrigation runoff as a result of being part of this project. The extent to which respondents reported feeling knowledgeable varied as a function of past participation. Prior to the project commencing, respondents that had not previously participated in projects reported feeling less knowledgeable compared to those that had involved in past projects. This difference persisted post the project's completion, whereby respondents that have participated in previous projects were still reporting higher levels of knowledge compared to those that had not.

Q4: On a scale of 1–7, how much of a negative impact do excess nutrient and irrigation runoff have on wetlands?

The responses did not change as a result of being part of the project. The post-scores were not significantly different from the pre-scores. This indicates that the project did not impact on the degree to which the farmers felt that excess nutrient and irrigation has a negative impact on wetlands. On average, they were neutral about the extent to which nutrient and irrigation have a negative impact on wetlands. The extent to which respondents felt that nutrient and irrigation runoff have a negative effect on wetlands did not vary as a function of past project participation prior to this project commencing, nor post the project completion.

It was noted that in the pre-project survey that farmers were unsure of the level of impact from excess nutrient and irrigation runoff. In the post-project survey, many of them remained neutral, saying it depended on the level of runoff and weather conditions. They also said they learned from the wetland evaluation publication that their local wetland was in good health for fish habitat. Some farmers from the LL area also said the large fish kill at the beginning of the year in the lagoon (due to natural causes) revealed that very high numbers of large barramundi could live there; therefore, it must be good. They connected the proof of the relatively good health of their wetland with their response to this question. This could indicate an increased connection to their wetland, even though their scores did not change as a result of the project.

Q5: On a scale of 1–7, how important are the wetlands to the health of the Great Barrier Reef?

The responses did not change as a result of being part of the project. The post-scores were not significantly different from the pre-scores. On average, they felt that wetlands are important to the health of the GBR, but this was not impacted by participating in the project. The extent to which respondents felt that wetlands are important to the health of the GBR did not vary as a function of past project participation, neither prior to this project commencing, nor post the project completion.

Q7: On a scale of 1–7, what do you think the quality of water in the lagoon (either HSL or LL) is currently like?

The responses changed as a result of being part of the project. The post-scores were significantly higher than the pre-scores. This indicates that the farmers felt the quality of the water in their lagoon increased over the life of the project and/or their understanding of water quality increased. At the end of the project, they felt that the water in the lagoon was of medium to high quality. The extent to which respondents rated the quality of the water in the lagoon did not vary as a function of past project participation, neither prior to this project commencing nor post the project completion.

Q8: Are cane farmers sufficiently acknowledged for their role in protecting local wetlands?

The responses did not change as a result of being part of the project. The post-scores were not significantly different from the pre-scores. This indicates that the extent to which the farmers feel acknowledged for their role in protecting local wetlands did not increase as over the life of the project. On average, they do not think that they are sufficiently acknowledged for their role in protecting local wetlands. The extent to which respondents felt acknowledged did not vary as a function of past project participation, neither prior to this project commencing nor post the project completion.

Responses were analysed in relation to age and farm size. Farm size did not affect any of their responses, but age affected two:

Prior to the project, younger respondents were significantly less likely to acknowledge the role farmers play in improving the health and function of local wetlands compared with older respondents. However, the influence of age was not evident in the post-project question.

The extent to which respondents felt that nutrient and irrigation runoff has a negative effect on wetlands did not vary as a function of age prior to this project commencing but did post the project completion. Younger participants felt that nutrient and irrigation have less of a negative impact compared with older participants.

Farmer responses did not vary according to location, with no evidence to suggest that any of the measures varied depending on the farm location ($p's > 0.069$).

Finally, farmer responses did not vary according to attendance at communication events ($p's > 0.091$), although with only three people indicating that they had attended a farm event, it was not possible to compare the results.

2.4 Final Practice Change Survey Results

2.4.1 Did The Project Lead To Long-Term Practice Changes?

Out of the 12 people that completed the survey, eight indicated that they would continue with the practice change. Out of these eight respondents, four (50%) suggested that they would not have made the change without being a part of the project.

Out of the eight respondents that indicated that they would continue with the practice change: six (75%) have or will recommend the change to other landholders.

One hundred per cent of the respondents (regardless of whether they indicated that they would make long-term changes or not) scored above the mid-point in response to the question 'Please rate your level of knowledge now (after participating in the trial) about the effects of nutrients and irrigation runoff on the wetlands?'

Only two respondents indicated that they did not have a clear understanding of the practice change and how making this change has resulted in improved water quality. It is worth noting that both of those respondents indicated that they were not intending on implementing long-term practice change as a result of being part of the project.

2.4.2 What Resources Did Respondents Report As Being Important?

The two highest-ranked resources that the farmers felt helped them achieve a high degree of knowledge were: BBIFMAC and Industry Groups. This was followed by their fellow farmers ('over the fence dissemination of information').

This was very similar to the rankings for the next question, which asked about the most important resources in terms of achieving a practice change. Again, the two highest-ranked sources were: BBIFMAC and Industry Groups.

With regards to achieving practice changes, the highest-ranked source was 'None of the above, we just got on and did it' with 100% of respondents scoring this option on or above the mid-point of the scale. Private Agronomists were also ranked as very important by the majority of the cane growers, and this was followed by fellow cane growers ('over the fence'). The lowest-ranked resources were consistently online resources and fact sheets.

2.4.3 What Impact/s Will The Practice Change Have?

Respondents were asked the extent to which they felt their practice change would have a positive or negative impact on a range of things. A low score would indicate a negative impact; a high score indicates a positive impact. Across all impacts, the responses were very positive. Not one single respondent indicated that they felt that their practice change would have a negative impact on any of the areas listed. Overall, however, the highest scores were given to the 'external' benefits (i.e. water quality, the health of the GBR, and local wetlands).

All of the respondents believed that they had the appropriate tools and resources required to implement practice changes after completing the program. From the 12 people that completed the survey, 100% of respondents answered at or above the mid-point of the scale with a mean score of 4.25, and the most common response was 5 'Very much increased my interest/excited to try new things'.

2.5 Dissolved Inorganic Nitrogen Results

The 14 farmers had their runoff monitored over two crop cycles, and 13 of the farmers investigated a practice change (11 changes were unique). The results of the practice changes were processed through the P2R Projector Tool, which provided modelled estimates of 2,582 kg dissolved inorganic nitrogen, 158 tonnes of soil and 1.907grams of pesticides saved. BBIFMAC reported an overall reduction in the average NO_3–N mg/l level in the runoff samples collected and analysed in the 2019 harvest season when compared with the 2018 harvest season.

Over the 2.5 years, changes in water quality and the wetland fauna were assessed. Water quality on the floodplain is largely dependent on rainfall and hydrology. This is a response common for most floodplains in the GBR. This means that altering the water quality conditions and hydrology can have dramatic and lasting impacts on the values and services expected for the wetlands on this floodplain. The wetlands on the lower Burdekin floodplain provide important habitat for a range of aquatic flora and fauna species, with many fish species, for example, requiring connection to downstream estuary areas to complete lifecycle stages (Waltham and Fixler 2017; Waltham et al. 2019). However, the wet season rainfall conditions experienced during this study resulted in limited, if any, possible connection with downstream estuaries, which was evidenced by the distinct lack of juvenile barramundi in Horseshoe Lagoon (HSL), Sheep Station creek and Lilliesmere creek. Recruitment of juvenile barramundi during this program period was observed in other creeks that have improved floodplain connectivity compared with those examined here. Dissolved oxygen is complex and varies over spatial-temporal scales. In both clear and turbid wetland sites, dissolved oxygen will reach critical levels, which increases the hypoxia exposure risk to fish—which under severe conditions will lead to fish kills (Perna et al. 2012). The hydrology of HSL is complex and further challenging without access to data on water extraction rates and groundwater interaction. During this study, the lagoon did seem to remain stable, with the exception of rainfall events where the lagoon rises and falls quickly.

Water quality in wetlands is strongly influenced by the presence of invasive aquatic weeds. As long as nutrient-rich water from the catchment continues to pass through the floodplain, the weeds will continue needing maintenance. There is an urgent need to ensure that any restoration projects on the floodplain develop a long-term strategy to keep up with the maintenance, or risk that the restoration site will return to a poor condition. Overall, this project has been an important first step in linking the data collected together and telling the connectivity story to cane farmers using science-based evidence (Waltham et al. 2019).

2.6 Discussion

2.6.1 Knowledge Attitudes, Skills and Aspirations (KASA) Surveys

Although limited by sample size, the application of an individualised CBSM strategy to participating cane farmers had a statistically significant effect on the (perceived) knowledge gained by cane farmers participating in the project. This included the role that they play in the improving the health and function of local wetlands and downstream areas, the effects of nutrient and irrigation runoff on the wetlands and the importance of the wetlands to the health of the GBR. Participating cane farmers also believed that they had increased their level of contribution to improving the health of the local wetlands over the course of the project and improved the quality of water in both HSL and LL. However, cane farmer attitudes towards the negative impacts that excess nutrients and irrigation runoff have on wetlands were not statistically changed by the project with farmers still reporting that they were undecided or only had a moderate amount of knowledge to support this response. Participants' responses to the acknowledgement that cane farmers receive for their role in protecting local wetlands were also statistically unchanged by the project with cane farmers reporting that they continued to believe that acknowledgement was low to very low.

While these findings support the use of a CBSM strategy to enhance practice change in cane farmers, it is clear that there are still some embedded attitudes to be overcome. These include an acknowledgement of the negative impacts on wetlands by excess nutrients and irrigation runoff, and the need for participants to be more widely acknowledged for their participation in practice change projects and in restoring the health of local wetlands.

Using the project experience, a Cane Farmer Practice Change Engagement for Wetland Health model (see Fig. 2.2) was developed to guide future projects looking to implement this strategy.

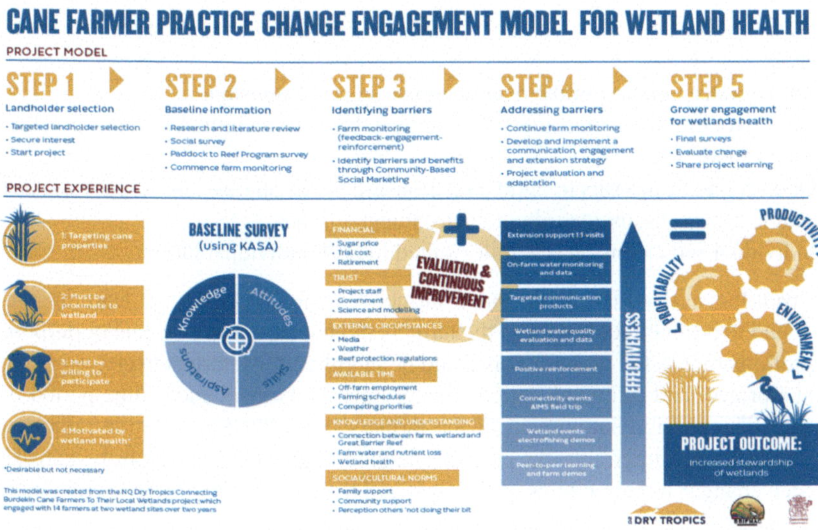

Fig. 2.2 Cane farmer practice change engagement model for wetland health

2.7 Practice Change

Fourteen farmers had their farm runoff monitored over two crops, 13 farmers investigated a practice change (11 changes were unique), and eight made a commitment to implement their new practice on suitable parts of their farm. The farmer who chose to undertake baseline monitoring for both years stated an intent to trial a new practice at the end of the project. Twelve farmers have increased (perceived) knowledge about their practices on water quality leaving their farm, and eight are confident that their new practice will improve it.

The project produced significant improvements to the farmers' (perceived) knowledge, understanding, and awareness of the connections between cane farming and wetlands. Farmers understood their role in the health and function of local wetlands and downstream areas and felt their level of contribution to improving local wetland health had increased because of the project. Knowledge about the effects of nutrients and irrigation runoff on the wetlands also increased.

Twelve of the 14 farmers featured in a publication including their photo, the changes made on their properties and eight made a public commitment to implement their new practice on other suitable parts of their farms.

The most effective of the methods trialled (to achieve reef water quality outcomes through increasing adoption of practice change and improving the ecological function of coastal wetland ecosystems) were:

- Farmers trialling a small-scale practice of their choice (typically strip trial) or receiving water monitoring data from standard practices, so they can identify areas where efficiencies can be made.

- Water quality and volume data collected—to compare the standard practice to the trial.
- Timely interpretation of the data with one-on-one extension.
- Targeted communication tools that address barriers identified through baseline social surveying.
- One-on-one farm visits to build a relationship based on trust and respect that can enable robust discussions about environmental issues.
- Individualised extension schedules suited to each individual farmer.

Given the wetland (location) based farmer selection, peer-to-peer learning was not found to be an effective tool as the farmers did not have a shared goal or purpose for participating in the project. This could be modified for future delivery.

The project had a limited sample size of only 14 farmers and a time period of 2 years. The public commitment of the farmers in the final publication indicated that over a longer timeframe, the project's methodology would successfully increase practice changes to larger areas of the farm and improve water quality leaving their farms.

This model of project delivery would work well in other agricultural commodities that are connected to downstream wetlands or waterways via runoff. Although relevant to the GBR catchments, this model could also be applied to other locations with altered land use and water bodies in close proximity.

2.8 Dissolved Inorganic Nitrogen

The wetland evaluation program could not directly attribute to changes on-farm and downstream, over the 2 years, due to the scale of the challenge (extensive floodplain with agricultural production), and highly variable rainfall patterns that were experienced across the floodplain during the 2018–2019 season. At the project scale, observable changes in the marine water quality conditions were not expected, given the volume of water from the floodplain and catchment during the wet season. However, the data generated, and the engagement with the community has conceivably started a trajectory of change in farming and raised awareness that floodplains, though modified, provide important values and services to the local region. Those farmers participating in the Meet the Scientists days (see Fig. 2.3) acknowledged the connection between land practices and impacts on local waterway quality and condition and offshore ecosystems. To this end, this project is the first step towards linking this information together and delivering the connectivity story to cane farmers using science-based evidence.

Fig. 2.3 Photo from the meet the scientists day

2.9 Conclusion

Fourteen sugarcane farmers were engaged in a broad suite of CBSM-informed strategies that led to changes in antecedents to behaviour.

CBSM was used to identify barriers and benefits to specific practice changes; develop an engagement strategy that utilises tools such as commitment, social norms, social diffusion, prompts, communication, incentives, and convenience; pilot the strategy with the 14 farmers at two wetland sites; and evaluate the effectiveness of the engagement strategy every six months.

The 13 practice changes made on the cane farms reduced dissolved inorganic nitrogen, soil and pesticides from entering the two wetland sites. The wetland monitoring program could not detect water quality changes in the wetlands and downstream areas that could be directly attributed to the reduced runoff from the farms.

A practice change engagement model for wetland health was developed from the learnings of the project and can be used to upscale the project to achieve greater outcomes in the adoption of practice changes for wetland health and improved water quality.

Acknowledgements **OGBR Reef Water Quality Program**—Billie Gordon, Megan Bickle and Tracy Schultz

BBIFMAC—Arwen Rickert, Dennis Stubbs and Gracie White

JCU—TropWater Team, Nathan Waltham NESP Tropical Water Quality Hub*

Behaviour Innovation—John Pickering and Jinny Hong

JCU—College of Business Law and Governance: Lynne Eagle and Rachel Hay https://nesptropical.edu.au/index.php/round-2-projects/project-2-1-3/

CSIRO—Nadine Marshall and Matt Curnock

NQ Dry Tropics staff.

*Nathan Waltham was supported by funding provided from the Australian Government's National Environmental Science Program Tropical Water Quality Research Hub—Project 3.3.2 https://nesptropical.edu.au/index.php/project-3-3-2/.

References

Adame MF, Virdis B, Lovelock CE (2010) Effect of geomorphological setting and rainfall on nutrient exchange in mangroves during tidal inundation. Mar Freshw Res 61(10):1197–1206

Adame MF, Reef R, Herrera-Silveira JA, Lovelock CE (2012) Sensitivity of dissolved organic carbon exchange and sediment bacteria to water quality in mangrove forests. Hydrobiologia 691(1):239–253

Beesley L, King AJ, Amtstaetter F, Koehn JD, Gawne B, Price A, Nielsen DL, Vilizzi L, Meredith SN (2012) Does flooding affect spatiotemporal variation of fish assemblages in temperate floodplain wetlands? Freshw Biol 57:2230–2246

Bryant CA, Brown KRM, McDermott RJ, Debate RD, Alfonso ML, Baldwin JA, Phillips LM (2009) Community-based prevention marketing: a new framework for health promotion interventions. In: DiClimente RJ, Crosby RA, Kegler M (eds) Emerging theories in health promotion practice and research. Wiley, San Francisco

Creighton C, Boon PI, Brookes JD, Sheaves M (2015) Repairing Australia's estuaries for improved fisheries production–what benefits, at what cost? Mar Freshw Res 66:493–507

Creighton C, Hobday AJ, Lockwood M, Pecl GT (2016) Adapting management of marine environments to a changing climate: a checklist to guide reform and assess progress. Ecosystems 19:187–219

Eagle L, Hay R, Farr M (2016) Harnessing the science of social marketing and behaviour change for improved water quality in the GBR: background review of literature. James Cook University, Townsville, Australia, p 110

Eberhard R, Thorburn P, Rolfe J, Taylor B, Ronan M, Weber T, Flint N, Kroon F, Brodie J, Waterhouse J, Silburn M, Bartley R, Davis A, Wilkinson S, Lewis S, Star M, Poggio M, Windle J, Marshall N, Hill R, Maclean K, Lyons P, Robinson C, Adame F, Selles A, Griffiths M, Gunn J, McCosker K (2017) Scientific consensus statement 2017: a synthesis of the science of land-based water quality impacts on the Great Barrier Reef, Chapter 4: management options and their effectiveness. State of Queensland

Gneezy U, Meier S, Rey-Biel P (2011) When and why incentives (don't) work to modify behavior. J Econ Perspect 25:191–209

Gopal B (2013) Future of wetlands in tropical and subtropical Asia, especially in the face of climate change. Aquat Sci 75:39–61

Greiner R, Gregg D (2011) Farmers' intrinsic motivations, barriers to the adoption of conservation practices and effectiveness of policy instruments: empirical evidence from northern Australia. Land Use Policy 28(1):257–265. https://doi.org/10.1016/j.landusepol.2010.06.006

Haldeman T, Turner JW (2009) Implementing a community-based social marketing program to increase recycling. Soc Mark Q 15(3):114–127. https://doi.org/10.1080/15245000903154618

Haynes D, Brodie J, Waterhouse J, Bainbridge Z, Bass D, Hart B (2007) Assessment of the water quality and ecosystem health of the Great Barrier Reef (Australia): conceptual models. Environ Manage 40(6):993–1003

Herbert ER, Boon P, Burgin AJ, Neubauer SC, Franklin RB, Ardón M, Hopfensperger KN, Lamers LP, Gell P (2015) A global perspective on wetland salinization: ecological consequences of a growing threat to freshwater wetlands. Ecosphere 6:1–43

Johnston CA, Detenbeck NE, Niemi GJ (1990) The cumulative effect of wetlands on stream water quality and quantity. A Landsc Approach Biogeochem 10(2):105–141

Johnston CA (1991) Sediment and nutrient retention by freshwater wetlands: effects on surface water quality. Crit Rev Environ Sci Technol 21(5–6):491–565

Litvin SY, Weinstein MP, Sheaves M, Nagelkerken I (2018) What makes nearshore habitats nurseries for nekton? an emerging view of the nursery role hypothesis. Estuaries Coasts 41:1539–1550

Mcleod E, Szuster B, Hinkel J, Tompkins EL, Marshall N, Downing T, Wongbusarakum S, Patwardhan A, Hamza M, Anderson C, Bharwani S, Hansen L, Rubinoff P (2016) Conservation organizations need to consider adaptive capacity: why local input matters. Conserv Lett. https://doi.org/10.1111/conl.12210

Nagelkerken I, Sheaves M, Baker R, Connolly RM (2015) The seascape nursery: a novel spatial approach to identify and manage nurseries for coastal marine fauna. Fish Fish 16:362–371

Neiger BL, Thackeray R, Barnes MD, McKenzie JF (2003) Positioning social marketing as a planning process for health education. Am J Health Stud 18(2/3):75–81

NQ Dry Tropics (2016) Burdekin region water quality improvement plan 2016. NQ Dry Tropics, Townsville

Perna CN, Cappo M, Pusey BJ, Burrows DW, Pearson RG (2012) Removal of aquatic weeds greatly enhances fish community richness and diversity: an example from the burdekin river floodplain, tropical Australia. River Res Appl 28:1093–1104

Schaffelke B, Collier C, Kroon F, Lough J, McKenzie, Ronan M, Uthicke S, Brodie J (2017) Scientific consensus statement 2017: a synthesis of the science of land-based water quality impacts on the Great Barrier Reef, Chapter 1: the condition of coastal and marine ecosystems of the Great Barrier Reef and their responses to water quality and disturbances. State of Queensland

Sheaves M, Brookes J, Coles R, Freckelton M, Groves P, Johnston R, Winberg P (2014) Repair and revitalisation of Australia' s tropical estuaries and coastal wetlands: opportunities and constraints for the reinstatement of lost function and productivity. Marine Policy 47:23–38

State of Queensland (2016a) Great Barrier Reef water science taskforce—final report. Prepared by the Great Barrier Reef water science taskforce, and the office of the Great Barrier Reef. Department of Environment and Heritage Protection, Brisbane

State of Queensland (2016b) Wetlands in the Great Barrier Reef catchments management strategy 2016–2021. State of Queensland

State of Queensland (2017) Reef 2050 water quality improvement plan 2017–2022. Published by the Reef Water Quality protection plan secretariat, 27 July 2018. https://www.reefplan.qld.gov.au/about/assets/reef-2050-water-quality-improvement-plan-2017-22.pdf

Steg L (2008) Normative concerns and environmental behavior. Int J Psychol 43(3–4):179

Waltham N, Fixler S (2017) Aerial herbicide spray to control invasive water hyacinth (*Eichhornia crassipes*): water quality concerns fronting fish occupying a tropical floodplain wetland. Trop Conserv Sci 10:1940082917741592. https://doi.org/10.1177/1940082917741592

Waltham NJ, Wegscheidl C, Smart JCR, Volders A, Hasan S, Waterhouse J (2017) Scoping land conversion options for high DIN risk, low-lying sugarcane, to alternative use for water quality improvement in wet tropics catchments. Reef and Rainforest Research Centre Limited, Report to the National Environmental Science Programme, Cairns, p 142

Waltham NJ, Burrows D, Wegscheidl C, Buelow C, Ronan M, Connolly N, Groves P, Audas DM, Creighton C, Sheaves M (2019) Lost floodplain wetland environments and efforts to restore connectivity, habitat, and water quality settings on the Great Barrier Reef. Front Mar Sci. https://doi.org/10.3389/fmars.2019.00071

Waters DK, Carroll C, Ellis R, Hateley L, McCloskey J, Packett R, Dougall C, Fentie B (2014) Modelling reductions of pollutant loads due to improved management practices in the Great Barrier Reef catchments-whole of GBR, vol 1. Department of Natural Resources and Mines, Technical Report. ISBN (978-1-7423-0999)

Waterhouse J, Schaffelke B, Bartley R, Eberhard R, Brodie J, Star M, Thorburn P, Rolfe J, Ronan M, Taylor B, Kroon F (2017) 2017 scientific consensus statement summary. State of Queensland.

Table 1. Type your title here. Obtain permission and include the acknowledgement required by the copyright holder if a table is being reproduced from another source

Waterhouse J, Duncanson P, Attard S, Bartley R, Bristow K, Brodie J, Coppo C, Davis A, Dougall C, Lewis S, Smith M, Star M, Roberts A, Wilkinson S (2016a) Spatial prioritisation and extension of information to manage sediment, nutrient and pesticide runoff in the Burdekin NRM region. A report for the Department of Environment and Heritage Protection, Prepared on behalf of NQ Dry Tropics, Townsville, Australia

Waterhouse J, Brodie J, Lewis S, Audas DM (2016b) Land-sea connectivity, ecohydrology and holistic management of the Great Barrier Reef and its catchments: time for a change. Ecohydrol Hydrobiol 16:45–57. https://doi.org/10.1016/j.ecohyd.2015.08.005

Weinstein MP, Litvin SY (2016) Macro-restoration of tidal wetlands: a whole estuary approach. Ecol Restor 34:27–38

Zedler JB (2016) What's new in adaptive management and restoration of coasts and estuaries? Estuaries Coasts 1–21

Laura Dunstan was the Program Manager of Waterways, Wetlands and Coasts at NQ Dry Tropics, the Natural Resource Management body for the Burdekin Dry Tropics in North Queensland. Laura recently began a new role as the Project Manager of Community Partnerships at the Great Barrier Reef Foundation. Laura has a Bachelor of Science (Marine Biology) and a Master of Education for Sustainability from James Cook University and over 15 years of experience in developing and implementing projects to protect and restore the Great Barrier Reef World Heritage Area through engaging the community and fostering sustainable behaviour changes.

Lisa Pulman was the Project Officer in the Waterways, Wetlands and Coasts Team at NQ Dry Tropics. Lisa is an environmental scientist living in working in Queensland whose work is inspired by the natural world and where humans fit within it. She is passionate about inspiring others on a daily basis and seeks real world change. Lisa is currently working in wetland research, and uses the lessons learned from the Burdekin cane farmers in her interactions with landholders.

Nathan Waltham is an aquatic ecologist with a deep interest in coastal landscape ecology and processes. His working career comprises positions in government, heavy industry and academia, working on a range of projects including fisheries sustainability, nutrient processes, landscape restoration, stormwater quality and quantity, geomorphic processes, erosion and sediment control, and contamination. Dr. Waltham works closely with a range of government, industry, NRM and community groups that are delivering on-ground restoration outcomes. His applied approach to research means he is able to evaluate the success of projects, which ensures that the maximum ecosystem services are reached.

Sue Sargent is a marine biologist and Natural Resource Management (NRM) consultant based in Queensland, Australia. An engagement and communications specialist, Sue works with the Queensland Government, non-government organisations, NRM and industry groups supporting behaviour change to improve water quality outcomes for the Great Barrier Reef (GBR). Sue is the Chair of the Burnett Local Marine Advisory Committee (LMAC) in the Southern GBR (one of twelve LMACs that provide advice to the Great Barrier Reef Marine Park Authority) and a representative on the Reef 2050 Advisory Committee.

Chapter 3
Waste not Want not: A Co-Created Food Waste Pilot

Jeawon Kim, Kathy Knox, and Sharyn Rundle-Thiele

Abstract Food waste costs the city, and hence rate payers, money and impacts the wider environment through food production, transport, waste collection and landfills generating methane which contributes to a climate change. Australian Councils are charged a kilogramme per household fee for waste collection, and food waste is the heaviest component in a general waste bin. Most household food waste is avoidable and one way to bring about change and reduce waste is to start with food wasted in the home. In response to this, Social Marketing @ Griffith piloted a food waste campaign in partnership with Redland City Council. Waste Not Want Not was co-created with community and evaluated using a controlled design following delivery of the two-week pilot program. The campaign consisted of two core components: household packs and a two-week interactive shopping centre display that featured interactions with volunteer staff, daily cooking demonstrations and a chef cook-off event. The campaign successfully decreased self-reported food waste and improved self-efficacy in the program group while no significant differences were observed in the control group. This chapter outlines the potential of social marketing to change food waste behaviour, outlining an engaging co-created program.

Keywords Co-creation · Households · Campaign · Food waste · Outcome evaluation · Social marketing

3.1 Behaviour Change

3.1.1 The Problem

On average, Australian households waste about $4,000 worth of food every year (Barometer 2018). Over 5 million tonnes of food ends up in landfill, and methane created during the process of food waste disposal contributes 8% of global warming (FAO 2019). Although food businesses and industry must bear some responsibility,

J. Kim (✉) · K. Knox · S. Rundle-Thiele
Griffith University, Brisbane, Australia
e-mail: jeawon.kim@griffith.edu.au

R. Hay et al. (eds.), *Broadening Cultural Horizons in Social Marketing*,
https://doi.org/10.1007/978-981-15-8517-3_3

approximately 40–50% of food waste happens at the consumer level (Statistics 2019). Council rates which pay for waste management reach up to AU$20 billion per year (ARCADIS 2019), and the costs to dispose of food waste are borne by ratepayers (householders). To put it simply, we create food waste on a daily basis, and then we pay taxes to remove our waste.

To date, few programs have been implemented in Australia and globally to reduce food waste, and application of social marketing to combat food waste appears limited (Barr et al. 2011). A recent review study identified 23 food waste studies that sought to reduce household food waste (Kim et al. 2019a, b). One of these studies identified as a social marketing study indicating there is potential for social marketing to be more broadly applied to combat household food waste. Examination of food waste studies in the Kim et al. (2019a, b) review indicated that only two had some form of consumer insight (Dai et al. 2015; Devaney and Davies 2017); the rest of the campaigns were designed solely based on expert opinion (Schanes et al. 2018). This portrays lack of understanding of what householders value, which can mean funds invested to lower food waste are not maximised. By centring program development on program user preferences, higher adoption rates for the program are expected and, in turn, positive program outcomes are anticipated.

In summary, while some progress in food waste reduction has occurred there is further room for improvement given that expert views differ from those of program users and social marketing remains underutilised in food waste reduction interventions. To apply social marketing to food waste reduction, a philosophical process—Co-Create, Build, Engage (CBE) was applied in this study. CBE ensures that people and their environments are placed at the heart of program development, implementation and evaluation (Rundle-Thiele et al. 2019). By covering each stage of the model, this chapter outlines how a social marketing food waste program titled Waste Not Want Not was designed, implemented, and evaluated.

3.2 Social Change

Behaviour change encompasses a wide range of approaches, including education, training, enforcement, infrastructure, technology, urban planning, community development, environment promotion, and social marketing (Tapp and Rundle-Thiele 2016). To date, a strong presence of information–education focus was noted in food waste programs (Bernstad et al. 2013; Dai et al. 2016; Rousta et al. 2015). For example, door-knocking campaigns have been applied to educate and inform individuals to reduce food waste (Bernstad et al. 2013; Dai et al. 2015). Education is an important component in behaviour change, and meta-analyses in other contexts indicate that education changed behaviour in 8% of the population (Snyder et al. 2004). However, education-focussed interventions assume that motivation is strong and assume that people will reduce food waste solely on the basis of being fully informed about the environmental benefits of doing so. While education remains an important aspect of behavioural change, alternative strategies are needed given that information provision is not a sufficient condition for behaviour to occur (Tapp and

Rundle-Thiele 2016). Thus, education and information provision used in isolation are unlikely to lead to strong program outcomes.

In 2017, the Nobel Memorial Prize in Economic Sciences was awarded to behavioural economist Richard Thaler. Since then, behavioural economics and 'nudge' have been popularised as tools that can be applied to deliver behavioural change. For example, Reisch and Thøgersen (2015) claim that nudges are promising tools for promoting a broad range of pro-environmental and sustainable consumption behaviours. However, actual outcomes of nudge interventions vary in different contexts (Cooper 2017; Ebert et al. 2017; Lehner et al. 2016) underpinning the difficulties of predicting efficiency in food waste campaigns. Despite this, attempts to use nudge is prominent in literature. For example, a recent study surveyed consumer preference on 13 nudging items for food waste reduction has discovered that German households were most interested in receiving tips on shopping via email (von Kameke and Fischer 2018). This study recommended future research to validate and build on this result.

On the other hand, social marketing 'offers' value by inviting people to engage with a program and assumes that existing levels of motivation and knowledge are insufficient to generate changes in behaviour without an external boost. Permission of volitional choice enables effective delivery of values to achieve the desired behaviour change (David et al. 2019; Dibb et al. 2013). By articulating the central role of considering consumers' needs and wants, social marketing is effective in changing behaviour in various contexts, including alcohol consumption among teenagers (Dietrich et al. 2016), Koala conservation (David et al. 2019) and food waste (Devaney and Davies 2017). However, as mentioned earlier, application and validation of use of social marketing in food waste is limited. Thus, this chapter will demonstrate the effectiveness of enacting social marketing in changing food waste behaviour.

3.3 The CBE Process

Three steps including CBE were adopted to ensure both people and the environment were considered in program development for a pilot food waste campaign (see Fig. 3.1, Rundle-Thiele et al. 2019).

According to this model, **Co-creation** ensures that social marketers gain insight to inform program build. Within the co-creation stage, understanding (i.e. segmentation) and community empowerment (i.e. stakeholder orientation) are features underpinning the co-creation process. This stage of the model assures all competition (i.e. previous and current program tactics and themes) are evaluated by the people and should involve identification of key theoretical constructs to inform program design. The second stage, **Build** involves the development of a social marketing program that ensures programs extend beyond communication campaigns that inform and raise awareness for the problem being tackled (e.g. reducing food waste). Programs are designed and built to ensure that all elements of the marketing mix (i.e. 4Ps) are encompassed.

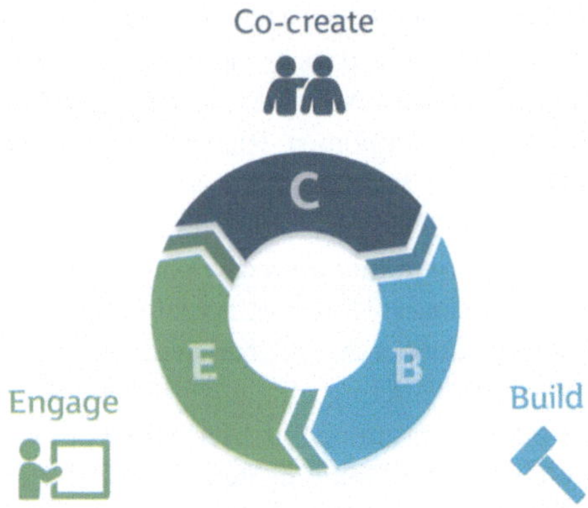

Fig. 3.1 The three step CBE process

The insight captured during the co-creation stage guides the program build giving clear strategic direction for the main approach needed to engage community in the program. In this stage, social marketers make sure they build a social marketing campaign to create 'exchange', thereby ensuring that people can voluntarily choose, which in turn delivers the desired health, social or environmental aims of the program (e.g. food waste reduction). This is where the true value of social marketing in the behaviour change landscape becomes clear. By delivering solutions that people value, and ideally they are willing to pay for, social marketing can deliver the outcomes sought. Importantly, given social marketing is based on voluntary exchange; we understand that the two parties are.

1. The people engaging in programs and
2. The civic and commercial institutions seeking to deliver outcome change.

Given the two parties come together to voluntarily transact, reactance is reduced. By voluntarily engaging in programs, participants not only benefit themselves, but also deliver benefits for the society within which they live.

The final stage, **Engage** allows social marketers to offer an exchange that delivers meaningful value. Within this stage, awareness of the program is raised to ensure community members know about the program and can choose to voluntarily engage with the program offering. The aim of the engagement stage can differ from increasing awareness to purchase and/or maintenance of the program. Recipients of the program could also vary from an individual to community and/or business. Since an ultimate aim of social marketing is in maintaining a desired behaviour, behavioural change is an on-going outcome of the CBE process.

3.4 The Program Background

In early 2016, Social Marketing @ Griffith partnered with Redland City Council as part of Council's Waste Reduction and Recycling Plan 2015–2020. The Council was charged a kilogramme per household fee for waste collection, and food waste was the heaviest component in a wheelie bin. Food waste was avoidable and one way to bring about change and reduce waste was to start with food wasted in the home. The City Council offered a modest budget to design and implement a community-level social marketing campaign to reduce household food waste behaviour. A formative research study was undertaken by Social Marketing @ Griffith over a twelve-month period. The formative research study included co-design and two online surveys. Formative research informed a desirable and acceptable social marketing food waste campaign for Redland City Council residents.

3.4.1 Co-Creation

A three step formative research program was conducted to gain insights, namely, (1) Co-design sessions ($n = 21$), (2) Online survey with the Australian population ($n = 414$) and a (3) Fridge audit with the Redland City Council residents ($n = 197$) (see Fig. 3.2).

The formative research study is described in detail in Kim et al. (2019a, b). The co-creation stage empowered Redland City Council householders to reveal their barriers to reducing their food waste, their opinions on the previous program tactics and to identify the foods they most commonly wasted. Over the series of studies, we gained the following insights:

Redland City Council householders:

- Perceived a lack of self-efficacy in cooking. Residents indicated they did not know what to do with the food that was left over in their fridges;
- Wanted a program that assisted them to understand how they could use the food that was already available in their fridge;
- Tended to waste a greater quantity of fruit and vegetables compared to other food groups.

These key insights guided program design. Through the co-creation stage, we learned that people often buy food following a recipe and frequently the amounts purchased for the recipe exceeded the amount they needed. This resulted in food left in the fridge. The fridge survey was employed in the co-creation stage to learn which foods were most commonly available in fridges in the Redlands City Council area to inform program build. Redland City Council residents had informed us that

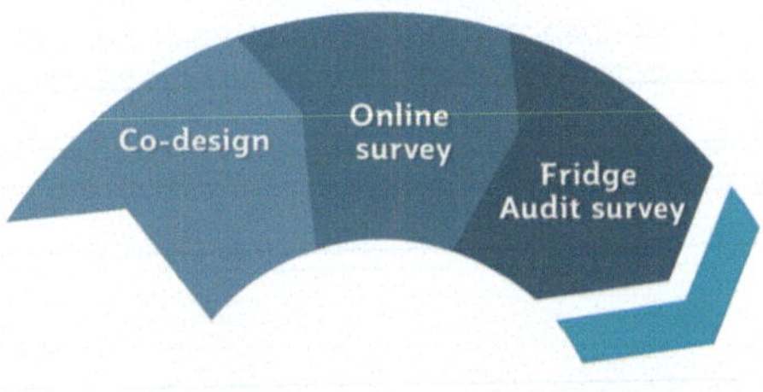

Fig. 3.2 Co-creation process for WNWN

targeting leftover reuse behaviour and creating a program that could assist them to understand how they could use the items left in their fridges may be an effective way to reduce household food waste.

3.4.2 Build

Based on insights captured from the co-creation stage, a social marketing pilot program that focussed on increasing self-efficacy in reusing leftovers and reducing the proportion of wasted fruit and vegetables was built in partnership with three local chefs (see Fig. 3.3). The Waste Not Want Not pilot campaign aimed to help residents understand how (1) what is already in the fridge can be turned into creative meals, (2) leftover food can be reused and (3) to avoid food spoilage through strategies including the use of shopping lists, menu planning and food storage tips (Kim et al. 2020).

Three local professional chefs created 16 leftover reuse recipes that consisted of food items that were commonly available in household fridges and therefore were most likely to be wasted (see Fig. 3.4 for two examples or visit 'https://www.wastenotwantnot-sm.com/' to download all for free).

The project team built a pilot campaign that featured daily cooking demonstrations and culminated in a Chef Cook-Off event to showcase the leftover reuse recipes. A two-week trial in one community in the Redland City Council area was feasible in terms of cost and time. We partnered with one local Shopping Centre located in the City Council area and nine local retailers who were located in close proximity to the shopping centre to provide additional support for the program. In addition, 18 volunteers were recruited from Griffith University ensuring that an interactive program atmosphere was delivered in the shopping centre over the two-week trial. By involving multiple partners into the program development phase, the program extended beyond a communication-only approach. For example, the shopping centre offered a space to locate the program (i.e. place); volunteers, chefs and the research

Fig. 3.3 Build process for WNWN

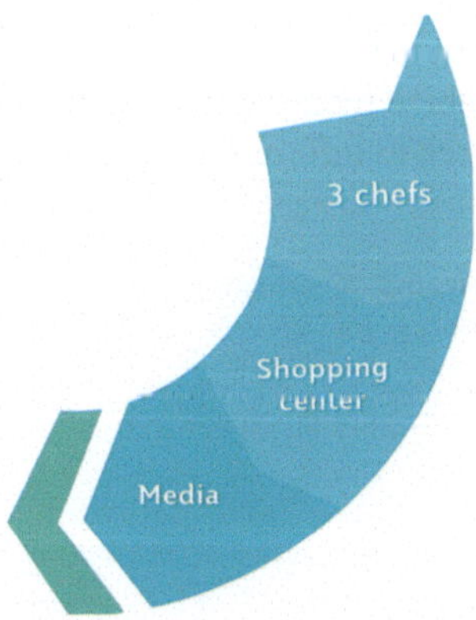

Fig. 3.4 Recipe cards: core product of WNWN

team were available to engage community members into a food waste conversation (people), and various supporting promotional elements were available to assist householders to minimise food waste (i.e. promotion). For example, price discounts at associated retailers, recipe cards, shopping planners, shopping bags and chopping boards were available. Recipe cards provided novel and innovative ways that household residents could use the contents that would be found in their fridges. Recipe names were linked to key barriers and motivations identified by residents during the co-creation stage.

Key motivations and barriers identified during co-creation included saving money and time, and lack of cooking inspiration. Hence, three promotional items were created to provide food waste prompts and reminders. A logo-printed reusable shopping bag, chopping board and magnetic shopping list notepad that could adhere to residents' fridges were key promotional items to support the food waste pilot program (see Fig. 3.5).

To extend awareness for Waste Not Want Not, a press release was issued by Griffith University. Additional communications included posters and leaflets, which were printed and distributed one week prior to the launch of the two-week shopping centre display. Extensive press coverage was received assisting the project team to raise awareness of food waste as an issue and use of leftovers as one means to combat food waste. For example, Channel 9 Queensland interviewed partner and daily cooking demonstrator chef Dominique Rizzo and the research team, and one

Fig. 3.5 Additional products

researcher was featured in the Australian Broadcasting Corporation Weekend News Breakfast program (see Fig. 3.6). Further to this, the research team visited a local radio station and promoted the program to the residents a week prior to the program delivery.

Additional promotion was undertaken by the research team and Redland City Council. For example, media platforms including the Council's Facebook page and local newspapers featured upcoming program details using the communication posters supplied by the research team (refer to top of Fig. 3.7). Posters were printed and located on advertising boards placed within the shopping centre (refer to bottom left of Fig. 3.7). Flyers containing key program details were handed out to shoppers prior to program delivery (refer to bottom right of Fig. 3.7), and these

Fig. 3.6 TV spots arising following press release

Fig. 3.7 Communication

were distributed via household packs to the 100 program households in the Waste Not Want Not evaluation study.

3.4.3 Engage

The Waste Not Want Not (WNWN) pilot social marketing campaign was delivered in early 2017 (see Fig. 3.8). During the Engage stage, the program focussed on delivering personal communication with residents. The interactive shopping centre display consisted of three main components: (1) daily food demonstrations delivered by chef Dominique Rizzo, (2) interaction with volunteers with a display fridge stocked with the top ten ingredients identified in the fridge survey undertaken during

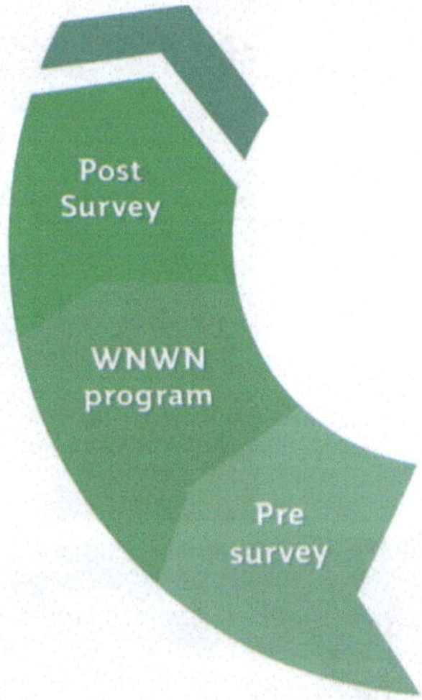

Fig. 3.8 Engage process of WNWN

the Co-Create stage and (3) a final chef cook-off event delivered by two local Chefs, namely Eugene Lee and Matthew Lee.

All residents surrounding the Stockland Cleveland Shopping Centre had access to the interactive shopping centre display over the two-week period. A mobile kitchen was installed in the middle of the shopping centre for the partnered local chef to demonstrate leftover reuse recipe daily (see Fig. 3.9) and for the final celebrity chef cook-off event.

The chef-showcased cooking tips and recipes were distributed each day to the shoppers between 9 a.m. and 3 p.m. Each day, members of the community were able to engage with the volunteers, watch and learn new cooking skills, taste and experience the novel recipes and take home the recipe card to try at home or share with family and friends.

Food samples were distributed to shoppers (see above Fig. 3.10), and the engagement fridge which contained the most commonly wasted and available food items in residents' fridges (see Fig. 3.11) was used as a discussion point. The food ingredients were supplied by partnering local businesses.

A Celebrity Chef Cook-Off (see Fig. 3.12) was planned as an additional means to engage residents assisting them to understand how they can make the most of the food they already have. The cook-off event was attended by a crowd of onlookers, and a judging panel was formed to cast their vote on the two speciality dishes. Recipe cards and food samples were distributed during the event. Considerable media attention was generated.

Fig. 3.9 Shopping centre display

3.5 Evaluation and Results

A repeated measures controlled outcome evaluation was undertaken for Waste Not Want Not. The follow-up survey also contained process evaluation measures. The results for each evaluation are discussed in the next section of the chapter.

Fig. 3.10 Daily food demonstrations

Fig. 3.11 Food samples and the engagement fridge

Fig. 3.12 The cook-off event

3.5.1 Process Evaluation. Program Reach and Overall Satisfaction

The process evaluation is reported in detail in Hodgkins et al. (2019). More than 5,000 food samples, 10,000 recipe cards and 300 program flyers were distributed over the two-week period of the program delivery. Over a half-million Australians were exposed to Waste Not Want Not through local and nationwide press coverage. Our partner, the Stockland Cleveland Shopping Centre experienced a 15% increase in foot traffic during program delivery. The overall satisfaction survey collected during the program revealed that almost 90% of the program recipients were satisfied with the program features. Moreover, the stakeholder survey showed that over 95% of partnered City Council staff members, chefs and the Shopping Centre staff members were very satisfied with the program delivery. Considering creation of mutual benefit between important parties is a key feature of social marketing; these results indicate potential for Waste Not Want Not to engage community in food waste conversations.

3.5.2 Evaluation Method 2. Pre-Post Survey

A repeated measures design was used to evaluate outcomes and assess the effectiveness of the program in changing households' self-efficacy and the proportion of wasted fruit and vegetables thrown away. Two groups were surveyed before and after the program: intervention households who received a household pack ($n = 91$) and control households ($n = 153$) who did not receive any Waste Not Want Not materials and were not exposed to the shopping centre display. Control households completing both studies were offered the chance to win one of ten \$50 retail vouchers. Data were gathered using a computer-assisted telephone survey by a market research agency. Validated scales were used to assess self-efficacy (Bandura 2010), and fruit and vegetable waste (Waste and Resources Action Programme 2007, 2008). For example, self-efficacy items measured the participant's level of confidence in planning meals, reusing leftover food and transforming leftovers into a different dish.

Independent sample t-tests were used to assess group differences for self-efficacy at baseline, and a chi-square test was used to examine the proportion of fruit and vegetables thrown away. Paired sample t-tests were conducted to determine the mean difference over time for the target and control groups for self-efficacy (see Fig. 3.13). Descriptive analysis on the scale 'throw away hardly any fruit and vegetables' was conducted to report the change in food waste over time for the program and control groups.

Overall, the survey results indicated that Waste Not Want Not achieved the desired positive behavioural change for residents in the local area. No statistically significant changes were observed in the control group, demonstrating an effect of participation in Waste Not Want Not. Taken together, this pilot campaign achieved its aims by successfully increasing cooking self-efficacy for the targeted audience and reducing the amount of fruit and vegetables discarded. The proportion of respondents reporting throwing hardly any fruit and vegetables out in the target (program) group increased by 41% with 44.5% reporting throwing hardly any fruit and vegetables out

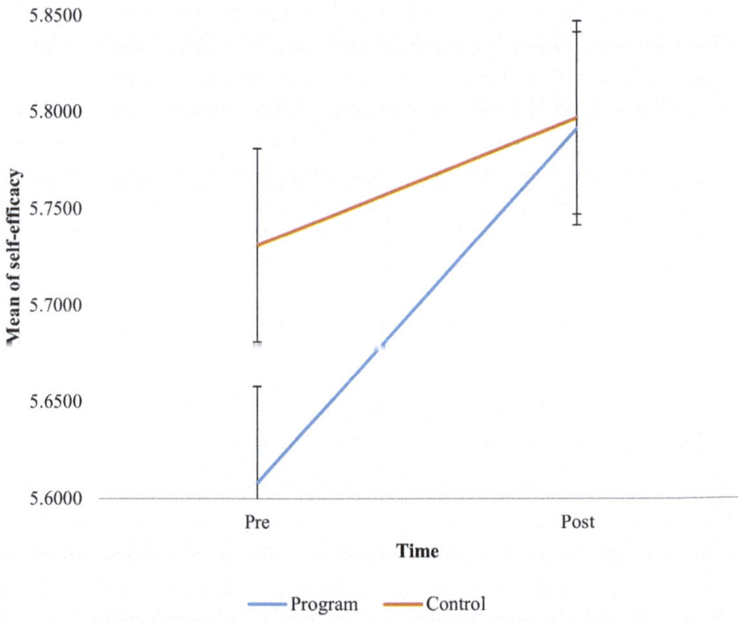

Fig. 3.13 Self-efficacy. Error bars represent standard error of the mean

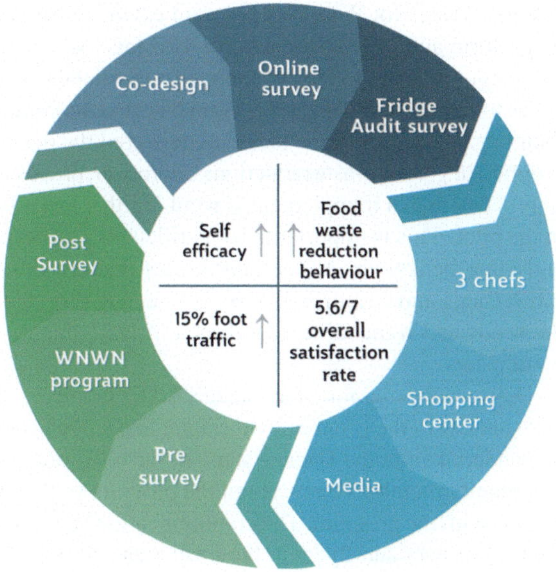

Fig. 3.14 Summary outcome of WNWN

at baseline compared to 62.6% reporting throwing hardly any fruit and vegetables out after WNWN (see Fig. 3.14).

In addition, 71.8% of respondents indicated that Waste Not Want Not encouraged them to think about food waste, and 31.8% reported that they reduced their food waste behaviour because of their exposure to the program. Approximately one half (49%) of respondents who had received the household pack had visited the shopping centre display and were able to rate the display. An approval rating of 5.6 (out of a maximum 7) was given to both the overall program and the shopping centre display indicating a high positive response. All components were perceived to be useful and were positively received. The outcome evaluation of the campaign from a social marketing benchmark perspective is detailed in Kim et al. (2020).

3.6 Discussion

The CBE framework delivers a procedural process demonstrating when activities were undertaken. CBE delivers a process that can be applied to develop, implement and evaluate a social marketing program. To ensure co-creation, we ran series of studies using mixed methods to gain insight to inform program build. Within the co-creation stage, three studies were conducted to expand our understanding of our target audience. As suggested by the model, all existing competitions (i.e. nine previous and current food waste programs) were evaluated by our target audience and we identified self-efficacy as a key theoretical construct to inform program design, implementation and evaluation. Through the Build stage, we involved multiple stakeholders to ensure that the Waste Not Want Not program extended beyond communication campaigns that inform and raise awareness. In this stage, we made sure that a social marketing campaign was built to create 'exchange'. Redland City Council residents voluntarily chose to engage with Waste Not Want Not, and the controlled evaluation demonstrates their engagement with the program delivered the desired environmental aims of the program (e.g. food waste reduction). Moreover, process evaluation identified areas for improvement that if implemented would further extend on program success identified in the pilot. This is important given criticisms may leveraged at policy, public health and other behaviour change approaches that are involuntary (French and Gordon 2019; Ranci and Arlotti 2019). By co-creating programs and building them to satisfy the needs and wants of residents, behavioural change can be achieved for program participants.

Delivering programs that engage is key to success in social marketing. Without engagement in the program, we cannot expect behavioural change to occur. Activities undertaken in the Engage stage of the CBE process focus attention on ensuring community members know about the program so that they can choose to voluntarily engage with the program. The communication activities which included press, placement of posters and flyers in the local area and use of social media assisted to gain awareness in the Redland City Council community. Through high engagement rates (e.g. 5,000 food samples trialled), Waste Not Want Not achieved the program aim of increasing self-efficacy in reusing leftovers and reducing fruit and vegetable waste behaviour. Insights gained during co-creation informed the design and development of the program and led to successful engagement of all stakeholders involved in the program. These key findings suggest that social marketing offers an additional behavioural change approach to reduce food waste behaviour.

Previous studies have called for strengthening stakeholder involvement to ensure that solutions identified by consumers can be implemented (Buyucek et al. 2016; Hodgkins et al. 2019). WNWN provides an example outlining how diverse stakeholders were involved in program design and implementation. For example, three chefs built a series of recipes focussed on contents most commonly found in Redland City Council resident fridges. Social marketing programs involve interdisciplinary collaboration (Dahl 2010; Grier and Bryant 2005; Lee and Kotler 2011; Truong 2014), and through this, residents came to see Chef Dominique Rizzo's daily cooking demonstrations and this delivered a 15% increase in foot traffic increase for project partner Stockland, which would have increased retail outlet turnover. Waste Not Want Not was supported by retailers in the shopping centre, some of whom supplied goods to the project at no cost. Stakeholder involvement delivered essential support and extended reach for the project team.

The current study was restricted to a downstream focus, focussing program efforts on households and fruit and vegetables, and this is considered to be a limitation. Previous researchers have argued that social marketers need to focus attention away from the individual, and they need to move upstream (Hoek and Jones 2011; Wymer 2011). In the context of organic food waste, an intervention for the upstream would be a policy to restrict food waste or the installation of a community waste station, while midstream efforts may involve retailers and restaurants in food waste reduction efforts. A broader system focus is recommended for future social marketing programs targeting food waste, given that food waste occurs right across the food supply chain. A field experiment is recommended to test whether a wider systemic intervention can deliver enhanced food waste reduction outcomes when compared to a downstream-only focussed intervention (such as WNWN).

3.7 Conclusions

This paper contributes to our understanding of how the CBE process can be applied to reduce food waste through a controlled, repeated measures assessment of an audience-targeted food waste pilot campaign. The WNWN program which focussed on leftover reuse effectively encouraged household food waste reduction behaviour and resulted in significant increases in self-efficacy for the intervention group, but not the control group. Behaviour change was achieved through delivering support tools that residents could use to reduce food waste.

References

ARCADIS (2019) Tackling Australia's food waste. Retrieved from https://www.environment.gov.au/system/files/pages/25e36a8c-3a9c-487c-a9cb-66ec15ba61d0/files/national-food-waste-baseline-final-assessment.pdf

Bandura A (2010) Self-efficacy. Corsini Encycl Psychol 1–3

Barometer FH (2018) Food waste findings. Retrieved from https://www.rabobank.com.au/food-waste-findings-from-financial-health-barometer/

Barr S, Gilg A, Shaw G (2011) Helping people make better choices: exploring the behaviour change agenda for environmental sustainability. Appl Geogr 31(2):712–720

Bernstad A, la Cour JJ, Aspegren A (2013) Door-stepping as a strategy for improved food waste recycling behaviour–evaluation of a full-scale experiment. Resour Conserv Recycl 73:94–103

Buyucek N, Kubacki K, Rundle-Thiele S, Pang B (2016) A systematic review of stakeholder involvement in social marketing interventions. Australas Mark J (AMJ) 24(1):8–19

Cooper EJ (2017) To nudge or not to nudge: promoting environmentally beneficial behaviors

Dahl S (2010) Current themes in social marketing research: text-mining the past five years. Soc Market Quart 16(2):128–136

Dai Y, Gordon M, Ye J, Xu D, Lin Z, Robinson N, Harder M (2015) Why doorstepping can increase household waste recycling. Resour Conserv Recycl 102:9–19

Dai Y, Lin Z, Li C, Xu D, Huang W, Harder M (2016) Information strategy failure: personal interaction success, in urban residential food waste segregation. J Cleaner Prod 134:298–309

David P, Rundle-Thiele S, Pang B, Knox K, Parkinson J, Hussenoeder F (2019) Engaging the dog owner community in the design of an effective koala aversion program. Soc Market Quart 25(1):55–68

Devaney L, Davies AR (2017) Disrupting household food consumption through experimental HomeLabs: Outcomes, connections, contexts. J Consum Cult 17(3):823–844

Dibb S, Carrigan M, Zainuddin N, Russell-Bennett R, Previte J (2013) The value of health and wellbeing: an empirical model of value creation in social marketing. Eur J Market

Dietrich T, Rundle-Thiele S, Schuster L, Drennan J, Russell-Bennett R, Leo C, Connor J (2016) Segmenting Australian high school students utilising a two-step cluster analysis: differential

effects following the game on know alcohol program. In: Rediscovering the Essentiality of Marketing. Springer, pp 413–414
Ebert P, Schubert T, Freibichler W (2017) Nudge Management: Wie Führungskräfte kluges Selbstmanagement anstoßen. Zeitschrift Fuehrung Und Organisation 86(2):82–86
FAO (2019) Food loss and food waste. Retrieved from https://www.fao.org/food-loss-and-food-waste/en/
French J, Gordon R (2019) Strategic social marketing: for behaviour and social change. SAGE Publications Limited
Grier S, Bryant CA (2005) Social marketing in public health. Annu Rev Public Health 26:319–339
Hodgkins S, Rundle-Thiele S, Knox K, Kim J (2019) Utilising stakeholder theory for social marketing process evaluation in a food waste context. J Soc Mark
Hoek J, Jones SC (2011) Regulation, public health and social marketing: a behaviour change trinity. J Soc Market 1(1):32–44
Kim J, Rundle-Thiele S, Knox K (2019) Systematic literature review of best practice in food waste reduction programs. J Soc Market
Kim J, Rundle-Thiele S, Knox K, Burke K, Bogomolova S (2019) Consumer perspectives on household food waste reduction campaigns. J Cleaner Prod 118608.
Kim J, Rundle-Thiele S, Knox K, Hodgkins S (2020) Outcome evaluation of an empirical study: food waste social marketing pilot. Soc Mark Q 1524500420918690
Lee NR, Kotler P (2011) Social marketing: influencing behaviors for good. Sage Publications
Lehner M, Mont O, Heiskanen E (2016) Nudging–A promising tool for sustainable consumption behaviour? J Cleaner Prod 134:166–177
Ranci C, Arlotti M (2019) Resistance to change. The problem of high non-take up in implementing policy innovations in the Italian long-term care system. Policy Soc 38(4):572–588
Reisch LA, Thøgersen J (2015) Research on sustainable consumption: Introduction and overview. Handbook of research on sustainable consumption, pp 1–16
Rousta K, Bolton K, Lundin M, Dahlén L (2015) Quantitative assessment of distance to collection point and improved sorting information on source separation of household waste. Waste Manage 40:22–30
Rundle-Thiele S, David P, Willmott T, Pang B, Eagle L, Hay R (2019) Social marketing theory development goals: an agenda to drive change. J Mark Manage 35(1–2):160–181
Schanes K, Dobernig K, Gözet B (2018) Food waste matters-A systematic review of household food waste practices and their policy implications. J Cleaner Prod 182:978–991
Snyder LB, Hamilton MA, Mitchell EW, Kiwanuka-Tondo J, Fleming-Milici F, Proctor D (2004) A meta-analysis of the effect of mediated health communication campaigns on behavior change in the United States. J Health Commun 9(S1):71–96
Statistics AB (2019) Waste account, Australia, experimental estimates, 2016–17. Retrieved from https://www.abs.gov.au/ausstats/abs@.nsf/mf/4602.0.55.005
Tapp A, Rundle-Thiele S (2016) Social marketing and multidisciplinary behaviour change. Key issues, interdisciplinary approaches and future directions, Beyond Behav Change, p 135
Truong VD (2014) Social marketing: a systematic review of research 1998–2012. Soc Market Quart 20(1):15–34
von Kameke C, Fischer D (2018) Preventing household food waste via nudging: an exploration of consumer perceptions. J Cleaner Prod 184:32–40
Waste and Resources Action Programme (2007) Survey of packaging with potential to reduce food thrown away at home report. Retrieved from https://www.wrap.org.uk/food-drink/business-food-waste/report/survey-packaging-potential-reduce-food-thrown-away-home-report
Waste and Resources Action Programme (2008) Helping consumers reduce fruit and vegetable waste: final report. Retrieved from https://www.wrap.org.uk/sites/files/wrap/WRAP%20RTL044-001%20Final%20report.pdf
Wymer W (2011) Developing more effective social marketing strategies. J Soc Market 1(1):17–31

Jeawon Kim is a Post-doc Research Fellow at Green Technology Center, Seoul and Waste Not Want Not campaign manager (Social Marketing @ Griffith) with experience in formative research and design, management, implementation, and evaluation of social marketing programs. Jeawon's PhD research focused on sustainability behaviours including leftover food reuse, reducing food waste, and increasing effective waste reduction behaviour from a social marketing perspective. Jeawon aims to challenge practice through scientific discovery. Her current research aims to facil-iate a global climate governance through an active collaboration with climate-finance resources to combat climate change.

Kathy Knox is a research psychologist with expertise in human development, attitude and health behaviour change across a range of settings. Kathy is a member of the Menzies Health Institute Queensland and Research Fellow with Social Marketing @ Griffith leading a rapidly developing program of empirical research on social behaviour change. Kathy's research focus is on communities and sustainability behaviours including health education and communication, reducing food waste and excess water use, increasing recycling and organ donation. Kathy has direct experience in designing and conducting community-based interventions and has evaluated local and national interventions to change awareness, knowledge and cultures of behaviour.

Sharyn Rundle-Thiele is Director, Social Marketing @ Griffith (GBS) and Editor-in-Chief, Journal of Social Marketing. Drawing on her commercial marketing background Sharyn's research focuses on applying marketing tools and techniques to change behaviour for the better. Research partners in 2018 include Defence Science and Technology Organisation, Australian Defence Force, Queensland Catholic Education Commission, Enhance Research, Department of Environment and Heritage Protection (Qld), Department of Health (Qld), and Redland City Council. Sharyn's research is published in more than 120 books, book chapters and journal papers.

Chapter 4
It Takes a Village: Co-creation and Co-design for Social Media Health Promotion

Brennan Linda, Annika Molenaar, Anouk Sherman, Shinyi Chin, Mike Reid, Helen Truby, and Tracy McCaffrey

Abstract Health promotion campaigns using Social Media have emerged as a potential way to change behaviour in a large, diverse group of people. However, successfully connecting with and engaging the target audience is still a key challenge faced by health professionals and social marketers. This chapter examines the use of co-creation and co-design processes as ways to engage young people in Social Media interventions. This method involves a 'Wicked Problems' studio (five-day workshop), where young adults (university students) across multidisciplinary areas and a variety of stakeholders (e.g. health organisations, government bodies, academics, etc.) collaborate to problem-solve. Using this method, we were able to understand what works and what does not work when it comes to co-creation with young adults. Towards a broader outlook, the results from this study will translate into practice through the development of a how-to guide on co-creation and co-design as a procedure for key stakeholders such as government bodies and health organisations.

Keywords Co-creation · Co-design · Social media · Healthy eating · Young adults · Obesity

---

B. Linda (✉) · S. Chin
School of Media & Communication, RMIT University, Melbourne, Australia
e-mail: linda.brennan@rmit.edu.au

A. Molenaar · A. Sherman · H. Truby · T. McCaffrey
Department of Nutrition, Dietetics & Food, Monash University, Notting Hill, Australia

M. Reid
Finance & Marketing, RMIT University School of Economics, Melbourne, Australia

H. Truby
School of Human Movement and Nutrition Sciences, The University of Queensland, Brisbane, Australia

R. Hay et al. (eds.), *Broadening Cultural Horizons in Social Marketing*,
https://doi.org/10.1007/978-981-15-8517-3_4

4.1 Introduction

Globally, 52% of adults aged 18 years and over were overweight or obese in 2016 (World Health Organization 2016). In young Australian adults, defined as aged 18–24 years old (National Health and Medical Research Council 2015), the prevalence of being over a healthy weight increased from 39% in 2014–2015 to 46% in 2017–2018 (Australian Bureau of Statistics 2019). Young Aboriginal and Torres Strait Islander Australians (hereafter referred to as Aboriginal people) have a higher prevalence, with 59% of young women well above a healthy weight (Australian Bureau of Statistics 2013). Overweight and obesity are well-established risk factors for the development of metabolic and psychological illnesses across the lifespan (Lobstein et al. 2004; Lim et al. 2012). The trajectory of weight gain is the fastest during young adulthood (Allman-Farinelli et al. 2008) much of which is driven by modifiable behaviours, particularly unhealthy food choices and behaviours combined with a lack of physical activity (Ashton et al. 2016; Ashton et al. 2017). Food choice-related behaviours of highest concern in this target group include (a) low fruit and vegetable intake (Menozzi et al. 2015; Chapman et al. 2016) and (b) high intake of sugar-sweetened beverages (Bleich et al. 2009; Australian Bureau of Statistics 2015; Sui et al. 2017) and as such provide a target for interventions.

Behavioural barriers to healthier eating for young adults include the food environment not having healthy food and beverage choices readily available, a perception of the relative expense of healthy foods, current social norms and peer influences that normalise and encourage consumption of unhealthy foods, lack of knowledge and skills to prepare healthy foods and lack of time and facilities to plan, shop and prepare healthy foods (Palermo et al. 2014; Paxton 2015; Munt et al. 2017). For some young Aboriginal Australians, an additional challenge is finding a balance between an easily available Westernised highly processed diet and obtaining access to a more traditional diet (Brimblecombe et al. 2014; Kerpan et al. 2015).

Young adulthood is a pivotal time where people are establishing behaviours and skills, creating an identity and establishing life-long values and habits (such as learning cooking skills, increasing nutrition knowledge) (Nelson et al. 2008) and these young adults are particularly vulnerable to food industry marketing (Freeman et al. 2016). However, there is an opportunity in young adulthood to create an environment that facilitates making healthier lifestyle choices, including broadening access to healthier food options, supporting physical activity and reduction of sedentary behaviour.

4.2 Current Challenges in Leveraging Social Media for Health Promotion

Social Media is defined as web-based communication channels dedicated to input from the community, interaction, social networking, content sharing and collaboration (Rouse 2016). Globally, Social Media use has reached 2.8 billion users and continues to grow (Hootsuite Media Inc. and We Are Social Ltd. 2017). Up to 55% of Australian young adults use Social Media more than five times a day with only 2% in one report stating they never use Social Media (Yellow 2018) with Facebook being the most commonly used platform (Greenwood et al. 2016; Sensis 2016). Reaching young people using traditional channels of health promotion and social marketing, such as television and radio, is increasingly challenging (Hebden et al. 2012). Social Media presents an opportunity to reach and engage young adults in a more targeted and effective way (Vaterlaus et al. 2015). Social Networking Sites (SNS) also provide a tool for engaging with young Aboriginal Australians where Social Media use is also widespread (Rice et al. 2016; Walker et al. 2019).

A significant advantage of using Social Media as a vehicle for health promotion messages is the opportunity to use customised messages for specific target subgroups in a way that has not previously been available to health practitioners (Dooley et al. 2013). Social marketers have long known that to impact and engage effectively with consumers requires prior segmentation of the target market in order to tailor specific messages and delivery platforms to population subgroups (Slater 1996). Consumer segmentation is increasingly being used as a strategy to improve effectiveness of social marketing campaigns (Moss et al. 2009), including for addressing obesity (Wills et al. 2015) and alcohol use (VicHealth 2013). Psycho-behavioural consumer segments, incorporating behaviours, attitudes, knowledge, opinions, goals and lifestyles (Maibach et al. 1996; Slater 1996) are more indicative of how and why people behave in the way they do and their likely reactions to interventions or communication efforts (Brennan et al. 2016). Moving beyond the 'one size fits all' approach" enables a focus on targeted message content and media channels, which may be critical to behaviour change (Kreuter et al. 2000; Kubacki et al. 2017).

Nutrition and health professionals, government, and non-government health organisations are attempting to leverage Social Media to reinforce nutrition-related behaviours during young adulthood (Capurro et al. 2014; Kite et al. 2016; Klassen et al. 2018b). Benefits of using Social Media for health promotion include increased reach and interaction (Moorhead et al. 2013). Despite this, health promotion campaigns run via Social Media have traditionally struggled to reach and engage with large numbers of people and have suffered from low engagement and high attrition rates (Thackeray et al. 2008; Waters et al. 2009; Maher et al. 2014; Lim et al. 2016a, b; Klassen et al. 2018b). These campaigns must also contend with sophisticated marketing campaigns of corporate brands and food industries (Mangold and Faulds 2009; Freeman et al. 2015). Regulation of Social Media content is lacking, and food industry is known to manipulate young adults' vulnerabilities using marketing tactics, including celebrity endorsements designed to sell an illusion of beauty and success from their products (Freeman et al. 2016).

4.3 Social Media Advertising and Influencing

Over 80% of Australian young adults are accessing Social Media platforms at least once per day, commonly during times of the day associated with choosing foods, such as first thing in the morning and at lunch (Yellow 2018). Corporate brands have been successful in using the features offered by Social Media to communicate with the public (Freeman et al. 2014). Social Media feeds can include both posts from brands the individuals follow as well as advertisements. Consumers are not only willingly engaging with brands, but also disseminating brand content to their friends, thereby increasing brand reach (Muntinga et al. 2011). Food industry brands selling Fast Moving Consumer Goods (FMCG) that are typically energy (calorie) dense and nutrient poor are known to target adolescents and young adults on Social Media, and this exposure to advertisements appears to influence their attitudes and intentions (Buchanan et al. 2017) and possibly behaviours (Fishbein et al. 2003). As young adults have a higher intake of sugar-sweetened beverages and are more likely to consume fast foods compared with other age groups (Australian Bureau of Statistics 2015), it is possible that the constant exposure to food brands via Social Media may be increasing or reinforcing these eating patterns.

Strategies used by those marketing on Social Media that may increase engagement include posting visually attractive content and linking posts to consumption cultures and contexts (Lim et al. 2016a, b; Klassen et al. 2018a). Alcohol and sexual health research has also found that consistency of posting and interaction between brands and users are associated with increased success (Burton et al. 2013; Veale et al. 2015). Social Media personalities, or 'Influencers', are Human Brands (Thomson 2006) and have been identified as being a strategic and powerful opportunity for product promotion (Uzunoğlu and Kip 2014). What makes these Influencers so successful appears to be their ability to engage with users on Social Media and develop a level of trust (Uzunoğlu and Kip 2014; Liu et al. 2015). Social Media Influencers have overtaken traditional celebrities in their ability to influence purchasing behaviour, as users find them more credible and relatable (Nielsen Company 2012) despite the relationship being parasocial in nature (Horton and Richard Wohl 1956). However, traditional celebrity figures still appear to have a strong influence on lifestyle behaviours and some are even seen as 'experts' in these areas (Hoffman and Tan 2015; Hoffman et al. 2017). Companies are utilising the influence of Social Media Influencers and celebrities to enhance their brand, and are using strategies used by Social Media Influencers in their own campaigns (Chapman 2012; Rayner 2012; Uzunoğlu and Kip 2014).

4.4 Particularities of Young Adults' Use of Social Media

Young adults may have multiple avatars (identities) that they experiment with as they expand into new social networks and determine who they want to be (Framroze 2017). Technology, including Social Media, facilitates playing with and trying on constructed identities to try and fit with both their ideal and aspirational 'self' (Molouki and Bartels 2017), in addition to a variety of social groups. These groups can be those that they aspire to belong to as well as those that they avoid because they do not want to belong to that group (Dwivedi et al. 2014). Many young adults have multiple Social Media accounts with multiple avatars; thus, different groups see different identities (Crozier 2016; Duncan 2016; Won 2016). For example, many young adults use a generic and sanitised persona on Facebook (i.e. Grandma might be watching) that they manage to create an impression appropriate for the different groups (Takahashi 2016).

To incentivise engagement with Social Media content, young adults want status, kudos or fame rather than 'things' or money (Crozier 2016). Thus, the pursuit of social currency is important to young adults (Forrester 2017), and Social Media may promote comparison to others (Vogel et al. 2014) and a drive to live for the approval of others and not for themselves (Won 2016). For example, 40% of young adults admit to feeling pressure to post only content that makes them look good, 39% posted content in order to gather likes and comments and 60% report feeling inadequate in comparison to their peers' Social Media personas (Won 2016). This social comparison and desire to appear a specific way has negative implications for both body image and disordered eating behaviours (Rounsefell et al. 2019).

While they are mindful of peer pressure, young adults remain motivated to fit in and conform with real and illusory social expectations (Solomon 2017). Conformity involves sufficient numbers of people participating (Blanton et al. 2008); however, young adults may be afraid of rejection by the social group and self-censor to align with subjective norms (Blanton et al. 2008). Since young people use personas themselves, they are hyper-alert to any lack of authenticity (Smith 2013; Won 2016). Connecting with young adults can be difficult if you are not part of their social circle. It may not be hard to become part of the social circle if you have something to say that they are interested in listening to (Perski et al. 2017). Social Media could be utilised to become part of the social circle as it is seen as a tool that is used to help people connect, is part of young adult's lives and does not exist in a separate online world (Smith 2013).

4.5 Engagement Means 'Look at Us' not Look at Me

Engagement, including engagement in Social Media content, has at least three dimensions: cognitive, affective and behavioural (Brodie et al. 2011). It is a precursor to outcomes such as affect, usage or responses to advertising. Social Media pro-engagement consists of: co-creation, positive contribution and consumption. Social Media anti-engagement consists of dormancy (not participating in Social Media), detachment (unfollowing a user or terminating a subscription), negative contributions and co-destruction (writing a public complaint) (Dolan et al. 2016; Hollebeek et al. 2016). Social Media engagement is participatory and reciprocal; it is a multiway interaction between and among an organisation and digital communities (Heldman et al. 2013).

Social marketing activities are unlikely to work unless they are personally relevant, engaging and provide a real benefit for the young adult (Brennan 2014). Young adults are willing to 'invest' in activity on behalf of others (Anderson and Masocha 2017), but any intervention has to be clearly beneficial to them and any short-term costs must be minimal and preferably sustainable. Too much pressure results in avoidance rather than positive engagement with the message (Brennan 2014). If you focus on the issue not the individual you can create autonomy and competency (Bentz et al. 2005), while focusing on 'others' can diffuse the fear and increase action (Brennan and Binney 2010). Emotional appeals will be more effective and more likely to engage on Facebook (Swani et al. 2013; Klassen et al. 2018a). However, positive emotions including humour (Field 2017) are more appealing than negative emotions such as fear, guilt and shame which can produce anxiety and maladaptive responses (Hastings et al. 2004).

Social Media is a way to connect to young adults who are always 'on-the-go'. Their phones are always in their hands, and the challenge is to hold their attention enough to engage with them. Upon reviewing both the health and marketing literature, it appears that an effective way to get young adults to engage is to get them to create something of their own; something that they have curated sufficiently to want to share with the world. Young people want to design their own spaces. All attempts to engage them in health messages need an element of co-design and co-creation attached to them (Heldman et al. 2013; Smith 2013; Crozier 2016).

4.6 Co-creation and Co-design in Health Promotion Campaigns

In order to engage young people in Social Media interventions, a process of co-design and co-creation is essential (Buyucek et al. 2016). Co-creation is the process where two or more people create something together collaboratively and in agreement with each other about desired outcomes (Hoyer et al. 2010). On the other hand, co-design is a form of design that actively involves a variety of stakeholders in the design process; its roots are embedded in theories of participatory design (Langley 2016).

Co-creation increases engagement with social marketing (Orazi et al. 2016). Young adults are resistant to messaging about the need to change unless it comes from their peers (Bentz et al. 2005). The message has to be empowering, pressuring and mobilising, not warning and informing, i.e. it is not an information gap it is a power gap (Bentz et al. 2005). Consumer-generated advertisements by peers trigger a process of identification which positively influences advertisement and brand evaluations (Orazi et al. 2016). Even with an unaffiliated audience, co-created advertisements appear to generate higher levels of engagement (Orazi et al. 2016); therefore, it is also worth spending advertisement space to show that 'this message has been created by…'. Furthermore, Consumer-Generated Advertising (CGA) or communication has a lower level of expectations in terms of production quality (Orazi et al. 2016), which is good for the budgetary issues of health organisations.

Social Media engagement is dialogic and conversational (Powell et al. 2011) as well as dynamic and iterative (Hollebeek et al. 2014). It is not monologic (Kent and Taylor 1998), which can sometimes be the approach of 'traditional' health promotion campaigns where health professionals are the 'experts' providing information or teaching the less-informed in a 'top-down' approach (i.e. assuming that information is the deficit to be addressed). When young adults are involved in co-creation and make the effort to help, they expect a response (Smith 2013), so it is important for public health professionals to have the capacity and readiness to put in time to engage and maintain engagement. Conversations ebb and flow, they require active listening and responding. Maintaining the conversation can be time consuming and expensive in a Social Media, social marketing setting.

When it comes to young adults and Social Media interventions, adopting a process of co-design and co-creation is essential (Buyucek et al. 2016). These processes enable young adults to participate directly in and contribute actively to the communication. A studio model where young university students participated in a five-day workshop across two weeks was developed as a method to engage young people in the co-design and co-creation of ideas, strategies and products utilising Social Media, with the aim to help communicate health to their peers. This co-creation workshop structure enables young adults to participate directly in a Social Media campaign's design, by contributing to the development of campaign messages (a bottom-up process). This includes determination of the best mix of media, and sharing these messages in real time with friends, family and networks (Vandelanotte et al. 2014). Using a co-creation approach with young adults draws on social marketing principles and allows the research team and stakeholders to connect more strongly with young adults, strongly integrate the 'voice of the customer' into change-related communications and interventions and have a larger societal impact (Greenhalgh et al. 2016).

4.7 Methodology

The term 'wicked problem' refers to a problem that is complex and difficult or impossible to solve. Here, the word 'wicked' signifies the problem's resistance to resolution, rather than evil. Rittel and Webber (Rittel and Webber 1973) contrasted 'wicked' problems with tame problems, clarifying that the former are less easily

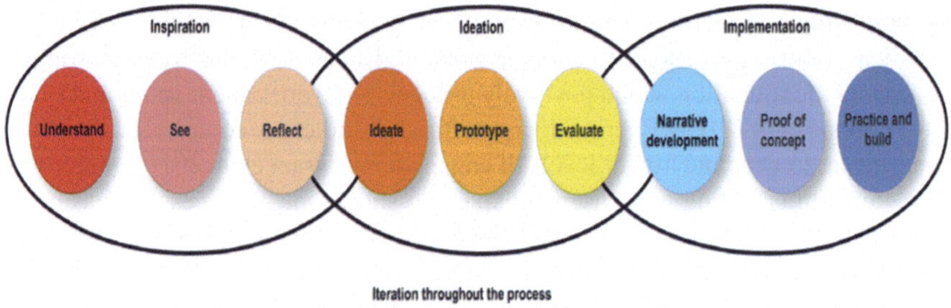

Fig. 4.1 User-centred design approach (adapted from IDEO)

defined and often socially complex, multicausal and highly resistant to resolution, such as that of obesity prevalence in many developed nations. Therefore, a process that encourages creativity, innovation and collaboration is proposed as a way to tackle such wicked problems.

Design thinking is a process in which problems are reconsidered through engaging in co-creation with those most connected to the problem to challenge assumptions and create positive, focused solutions (van de Grift and Kroeze 2016). In principle, it involves three steps—inspiration, ideation and implementation. This process is a user-centred approach for tackling complex wicked problems that do not have clearly identifiable solutions and where the user is instrumental in resolving the issue. Given the complex nature of the environment, particularly the obesogenic nature of the environment, and the need to use a scientific evidence base for decision-making, our team added an additional element to the process—that of information gathering—to the inspiration phase (see Fig. 4.1). Design thinking has been promoted as a problem-solving method for wicked problems (Buchanan 1992). This approach represents a collaborative co-creation process of knowledge creation and iterative learning towards innovation (Hamby et al. 2019). The process involves gathering together a range of different people and perspectives, particularly those who will be the 'end user' of the outcome in the brainstorming process, and exploring and experimenting with possible ideas and solutions (Brown 2008).

Our five-day Wicked Problem Studio workshop (WPS) was conducted as part of a university undergraduate course at the School of Media and Communication, RMIT University (Australia). This studio-based workshop ran as a five-day intensive course across a two-week period in November 2017. The course's title was 'Wicked problems: innovative solutions' and had a focus on the issue of obesity. The study is connected to a larger investigation of how social marketing particularly through Social Media can be used to communicate obesity prevention strategies to young adults—the Communicating Health study (Lombard et al. 2018).

4.7.1 Sample

A total of 54 students, aged within the target demographic (18–24 years), enrolled in WPS. As the course was an elective option, students from any disciplinary background at RMIT University were able to take part. Table 4.1 provides a breakdown of the number of students who participated, their institutions and disciplinary backgrounds. 40.7% (N = 22) of the cohort consisted of Advertising students, which was not unexpected, as the course was administered by the Advertising department of the School of Media and Communication (RMIT University). The cohort also included students from Business, Social Science, Applied Science, Arts and Design backgrounds (Table 4.1). Additionally, as the 'wicked problem' was obesity, students (n = 6) studying nutrition from Monash University were invited to participate as volunteers and without course credit advantage to provide subject expertise to the group, while RMIT students gained course credit for their participation.

The allocation of students into teams of 6–7 was partially random. There was an emphasis to ensure each team consisted of students from multidisciplinary backgrounds. For example, a team would have two Advertising students, two Business students, one Psychology student, one Design student and one Nutrition student. From the 54 students, eight teams were established—six teams with seven students each and two teams with six students each. Teams were announced on day one of WPS, and students were to work in the same teams throughout the five days of workshops.

Table 4.1 Breakdown of students' discipline background

Institution	Undergraduate degree	N
RMIT university	Bachelor of social science (Psychology)	4
	Bachelor of applied science (Construction management)	1
	Bachelor of business (Accountancy)	3
	Bachelor of business (Economics and finance)	5
	Bachelor of business (Management)	1
	Bachelor of business (Human resource management)	1
	Bachelor of business (Financial planning)	1
	Bachelor of business (International business)	1
	Bachelor of communication (Professional communication)	2
	Bachelor of communication (Advertising)	22
	Bachelor of design (Communication design)	3
	Bachelor of design (Digital media)	3
	Bachelor of arts (Music industry)	1
Monash university	Bachelor of nutrition science	6

4.7.2 Process

The stages of WPS were

1. **The Brief**—run by the course coordinator with presentations from eight academic experts (refer to Sect. 4.2.5 for details)
2. **Inspiration**—via a research walk through Melbourne CBD, observations, interviewing the public and online questionnaire. A researcher or academic accompanied each group (refer to Sect. 4.2.4 for details)
3. **Ideation**—team brainstorming (refer to Sect. 4.2.4 for details)
4. **Prototyping**—testing ideas with the public (refer to Sect. 4.2.4 for details)
5. **Implementation/Presentation**—team pitches of their ideas to an assessing panel (refer to Sect. 4.2.6 for details).

4.7.3 WPS Schedule

The WPS stages were implemented across the five workshop days as follows:

Day 1	Course introduction, assessment brief and Wicked Problem brief Team formation and icebreaker activities Team lunch and getting to know the Communicating Health mentors Expert speakers on obesity, body positivity, behaviours and indigenous health Debrief for the following days
Day 2	Expert speakers on design methodology, ethnography, empathy and Social Media Inspiration and research walk from RMIT to the Community Hub, Docklands Teams determine their own route Team lunch along the way Converge at Vic Harbour Library for teams to share insights and findings
Day 3	Brainstorming preparation and students given materials Team brainstorming and ideation in separate rooms Team lunch Continued ideation or fast tracking to prototyping upon approval Teams share their progress and offer feedback to other groups
Day 4	Students briefed on prototyping Begin building and live-testing prototypes on-site Team lunch Progress sharing and feedback to other groups
Day 5	Students briefed on pitching process and due diligence testing/business forecasting Each team work on developing a slide deck and preparing a 20 min pitch Team lunch Team pitch presentation to the expert panel and feedback given on their ideas Prizes awarded for different criteria, such as teamwork and innovation, and overall winning team received a $250 cash prize

4.7.4 Resources

RMIT University was the 'basecamp' during the five days. A large classroom was utilised throughout that period where students were given initial briefing about the wicked problem of obesity, as well as information and guidance on design thinking. The layout of the classroom allowed 7–8 students to sit together at large tables, as opposed to the standard lecture-style theatre. The classroom had a large projector screen at the front, several smaller TV screens and also multiple medium-sized white-boards spread out around the classroom. Eight smaller classrooms were utilised on day 3 to allow each team to work in their own spaces. These classrooms were also equipped with tables, whiteboards and AV equipment, allowing students to brainstorm ideas as well as practise their pitches. A traditional lecture theatre was used on the last day of the workshop where each team pitched their ideas to a panel of experts with the rest of the cohort as an audience.

Additionally, off-site locations were utilised. Community Hub at The Dock is a community facility that provides a range of services and facilities to the public. Students converged at Community Hub after the inspiration and research walk and shared their insights with each other on day 2 of WPS. Students were encouraged to gain inspiration during their walk from RMIT University to Victoria Harbour (see map in Fig. 4.2). The Exchange at Knowledge Market was also used during day 4 of the workshop, where students built and tested their prototypes. The Exchange is a collaboration between the developer Lendlease and RMIT University, where

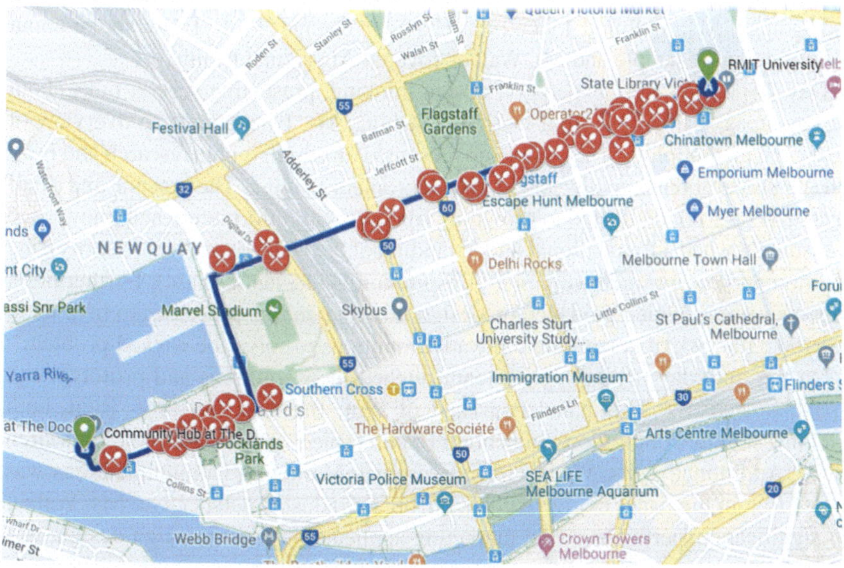

Fig. 4.2 Map of Inspiration and research walk (RMIT to Community Hub at The Dock) with food outlets along the way highlighted in red

new creative projects and partnerships are fostered as public workshops, exhibitions, forums and community events are curated and hosted for the community.

Students used their digital devices (e.g. laptops, tablets and smartphones) for their own research, information gathering and content creation. They also had access to whiteboards, butcher's paper, sticky notes and markers for brainstorming ideas. The large classroom AV equipment was utilised by the unit coordinator and speakers for information briefings and presentations to the students. Students were given access to a computer and projector screen for their pitches at the end of the workshop.

Students were provided with a course guide, lecture slides, field guide to human-centred design, workbook and marking rubric. A purpose-built website was used to disseminate information such as teams and schedules, and additional resources from the experts for self-led research.

4.7.5 Experts

A number of experts in the fields of obesity, nutrition, advertising/communication and design thinking were invited to speak to the students (Table 4.2), in particular, during days 1 and 2. On the first day of WPS, students were briefed on the wicked problem of obesity. As one of the chief investigators of Communicating Health, Prof. Linda Brennan also provided a brief overview of the research project, its goals and aims. Day 1 speakers focused on the topic of health and obesity, with each speaker covering their niche area (Fig. 4.1—Inspiration/Understand phase). Dr. Previte focused on body image, Ms. Willmott looked at behaviour change, Prof. Truby spoke about obesity in young people and Mr. Walker covered Aboriginal health.

Day 2 speakers were experts on design thinking, presenting several talks about the key areas of this method of learning/teaching. Ms. Ivanka gave an overview of the design methodology, Dr. Sumartojo presented on ethnographic research and Prof. Chester spoke about empathy and its role in problem-solving. The first one and a half days of WPS were dedicated to providing students with information and knowledge to ensure they had a foundational understanding of the wicked problem (obesity), the situation and environment (Fig. 4.1—Inspiration/Understand phase). Furthermore, students were provided guidance on design thinking and its processes to help them undertake the task of generating innovative solutions to solve the wicked problem.

On the final day of WPS, each team pitched their solutions and prototypes to a panel of experts. This panel consisted of Prof. Brennan and Prof. Reid, both chief investigators of the Communicating Health project, and Dr. Klassen, a research fellow. Additionally, Ms. Walker from Evergreen Marketing Communications was invited to join the panel to give feedback from an industry perspective. Ms. Walker has significant expertise in social marketing applications and campaign design.

Table 4.2 List of experts invited to participate in WPS

Day 1 speakers—obesity, social media, marketing, nutrition and Aboriginal culture and engagement experts	Prof. Linda Brennan* (Professor of Advertising, School of Media and Communication, RMIT University) Dr. Josephine Previte (Senior Lecturer, Department of Marketing, University of Queensland) Ms. Taylor Willmott (Research Assistant, Social marketing @ Griffith, Griffith Business School) Prof. Helen Truby* (Professor of Nutrition, Department of Nutrition, Dietetics and Food, Monash University) Mr. Troy Walker (Research officer, Department of Nutrition, Dietetics and Food, Monash University)
Day 2 speakers—ethnographic, design thinking and psychology experts	Ms. Tania Ivanka (Associate Lecturer, School of Communication Design, RMIT University) Dr. Shanti Sumartojo (Senior Research Fellow, School of Media and Communication, RMIT University) Prof. Andrea Chester (Deputy Vice Chancellor, College of Design and Social Context, RMIT University)
Day 5—panel members for pitch feedback	Prof. Linda Brennan* (Professor of Advertising, School of Media and Communication, RMIT University) Prof. Mike Reid* (Professor of Marketing, Department of Finance, Economics and Marketing, RMIT University) Dr. Karen Klassen* (Research Fellow, Department of Nutrition, Dietetics and Food, Monash University) Ms. Gill Walker (Founder and Managing Director, Evergreen Advertising and Marketing)

*Also chief investigators of Communicating Health, an NHMRC-funded research project (Grant number: GNT1115496)

4.7.6 *Assessments*

Throughout the course of the five days, students were required to undertake three assessment tasks:

1. Individual learning contract (Due day 2; contribution to assessment 20%)
    - Students were asked to outline:
    - Why they chose to participate in the WPS
    - What three skills they hoped to achieve/improve from participation
    - Their plan on how they would achieve those skills, such as what activities would be beneficial

- Methods by which they would prove the obtainment/improvement of those skills, such as video records
- They were encouraged to present this assessment as an infographic.

2. Multidisciplinary team presentation (Due day 5; contribution to assessment 50%)
   The team pitch presentation and slide deck were assessed.
   Students were asked to show that their prototype and business model:

   - Answered the problem
   - Could be successfully rolled out
   - Would have the desired impact and be low risk
   - Be implementable by their team, showing they have the requisite expertise, resources, skills and collaborative knowledge.

3. Reflective video (Due four days post-WPS; contribution to assessment 30%)
   Students were asked to design and film a two-minute video in which they described their role in the multidisciplinary team, what they got out of participating and the main skills they gained. They were guided to frame this within:

   - Context: their course and team
   - Learning reflections: identify key learnings and insights
   - Future application: discussing how they would transfer new knowledge to future projects and professional life.

4.7.7 Research

Research conducted over the course of WPS included observation (photos, videos, and taking notes), and interviews with stakeholders, guest speakers and with each student group. Guest speakers were interviewed once, usually on the day of their presentation, and stakeholders were interviewed once or twice at varying time points during WPS, and two took place two months later. Student groups were interviewed on days 3 and 5 of WPS. Post-WPS research utilised the first and third student assessments to record, assess and identify patterns and outcomes. All research participants signed a consent form prior to being interviewed or observed. This project was approved by Royal Melbourne Institute of Technology College Human Ethics Advisory Network (Project number: 21175) and Monash University Human Research Ethics Committee (Project number: 12862).

4.8 Findings and Discussion

4.8.1 Reflection on the Process

From the WPS, the teams produced eight co-created ideas that they believed could help tackle the wicked problem of obesity through healthy eating. These ideas consisted mainly of mobile applications or plug-ins for existing applications or programmes that often promoted health beyond healthy eating. Their ideas highlighted the need to talk about health and healthy eating with a positive tone, incorporating the benefits of health-enhancing behaviours, particularly those that are short term, rather than the longer term consequences such as disease prevention. Through prototyping with others in their target age group, the teams created messages they believed were appropriate, interesting and would connect with the needs and values of their age group.

Most participants regarded the multidisciplinary nature of WPS as one of the highlights. Working in teams with people of varying educational backgrounds allowed for a broad perspective of obesity as a complex, whole of society issue and an understanding of diverse methods that may be required to reduce its prevalence. At times, these differing viewpoints resulted in disagreements, however teams were able to negotiate a compromise and find a way forward. Unique sets of skills from different disciplines complemented each other within each team. When there were multiple people with similar skill sets within one team, participants found themselves having to adjust and take on a role they normally would not. Roles were not specified in the brief, so students decided throughout the process what each team member would be responsible for. This was seen as an important learning curve and pushed participants out of their comfort zone which sometimes resulted in the group being more innovative as a result. Similarly, teams that did not have members from a wider range of disciplines believed they would have benefited from a broader perspective. Many participants stated that working and troubleshooting within a multidisciplinary team strengthened their empathy and ability to give and receive constructive criticism and apply this feedback. For university students, the multidisciplinary nature of the studio was unlike anything they had done in university thus far. They enjoyed being part of a 'real-world' experience and believed it would be beneficial for their future careers in which working with a range of people would be required.

The first day of WPS, 'The Brief' which involved expert speakers providing information and context on the wicked problem, was considered by participants to be an information overload experience and hindered their ability to recall the material for future application in the studio. Participants were not expecting lecture-style information dissemination and some were overwhelmed and found it hard to consolidate their thoughts with many concepts being presented in a short time span. This was particularly the case as obesity and nutrition were relatively unknown topics to many participants with some participants stating they could not take in enough information to improve their understanding on the issues. Participants suggested that an interactive session would have increased their engagement in the topic area, allowing them to actively listen, be inspired and start brainstorming potential solutions. Suggestions included, rather than having experts give lecture-style presentations, each team could have an expert mentoring or working with the team for part of the studio.

The inspiration and research walk and prototyping with the general public provided additional opportunities to experience the 'real-world' applicability of this wicked problem from their target audience. The participants saw benefit in hearing other people's perspectives on both the wicked problem, and later in the prototyping stage, on their campaign idea. Some participants believed it would have been more beneficial to go to a location where there were more people within their target audience and for not all teams to go to the same location and talk to the same people. Also highlighted by some participants was the ambiguity of tasks to be completed on the research walk and some suggested they would have appreciated more guidance on what was expected of them during this stage. Interviewing others built their confidence and allowed them to trial ethnographic interviewing techniques, which they had learnt as part of the WPS process. This built their empathetic listening skills, and many participants highlighted empathy as an important skill gained throughout WPS. Working in multidisciplinary teams required a level of understanding of their teammate's perspectives and the ability to compromise in order to establish agreement on both the process of ideation, prototyping and iteration and their final campaign implementation. Interviewing and working with their target audience on prototyping their ideas was useful as it required them to take how their audience feels into deeper consideration when creating ideas and was interesting as they received first-hand feedback on their campaigns. However, daily debriefs to the whole group to share their findings, progress and gain feedback, manifested in groups adopting other groups' ideas into their own solution, resulting in some convergence of ideas and similarities in the final campaigns.

The perceived short time frame of the studio overall was regarded by some participants as both a challenging and rewarding aspect of the studio. In both the reflective videos and interviews, participants described the studio as an intense week that pushed them to get to know their team, research and develop a campaign in a time frame in which they were not accustomed to working. Idea generation was required to be as succinct and fast paced as possible, with a limited window to decide on an idea and refine this before prototyping. This required them to be efficient rather than labouring the details of a whole subject area such as obesity. This sometimes hindered their ability to develop new skills as they believed they were best to stick within their comfort zone due to the need for efficiency or did not have a chance to reflect before moving on to the next stage. However, the time frame was applicable to real life circumstances where they may have limited time to complete a brief. Participants stated they built skills in organisation, negotiation and adaptability to overcome the time constraints. However, there were some participants with a differing viewpoint, stating that working on one task for multiple hours was not conducive to productivity due to their short attention span. These participants also felt they had too much time to complete the overall campaign development, and it was not necessary for the five-day studio to be spaced across two weeks.

4.8.2 Benefits of Wicked Problem Studio

The use of co-creation in health messaging design has the potential to engage young adults and encourage behaviour change. The involvement of the target audience in the design of campaigns directed towards them created a sense of empowerment and ownership of the behaviour the campaign is targeting. There is a positive impact when involving young adults in campaign development, particularly when the target behaviour is not in an area of interest or expertise for participants. Participants in WPS felt invested in the message, and for some participants, there was an observed change in their attitude towards obesity and healthy eating. Some participants reflected that they had changed their diet since, or healthy eating was something they were now going to pursue in their personal life. Young adults respond well to calls for participation and take their responsibilities for others quite seriously. The participants were proud of their campaigns and achievements, and believed that the unique and creative ideas would not have been produced without the knowledge and insights from their multidisciplinary group. There was a sense of ease within many teams in sharing their own ideas, being able to build fast rapport with a group of people with the same end goal but potentially differing values and ideals, which makes for interesting conversations and unique ideas.

Participants stated in their reflective videos that WPS was a beneficial experience and were satisfied that they participated. Benefits included the development of transferrable research and design skills and increased knowledge on obesity, nutrition, human-centred design process and ethnography. Many participants stated they were able to achieve what they had hoped to learn as indicated in their individual learning contracts. This included skills such as improved communication, researching, critical thinking, ability to collaborate effectively, emotional intelligence and confidence in presenting their ideas. Many also mentioned an improvement in their ability to listen to others empathically, developed through the research walk and prototyping phases. The process of working in a multidisciplinary team and having to navigate different opinions, values and work processes also enhanced their empathy and adaptability. Working in a team with people they did not know, without a defined leader, required many to take on leadership roles, often unexpectedly, which, along with interviewing and presenting, increased their confidence.

There is willingness by young adults, including those that participated in the WPS to connect and help; so when asked for engagement with a manageable scope and specific tasks, they will try to help. The WPS studio allowed participants to think creatively and develop ideas they believed were innovative and not something they would have achieved without their multidisciplinary group. Creativity is important to young adults and allowing them to make their own campaigns and artefacts by giving them creative licence is sometimes the only way to get them to engage (Smith 2013; Crozier 2016). Young adults like to create their own content, which is evident on Social Media and through the participants of WPS who enjoyed the creative and hands-on aspects of co-creation.

4.8.3 Cautions

It is important to reflect on the process both with stakeholders and researchers, but importantly with the participants who are the target audience for the outcome and involved in the co-creation process. Participant reflections highlighted important strengths of the process and areas for improvement. Reflection throughout the process at different stages may also be necessary, as the post-WPS personal video reflections were generally more positive than the team interviews throughout the studio. This may be due to the mid-WPS interviews and observations being with a group or because the participants were more recently affected by the negative aspects of the studio compared to their individual reflections.

The reflections highlighted the importance of making the studio as interactive and hands-on as feasible, as didactic information dissemination was viewed as ineffective and was not conducive to inspiration or creative idea generation. There is also the potential to provide too much information in the Brief as part of the inspiration phase, which may overwhelm and hinder the participant's ability to view the issue and potential solution through his or her own lens. It is of importance to find a balance between providing enough information to participants, so they are aware of the broad issue and the requirements of the studio, and providing too much information that manipulates their decision-making.

The requirement for teams to present their progress and ideas to each other throughout the WPS resulted in a convergence of ideas among teams. This represents the desire for young adults to conform and be accepted by their peers and is therefore a difficulty in this age group (Erikson 1994). Groupthink, or the desire to conform to create harmony, among young adults allows for morale to be established, however may be at the expense of critical thinking (Turner and Pratkanis 1998). This is an important consideration for future use of co-creation studios as there were teams who abandoned unique and creative ideas, in favour of incorporating ideas other teams presented.

Health promotion initiatives need to be designed to ensure that the right people are participating in the co-creation process. Health promotion programmes need to increase involvement by their target audience which may involve separate initiatives or strategies for different target groups (Orazi et al. 2016). Wherever the target is located, we need to take the message to where young people are most engaged. We, as researchers, cannot afford to have the message broadcast through our most familiar channels when these channels do not reach our target audience. We also need to ensure that we understand how best to convey these messages on these channels. Creating a variety of artefacts that meet the audience's communication needs is an important part of the Social Media and digital communication strategy.

4.8.4 Directions for Future Research

The use of co-creation in social marketing may be the key to successful engagement of people in campaigns, however, remains infrequently used in the health and nutrition space. Future research in health and nutrition could incorporate these co-creation and other social marketing techniques to enhance patient-centred care and develop health messaging that resonates with their audience. The success of the multidisciplinary nature of WPS highlights the importance of incorporating different disciplines within the design and development of health messaging and future research studies. This was made possible by cross-university collaboration, which was an important element that brought the academic team together as part of a larger research programme. As WPS included young adult university students, their experiences may not be applicable to other age groups or those outside of a university course. Therefore, it is important to conduct co-creation health research in different age and demographic groups or segments. The use of psycho-behavioural consumer segmentation for both the target population of the campaign and the co-creation participants could be of benefit beyond segmenting based on broad age groups or other demographics.

Health and nutrition could benefit and has much to learn from using psycho-behavioural consumer segmentation when designing and applying interventions and campaigns. There is also the potential to gain a deeper understanding of important values of a target population by allowing participants in a co-creation studio to design a campaign based on a health or social topic they believe is important for them to focus on. In addition, exploring how Social Media Influencers are able to engage and retain their audience's attention and gain authenticity could provide useful strategies to incorporate in future research. Health promotion campaigns could utilise these learnings in their development of campaigns, as well as the potential to engage Influencers with like values and an appropriate audience, in the design or dissemination of health promotion messaging.

4.9 Conclusions

Co-creation, empowerment and enabling participation are key to the successful engagement of young people with Social Media and health messaging. We know that young people are creating and adapting who they will be in their adult futures in the form of Social Media personas. We know that they want to conform to societal expectations, but also want to have a say in what those expectations are. Consequently, health promoters can leverage these motivations and design programmes that enable young people to participate in creating their own futures.

For health promotion to be successful, it is necessary to understand the language that resonates with the target audience and communicate messages that are meaningful to them, so you can access them as if you were in their social circle. This involves talking to your audience and gathering an understanding of their behaviours, values and needs, which is where co-creation and co-design may be beneficial. Through co-creation, we were able to understand what works and what does not work when it comes to co-creation of Social Media interventions aiming to encourage young adults to make healthier lifestyle choices.

Young adults may prefer messages that are positive, whereas deficit language and the health risks of their eating behaviours may not be appropriate. Food and eating should be more about well-being, how food makes you feel, how you can eat affordably, learn to budget and provide solutions to their issues, such as recipe ideas. Thinking should extend beyond instructions of what they should be doing to how the desired behaviours could make them feel and how you can make it easy to eat this way. Co-creation helps with this understanding of why, how and what prompts young adults to engage with and respond to health messages.

As using Social Media for health research is relatively new and rapidly evolving, we are still learning which channels and types of messages to use to reach our target audience, how to evaluate audience engagement and the effectiveness of campaign messages. There is a need to use platforms that your audience frequents, not necessarily those that are traditionally used. Different messaging styles and communication strategies perform better on different platforms and therefore there is a need for multiple artefacts for a health promotion campaign tailored to different platforms. Campaigns in health and nutrition could benefit from greater application of social marketing principles, particularly psycho-behavioural consumer segmentation. One size does not fit all, and with Social Media, there is the unique ability to target audiences and influence behaviours based on more than standard demographics, by specifically taking into consideration interests and values.

Social Media remains a challenging place to promote health among the enormity of often misinformed and sometimes damaging information about food, health and weight loss. As Social Media is unlikely to reduce in its momentum of popularity, it is important for evidence-based messages to have a strong and clear voice among the noise. However, we do know that young people respond well to calls for participation and take their responsibilities for others quite seriously. Social Media in this context is therefore an ideal mechanism by which to generate participation and engagement. Co-creation provides that opportunity for health professionals and social marketers to increase the relevance, reach of the messages, and gain traction against the breadth of information on Social Media.

4.9.1 Implications for Key Stakeholders Such as Government Bodies and Health Organisations

The findings and learnings from the WPS will translate into practice for the use of co-creation and co-design. Participant reflections highlighted the importance and benefit of multidisciplinary perspectives in a co-creation exercise. The user-centred design approach and interaction with other individuals within the target audience was important to create a well-rounded and well thought out product and messaging strategy. Importantly, it is essential for reflection with participants throughout the process to learn and adapt from experience. The refined methods based on reflection from this WPS provide an example and how-to for using co-creation to tackle an issue as complex as obesity with young adults that could be utilised by key stakeholders to inspire social change. Importantly, the whole village will have to be involved in the design process—from macro level government and social marketing agencies to the micro level with affected individuals.

Acknowledgements This research was funded by National Health and Medical Research Council 'Communicating Health: optimising engagement and retention using Social Media' (GNT1115496).

The authors would like to acknowledge the support of Ms. Kerin Elsum, the course coordinator of the Wicked Problems Studio, without whom this research could not have taken place.

References

Allman-Farinelli MA, Chey T, Bauman AE, Gill T, James WPT (2008) Age, period and birth cohort effects on prevalence of overweight and obesity in Australian adults from 1990 to 2000. Eur J Clin Nutr 62(7):898–907

Anderson H, Masocha S (2017) Bottom-up and inside-out: using participatory methodology to measure the social impact of a radio engagement project with young people of refugee experience. Commun Res Pract 3(1):60–73

Ashton LM, Hutchesson MJ, Rollo ME, Morgan PJ, Collins CE (2016) Motivators and barriers to engaging in healthy eating and physical activity. A cross-sectional survey in young adult men. Am J Men's Health 11(2):330–343

Ashton LM, Morgan PJ, Hutchesson MJ, Rollo ME, Collins CE (2017) Feasibility and preliminary efficacy of the 'HEYMAN' healthy lifestyle program for young men: a pilot randomised controlled trial. Nutr J 16(1):2

Australian Bureau of Statistics (2013) 4727.0.55.001—Australian aboriginal and torres strait islander health survey: first results, Australia, 2012–2013. Australian Bureau of Statistics. https://www.abs.gov.au/ausstats/abs@.nsf/mf/4727.0.55.001

Australian Bureau of Statistics (2015) Consumption of sweetened beverages. 4364.0.55.007—Australian health survey: nutrition first results—foods and nutrients. Canberra, Australian Bureau of Statistics

Australian Bureau of Statistics (2019) 4364.0.55.001—national health survey: first results, 2017–2018, 24 September 2019. https://www.abs.gov.au/ausstats/abs@.nsf/Lookup/by%20Subject/4364.0.55.001~2017-18~Main%20Features~Overweight%20and%20obesity~90

Bentz JW, Dorfman L, Denniston R, Novelli W (2005) Opportunities for social change through upstream partnerships. Soc Mark Q 11(3–4):17–25

Blanton H, Köblitz A, McCaul KD (2008) Misperceptions about norm misperceptions: descriptive, injunctive, and affective 'Social Norming' efforts to change health behaviors. Soc Pers Psychol Compass 2(3):1379–1399

Bleich SN, Wang YC, Wang Y, Gortmaker SL (2009) Increasing consumption of sugar-sweetened beverages among US adults: 1988–1994 to 1999–2004. Am J Clin Nutr 89(1):372–381

Brennan L (2014) Social marketing and behaviour change: models, theory and applications. Edward Elgar, Cheltenham

Brennan L, Binney W (2010) Fear, guilt, and shame appeals in social marketing. J Bus Res 63(2):140–146

Brennan L, Previte J, Fry ML (2016) Social marketing's consumer myopia Applying a behavioural ecological model to address wicked problems. J Soc Mark 6(3):219–239
Brimblecombe J, Maypilama E, Colles S, Scarlett M, Dhurrkay JG, Ritchie J, O'Dea K (2014) Factors influencing food choice in an Australian Aboriginal community. Qual Health Res 24(3):387–400
Brodie RJ, Hollebeek LD, Jurić B, Ilić A (2011) Customer engagement: conceptual domain, fundamental propositions, and implications for research. J Serv Res 14(3):252–271
Brown T (2008) Design thinking. Harvard Bus Rev 86(6):84
Buchanan L, Kelly B, Yeatman H (2017) Exposure to digital marketing enhances young adults' interest in energy drinks: an exploratory investigation. PLoS ONE 12(2):e0171226
Buchanan R (1992) Wicked problems in design thinking. Des Issues 8(2):5–21
Burton S, Dadich A, Soboleva A (2013) Competing voices: marketing and counter-marketing alcohol on Twitter. J Nonprofit Public Sect Mark 25(2):186–209
Buyucek N, Kubacki K, Rundle-Thiele S, Pang B (2016) A systematic review of stakeholder involvement in social marketing interventions. Australas Mark J (AMJ) 24(1):8–19
Capurro D, Cole K, Echavarria MI, Joe J, Neogi T, Turner AM (2014) The use of social networking sites for public health practice and research: a systematic review. J Med Internet Res 16(3):e79
Chapman K, Havill M, Watson WL, Wellard L, Hughes C, Bauman A, Allman-Farinelli M (2016) Time to address continued poor vegetable intake in Australia for prevention of chronic disease. Appetite 107:295–302
Chapman S (2012) Does celebrity involvement in public health campaigns deliver long term benefit? Yes. BMJ: Br Med J 345:e6364
Crozier M (2016) Engaging young people online, 18 Sept 2017. http://www.bangthetable.com/engaging-young-people/
Dolan R, Conduit J, Fahy J, Goodman S (2016) Social media engagement behaviour: a uses and gratifications perspective. J Strat Mark 24(3–4):261–277
Dooley JA, Jones SC, Iverson D (2013) Social marketing and health promotion communication: from traditional marketing to Web 2.0. Cases Public Health Comm Mark 7:120–139
Duncan F (2016) So long social media: the kids are opting out of the online public square. Conversation, 18 Sept 2017. https://theconversation.com/so-long-social-media-the-kids-are-opting-out-of-the-online-public-square-53274
Dwivedi A, McDonald RE, Johnson LW (2014) The impact of a celebrity endorser's credibility on consumer self-brand connection and brand evaluation. J Brand Manag 21(7–8):559–578
Erikson EH (1994) Identity: youth and crisis, WW Norton & Company
Field P (2017) ROI in 2017: effectiveness in the digital age. WARC Trends, 18 Sept 2017. http://content.warc.com/warc-toolkit-2017-of-marketing-trends-effectiveness-in-the-digital-age
Fishbein M, Hennessy M, Yzer M, Douglas J (2003) Can we explain why some people do and some people do not act on their intentions? Psychol Health Med 8(1):3–18
Forrester I (2017) Content in 2017: how video is transforming social marketing. WARC Trends, 20 Sept 2017. http://content.warc.com/warc-toolkit-2017-of-marketing-trends-how-video-is-transforming-social-marketing
Framroze, M. (2017). Self-spectacle online: the construction and representation of identity in contemporary digital culture. Doctor of Philosophy in Education Ph.D., University of California, Los Angeles
Freeman B, Kelly B, Baur L, Chapman K, Chapman S, Gill T, King L (2014) Digital junk: food and beverage marketing on Facebook. Am J Public Health 104(12):e56–e64
Freeman B, Kelly B, Vandevijvere S, Baur L (2016) Young adults: beloved by food and drink marketers and forgotten by public health? Health Promot Int 31(4):954–961
Freeman B, Potente S, Rock V, McIver J (2015) Social media campaigns that make a difference: what can public health learn from the corporate sector and other social change marketers? Public Health Res Pract 25(2)

Greenhalgh T, Jackson C, Shaw S, Janamian T (2016) Achieving research impact through co-creation in community-based health services: literature review and case study. Milbank Q 94(2):392–429
Greenwood S, Perrin A, Duggan M (2016) Social media update. PEW Internet Research http://www.pewinternet.org/2016/11/11/social-media-update-2016/
Hamby A, Pierce M, Daniloski K (2019) 9 co-creating social change using human-centred design. Macro-Social Marketing Insights: Systems Thinking for Wicked Problems, Routledge.
Hastings G, Stead M, Webb J (2004) Fear appeals in social marketing: Strategic and ethical reasons for concern. Psychol Mark 21(11):961–986
Hebden L, Chey T, Allman-Farinelli M (2012) Lifestyle intervention for preventing weight gain in young adults: a systematic review and meta-analysis of RCTs. Obes Rev 13(8):692–710
Heldman AB, Schindelar J, Weaver JB III (2013) Social media engagement and public health communication: implications for public health organizations being truly "social". Public Health Rev 35(1):1–18
Hoffman SJ, Mansoor Y, Natt N, Sritharan L, Belluz J, Caulfield T, Freedhoff Y, Lavis JN, Sharma AM (2017) Celebrities' impact on health-related knowledge, attitudes, behaviors, and status outcomes: protocol for a systematic review, meta-analysis, and meta-regression analysis. Syst Rev 6(1):1–13
Hoffman SJ, Tan C (2015) Biological, psychological and social processes that explain celebrities' influence on patients' health-related behaviors. Arch Public Health 73(1):3
Hollebeek LD, Conduit J, Brodie RJ (2016) Strategic drivers, anticipated and unanticipated outcomes of customer engagement. J Mark Manag 32(5–6):393–398
Hollebeek LD, Glynn MS, Brodie RJ (2014) Consumer brand engagement in social media: conceptualization, scale development and validation. J Interact Mark 28(2):149–165
Hootsuite Media Inc., We Are Social Ltd. (2017) Digital in 2017 Global Overview, 20 March 2017. https://www.slideshare.net/wearesocialsg/digital-in-2017-global-overview
Horton D, Richard Wohl R (1956) Mass communication and para-social interaction: observations on intimacy at a distance. Psychiatry 19(3):215–229
Hoyer WD, Chandy R, Dorotic M, Krafft M, Singh SS (2010) Consumer cocreation in new product development. J Serv Res 13(3):283–296
Kent ML, Taylor M (1998) Building dialogic relationships through the world wide web. Public Relat Rev 24(3):321–334
Kerpan ST, Humbert ML, Henry CJ (2015) Determinants of diet for urban Aboriginal youth: implications for health promotion. Health Promot Pract 16(3):392–400
Kite J, Foley BC, Grunseit AC, Freeman B (2016) Please like me: facebook and public health communication. PloS one 11(9):e0162765
Klassen KM, Borleis ES, Brennan L, Reid M, McCaffrey TA, Lim MS (2018a) What people "Like": analysis of social media strategies used by food industry brands, lifestyle brands, and health promotion organizations on facebook and instagram. J Med Internet Res 20(6):e10227
Klassen KM, Douglass CH, Brennan L, Truby H, Lim MS (2018b) Social media use for nutrition outcomes in young adults: a mixed-methods systematic review. Int J Behav Nutr Phys Act 15(1):70
Kreuter MW, Oswald DL, Bull FC, Clark EM (2000) Are tailored health education materials always more effective than non-tailored materials? Health Educ Res 15(3):305–315
Kubacki K, Dietrich T, Rundle-Thiele S (2017) Segmentation in social marketing: why we should do it more often that we currently do. Segmentation in Social Marketing, Springer, pp 1–6
Langley J (2016) Participatory design: co-creationl co-productionl co-design combining imaging and knowledge. Knowledge Utilisation Colloquium, Llandudno, Wales, 29/06/2016 - 01/07/2016.
Lim MS, Wright CJ, Carrotte ER, Pedrana AE (2016) Reach, engagement, and effectiveness: a systematic review of evaluation methodologies used in health promotion via social networking sites. Health Promot J Austr 27(3):187-197
Lim MSC, Hare JD, Carrotte ER, Dietze PM (2016) An investigation of strategies used in alcohol brand marketing and alcohol-related health promotion on Facebook. Digit Health 2:1–7

Lim SS, Vos T, Flaxman AD, Danaei G, Shibuya K, Adair-Rohani H, AlMazroa MA, Amann M, Anderson HR, Andrews KG (2012) A comparative risk assessment of burden of disease and injury attributable to 67 risk factors and risk factor clusters in 21 regions, 1990–2010: a systematic analysis for the global burden of disease study 2010. Lancet 380(9859):2224–2260

Liu S, Jiang C, Lin Z, Ding Y, Duan R, Xu Z (2015) Identifying effective influencers based on trust for electronic word-of-mouth marketing: a domain-aware approach. Inf Sci 306:34–52

Lobstein T, Baur L, Uauy R for the IASO International Obesity TaskForce (2004) Obesity in children and young people: a crisis in public health. Obes Rev 5(Suppl 1):4–104

Lombard C, Brennan L, Reid M, Klassen KM, Palermo C, Walker T, Lim MS, Dean M, McCaffrey TA, Truby H (2018) Communicating health—optimising young adults' engagement with health messages using social media: study protocol. Nutr Diet 75(5):509–519

Maher CA, Lewis LK, Ferrar K, Marshall S, De Bourdeaudhuij I, Vandelanotte C (2014) Are health behavior change interventions that use online social networks effective? A systematic review. J Med Internet Res 16(2):e40

Maibach EW, Maxfield A, Ladin K, Slater M (1996) Translating health psychology into effective health communication: the American healthstyles audience segmentation project. J Health Psychol 1(3):261–277

Mangold GW, Faulds DJ (2009) Social media: the new hybrid element of the promotion mix. Bus Horiz 52(4):357–365

Menozzi D, Sogari G, Mora C (2015) Explaining vegetable consumption among young adults: an application of the theory of planned behaviour. Nutrients 7(9):7633–7650

Molouki S, Bartels DM (2017) Personal change and the continuity of the self. Cogn Psychol 93(Supplement C):1–17

Moorhead SA, Hazlett DE, Harrison L, Carroll JK, Irwin A, Hoving C (2013) A new dimension of health care: systematic review of the uses, benefits, and limitations of social media for health communication. J Med Internet Res 15(4):e85

Moss HB, Kirby SD, Donodeo F (2009) Characterizing and reaching high-risk drinkers using audience segmentation. Alcohol: Clin Exp Res 33(8):1336–1345

Munt A, Partridge S, Allman-Farinelli M (2017) The barriers and enablers of healthy eating among young adults: a missing piece of the obesity puzzle: a scoping review. Obes Rev 18(1):1–17

Muntinga DG, Moorman M, Smit EG (2011) Introducing COBRAs: exploring motivations for brand-related social media use. Int J Advert 30(1):13–46

National Health and Medical Research Council (2015) Targeted call for research into engaging and retaining young adults in interventions to improve eating behaviours and health outcomes (Preventing Obesity in 18–24 year olds TCR), 1 July 2015. https://www.nhmrc.gov.au/grants-funding/apply-funding/targeted-and-urgent-calls-research/preventing-obesity-18-24-year-olds-t

Nelson MC, Story M, Larson NI, Neumark-Sztainer D, Lytle LA (2008) Emerging adulthood and college-aged youth: an overlooked age for weight-related behavior change. Obesity 16(10):2205–2211

Nielsen Company (2012) Global trust in advertising and brand messages. https://www.nielsen.com/us/en/insights/report/2012/global-trust-in-advertising-and-brand-messages-2/

Orazi DC, Bove LL, Lei J (2016) Empowering social change through advertising co-creation: the roles of source disclosure, sympathy and personal involvement. Int J Advert 35(1):149–166

Palermo C, Perera-Schulz D, Kannan A, Truby H, Shiell A, Emilda S, Quenette S (2014) Development and evaluation of an iPad App for measuring the cost of a nutritious diet. JMIR Mhealth Uhealth 2(4):e50

Paxton SJ, McLean, SA (2015) Peer interactions and relationships—risk factors for eating disorders. Encyclopedia of Feeding and Eating Disorders, Springer

Perski O, Blandford A, West R, Michie S (2017) Conceptualising engagement with digital behaviour change interventions: a systematic review using principles from critical interpretive synthesis. Transl Behav Med 7(2):254–267

Powell GR, Groves SW, Dimos J (2011) ROI of social media: how to improve the return on your social marketing investment. Wiley, New York

Rayner G (2012) Does celebrity involvement in public health campaigns deliver long term benefit? No. BMJ: Br Med J 25;345:e6362
Rice ES, Haynes E, Royce P, Thompson SC (2016) Social media and digital technology use among Indigenous young people in Australia: a literature review. Int J Equity Health 15:81
Rittel WJ, Webber MM (1973) Dilemmas in a general theory of planning. Policy Sci 4(2):155–169
Rounsefell K, Gibson S, McLean S, Blair M, Molenaar A, Brennan L, Truby H, McCaffrey TA (2019) Social media, body image and food choices in healthy young adults: a mixed methods systematic review. Nutr Diet 77(1):19-40
Rouse M (2016) Definition: social media, 24 Aug 2017. http://whatis.techtarget.com/definition/social-media
Sensis (2016) Sensis social media report 2016
Slater MD (1996) Theory and method in health audience segmentation. J Health Commun 1(3):267–283
Smith B (2013) Tips for youth engagement on social media, 18 Sept 2017. http://www.buildsocialconsulting.com/youth-engagement-social-media/
Solomon MR (2017) Consumer behavior: buying, having, and being, global edition. Pearson
Sui ZX, Wong WK, Louie JCY, Rangan A (2017) Discretionary food and beverage consumption and its association with demographic characteristics, weight status, and fruit and vegetable intakes in Australian adults. Public Health Nutr 20(2):274–281
Swani K, Milne G, Brown BP (2013) Spreading the word through likes on Facebook: evaluating the message strategy effectiveness of Fortune 500 companies. J Res Interact Mark 7(4):269–294
Takahashi T (2016) Creating the self in the digital age: young people and mobile social media. The good life in Asia's digital 21st Century. Digital Asia Hub, Hong Kong, pp 44–50
Thackeray R, Neiger BL, Hanson CL, McKenzie JF (2008) Enhancing promotional strategies within social marketing programs: use of Web 2.0 social media. Health Promot Pract 9(4):338–343
Thomson M (2006) Human brands: investigating antecedents to consumers' strong attachments to celebrities. J Mark 70(3):104–119
Turner ME, Pratkanis AR (1998) Twenty-five years of groupthink theory and research: lessons from the evaluation of a theory. Organ Behav Hum Decis Process 73(2–3):105–115
Uzunoğlu E, Kip SM (2014) Brand communication through digital influencers: leveraging blogger engagement. Int J Inf Manage 34(5):592–602
van de Grift TC, Kroeze R (2016) Design thinking as a tool for interdisciplinary education in health care. Acad Med 91(9):1234–1238
Vandelanotte C, Kirwan M, Rebar A, Alley S, Short C, Fallon L, Buzza G, Schoeppe S, Maher C, Duncan MJ (2014) Examining the use of evidence-based and social media supported tools in freely accessible physical activity intervention websites. Int J Behav Nutr Phys Act 11(1):105
Vaterlaus JM, Patten EV, Roche C, Young JA (2015) #Gettinghealthy: the perceived influence of social media on young adult health behaviors. Comp Human Behav 45:151–157
Veale HJ, Sacks-Davis R, Weaver ER, Pedrana AE, Stoove MA, Hellard ME (2015) The use of social networking platforms for sexual health promotion: identifying key strategies for successful user engagement. BMC Public Health 15:85
VicHealth (2013) Drinking-related lifestyles: exploring the role of alcohol in Victorians' lives. Melbourne, Australia, VicHealth
Vogel EA, Rose JP, Roberts LR, Eckles K (2014) Social comparison, social media, and self-esteem. Psychol Popul Media Cult 3(4):206
Walker T, Palermo C, Klassen K (2019) Considering the impact of social media on contemporary improvement of Australian aboriginal health: scoping review. JMIR Public Health Surveill 5(1):e11573
Waters RD, Burnett E, Lamm A, Lucas J (2009) Engaging stakeholders through social networking: how nonprofit organizations are using facebook. Public Relat Rev 35(2):102–106
Wills J, Crichton N, Lorenc A, Kelly M (2015) Using population segmentation to inform local obesity strategy in England. Health Promot Int 30(3):658–666

Won M (2016) Contextual communities: why young people are straying from traditional social media in search of more authentic connections. Soc Media Today, 4 Sept 2017. http://www.socialmediatoday.com/social-networks/contextual-communities-why-young-people-are-straying-traditional-social-media-search
World Health Organization (2016) Obesity and overweight, 24 Aug 2017. http://www.who.int/mediacentre/factsheets/fs311/en/
Yellow (2018) Yellow social media report 2018—consumers

Brennan Linda is a Professor at the School of Media and Communication at RMIT University. Her research interests are social and government marketing and especially the influence of marketing communications and advertising on behaviour.

Annika Molenaar is a Nutrition Scientist and Research Officer in the Department of Nutrition, Dietetics and Food at Monash University. Her research is currently focused on health promotion and the delivery of appropriate and engaging messages about nutrition and health through the NHMRC funded Communicating Health project (GNT1115496). She has experience in both qualitative and quantitative methods working on projects for both food industry and government.

Anouk Sherman is a Clinical Nutritionist based in Melbourne. She treats patients for a range of conditions and is passionate about nutrition as preventative medicine. As well as private consultations, Anouk is passionate about improving health outcomes on a broader scale by speaking to groups, from kindergarten children to corporates to retirees. Prior to becoming a healthcare practitioner, Anouk had a career in media and marketing, with extensive experience in scientific publishing whilst living in Amsterdam. Alongside her nutrition practise, Anouk assists various research projects at RMIT and Monash universities, with a focus on health promotion.

Shinyi Chin is a doctoral candidate at the School of Media and Communication at RMIT University. Her research interests are digital marketing and social marketing with a focus on promoting and encouraging positive behaviour change by furthering our understanding of how and why people use digital technologies.

Mike Reid is a Professor in RMIT University's School of Economics, Finance and Marketing. He has both commercial and academic experience in qualitative and quantitative marketing research and has been awarded a number of competitive grants including NHMRC, ARC Discovery, VicHealth Partnership, Sustainability CRC and VicHealth Innovation grants. His research encompasses social marketing areas of food literacy, consumer food choice, obesity and behaviour change, alcohol consumption and harm reduction, and integrated marketing communications.

Helen Truby is a Professor and Chair in Human Nutrition at the University of Queensland and at the time of writing was a Professor at Monash University Department of Nutrition, Dietetics and Food. She has both clinical and academic experience in quantitative research and translation of findings into practice. She has been awarded a number of competitive grants including NHMRC, Department of Health, Cancer Council and various philanthropic trusts. Her research encompasses how to use food as medicine in terms of supporting healthy food choices for the population and behaviour change.

Tracy McCaffrey is a Senior Lecturer in the Department of Nutrition, Dietetics and Food at Monash University, as a co-investigator and project co-ordinator for Communicating Health (NHMRC GNT1115496). She is a Registered Nutritionist (UK), with a research focus on the assessment of food and nutrient intakes in populations and the development of evidence-based dietary guidelines for healthy eating using quantitative and qualitative methods. She has advised government departments in Australia, New Zealand, UK and Ireland. Dr. McCaffrey is a Director of the Australian Nutrition Trust Fund, a not-for-profit organisation, aimed at advancing nutrition science research in Australia.

Part II
Health

A range of healthcare-related social marketing issues are discussed under the health theme.

It *Takes A Village: Co-creation and Co-design for Social Media Health Promotion*, is another co-creation case study that shows the use of co-creation and co-design processes in order to engage young people in social media interventions. This method involves a 'Wicked Problems' studio (5-day workshops), where young adults (university students) across multi-disciplinary areas and a variety of stakeholders (i.e. health organisations, government bodies, academics etc.) collaborate to problem-solve . This studio approach of co-creation and co-design delivers a deeper understanding of effective messages and content creation for social media platforms designed to connect with young adults. The chapter segues into the next theme—healthcare.

The *Male engagement through increased awareness for men's Counselling—A Case Study of Pulih Foundation* chapter focuses specifically on the efforts undertaken in creating campaigns aimed specifically at the prevention of violence against women in urban areas, focusing on awareness raising in help-seeking behaviour in men. One of the outcomes expected from the programme is there will be increased awareness, change of attitude and behaviour of men towards this issue—to include men to prevent violence against women. The main purpose of this case study is to gain awareness of help-seeking behaviour in men by having psychological counselling. The indicator is the increase of men who do counselling in Pulih Foundation.

The *Behavioural change regarding intention indication whether to donate organs: The case of 'Share Your Value Project'* chapter discusses Japan, in case where opportunities to declare one's intent for organ donation exist but people are not aware of them. Therefore, in many critical situations, family members must bear the mental burden to make the decision in a limited time.

The role of coordinated effort in successful healthcare campaign is illustrated in *Mental Health, Alcohol and Other Drugs: After Hours Crisis Communication and Marketing Initiative*, case study wherein communities living within the Murrumbidgee area of New South Wales, Australia, identified issues with mental health, alcohol and other drugs as being a specific concern for the region. In response, the Murrumbidgee Mental Health Drug and Alcohol Alliance (MMHDAA) together with the Murrumbidgee Primary Health Network (MPHN) initiated the development of a marketing and communications initiative to improve access to information at the time of a mental health crisis. A high-profile, integrated marketing campaign was launched in late 2017, and incorporated the elements that equipped all stakeholders to actively support long term increases in help-seeking behaviour. The marketing and communications programme is part of a wider set of comprehensive regional initiatives aimed at addressing these identified needs.

The *Make it Count 2018* case presents another example of an organ donation campaign using social media and smart technologies. This project aimed to raise awareness and increase Australian Organ Donor registrations at the 2018 Brisbane Ekka, in Brisbane, Queensland, Australia. Researchers conducted targeted awareness and registration activities, concurrent with The Ekka, featuring a unique smart device application named the 'DonateLife App', to increase Australians' awareness of donation and directly facilitate online donor registrations.

Chapter 5
Male Engagement Through Increased Awareness for Men's Counselling: A Case Study of Pulih Foundation

Nirmala Ika Kusumaningrum and Irma S. Martam

Abstract Pulih Foundation's lengthy involvement in assisting victims of gender-based violence supports the need for a gender transformation approach to help reduce violence against women and children. This approach uses the socio-ecological framework where violent prevention efforts must be conducted at the individual, family, community, and country level. To make this happen, we participated in the MenCare+ (2013–2015) program and the Prevention+ (2016–2018) program, as well as making our own independent efforts to share campaigns through various media. This paper specifically focuses on the efforts undertaken to create campaigns aimed directly at the prevention of violence against women in urban areas and on the efforts to raise awareness in help-seeking behaviour in men. One of the outcomes expected from the program is that there will be an increased awareness, and a change of attitude and behaviour of men towards preventing male violence against women. The main purpose of this case study is to raise awareness of help-seeking behaviour in men by having psychological counselling. The indicator is the increase of males who participate in counselling in the Pulih Foundation (Yayasan Pulih 2015).

Keywords Masculinity · Men · Women · Gender-based violence · Health and wellbeing · Charity and not-for profit

Note The definition of gender for this paper refers to the roles and responsibilities of men and women that are created in our families, our societies, and our cultures. The concept of gender also includes the expectations held about the characteristics, aptitudes, and likely behaviours of both women and men (femininity and masculinity). Gender roles and expectations are learned. They can change over time and they vary within and between cultures. Systems of social differentiation, such as political status, class, ethnicity, physical and mental disability, age and more, modify gender roles (UNESCO 2003). Gender transformative approaches are programs or interventions that create opportunity for individuals to actively challenge gender norms, promote positions of social and political influence of women in communities and address power inequities between persons of different genders, going beyond just including

N. I. Kusumaningrum (✉) · I. S. Martam
Clinical Psychologist University of Indonesia, Depok, Indonesia
e-mail: nirmala.ika.k@gmail.com

R. Hay et al. (eds.), *Broadening Cultural Horizons in Social Marketing*,
https://doi.org/10.1007/978-981-15-8517-3_5

women as participants http://www.healthcommcapacity.org/wp-content/uploads/2014/08/Gender-Transformative-Approaches-An-HC3-Research-Primer.pdf.

5.1 Introduction

The violence rate against women and children in Indonesia is high. The results of Komnas Perempuan's CATAHU data in 2017 show 348,446 cases of which 70–80% occurred in the private environment (Catatan Tahunan 2018). One of the causes of this violence is the existence of a patriarchal hegemony and deeply rooted social norms that view men as dominant figures who have powers beyond women and children, who occupy subordinate roles. Patriarchal hegemony and social norms that are communicated through education and through nurture have passed along generations and have become part of existing rules and policies.

Men also experience a burden under the masculine and dominant role expected of them and as such, due to a lack of knowledge, experience, and skills to manage the situation, tend to use violence as their catalyst for anger and frustration. Yet when further explored, men are in fact, not comfortable with their displays of violence. Men themselves need to be understood and they need help to be able to manage themselves well to foster better relationships with spouses or people around them.

The Pulih Foundation, with a background in psychosocial services, understands how the process of internalising cultural values is transmitted through education or family value to the individual. Sometimes, the experience leaves a personal trauma and requires a recovery process that requires psychological counselling. However, psychological counselling is not common practice in Indonesia, especially for men. Seeking help tends to be seen as a sign of weakness, which men through learned culture, are not allowed to show. Therefore, introducing new 'help-seeking behaviours' amongst men presents a challenge due to the cultural norms within which, the men have grown up.

5.2 The Social Marketing Intervention Mix

The social marketing intervention for this program has been directed towards changing the behaviour of men—to assist them to seek for help (through male counselling) when they are aware that they have a problem that may lead to an inappropriate response in a relationship (e.g. the use of violence). To achieve this awareness was not easy, considering the social norms attached to seeking help, i.e. it may cause them to be seen as weak or not 'man' enough (which becomes a psychological cost for them).

5.3 Behaviour (Social Change Goals)

Since 2002, the Pulih Foundation has counselled victims of gender-based violence, especially victims of violence against women and children in private environments. In doing this, the Pulih Foundation uses a psychosocial approach, in which the client is seen not only as an individual but also as part of his social circle. Based on experience, the foundation has found that both men and women need to be included in the prevention of violence against women and children because men and women should share gender education in order to understand each other. Many men also disagree with violence and the dominating roles attached to it. The issues related to the gender constructs attached to men, and the development of society today requires a more open view of both men and women in terms of the gender constructs that they each have implanted over their lifetime.

With that in mind, the Pulih Foundation welcomes the WPF Rutgers offer to work with other partners in Indonesia to participate in the MenCare+ program in 2013–2015 and then continue with the Prevention+ program from 2016 to 2018. Both programs aim to assist men to reduce violence against women and children through various approaches. The MenCare+ program is a program implemented in Brazil, South Africa, Rwanda, and Indonesia (Jakarta, Lampung, Yogyakarta and East Java) and the Prevention+ Program as a continuation of MenCare+ is implemented in Pakistan, Rwanda, Uganda, MENA, and Indonesia (Lampung and Yogyakarta).

MenCare+ and Prevention+ attempt to generate gender transformation, with a socio-ecological approach, where men are seen as part of society and social system so that the approach is actioned at an individual, family, community, culture, and country level. The socio-ecological perspective encompasses context in the broadest sense of the world, to include physical, social, cultural, and historical aspects of context (including trends at the local and global level such as globalisation, urbanisation, and large-scale environmental change) as well as attributes and behaviours of persons within (McMichael 2002). This approach is in accordance with the psychosocial approach that the Pulih Foundation has enacted. This multi-level approach in Prevention + program (individual, community, institutional, and government) has become crucial, complementing all the work and advocacy of feminist groups in order to transform the intersecting social and structural factors that allow Gender-Based Violence (GBV) to persist. Another component of the program is engaging young and adult men to be part of the solution, become partners/advocates for change together with young and adult women to challenge and transform harmful gender norms and practices.

Upstream marketing conducted through advocacy of certain policy such as a mandatory attendance to anti-natal care in community health centre for fathers aims to change the behaviour of fathers and fathers to be. Change agents such as midwives, primary health carers, teachers, and religious leaders also need to be involved and persuaded that their role is important to influence the fathers and fathers to be, to change their behaviour towards child rearing. The combination of different approaches utilised at different levels, promotes the behaviour change of the father to be and of new fathers on child-rearing behaviour.

For the Prevention+ program, the Pulih Foundation has developed a strategy through health and education sectors with activities that mobilise communities through strengthening the capacity of individuals (such as counselling) and institutions (training and workshops), public education for empowerment (women), and awareness (men) in a form of seminars, lectures and of discussion forums. Media campaigns promoting the values of cooperation, healthy relations, gender equality, non-violent nurturing, and the value of positive masculinity are also used.

The project, together with its partners, developed a male counselling program that is in line with local conditions based on a module that was developed by Rutgers Netherlands (Rutgers 2016–2018). This counselling program was originally an attempt to make behavioural changes in men who committed domestic violence, but later in their use, developed into counselling for men in general in order to prevent further and recurrent violence.

5.3.1 Customer Orientation

An effort towards male inclusion through campaigns and counselling promotes values that are more positive for women and men, both children and adults, in attempt to minimise the risk and impact of violence. However, many men simply do not understand that their behaviour—because of social norms, culture, and religion—is an act of violence that hurts their spouse and/or child. In general, both women and men do not like to be in conflict situations. In particular, women victims of violence do not want to be separated from their spouses. The experience shared by victims of violence in 2002 highlighted that the majority of women victims of violence did not want a divorce, but they expect their partners to change so they could return to the household. The program aims to manifest synergy between women and men.

5.3.2 Theory

The dynamics and connection behind the approach between activities that have been conducted so far in Indonesia use the Theory of Change (Anderson 2006), the Socio-ecological Model (Sword 2001) and Systems Thinking (Kelly 2012), which saw the intervention create change and increase male clients attending the clinic.

5.3.3 *Insight*

This case study is based on the Pulih Foundation experience (Internal Evaluation Report and Report from the Pulih Counselling Center), Annual report for MenCare+ and Prevention+ (Pulih 2018).

Insights based on the results of the two programs include:

(1) men want to be involved in parenting but they do not understand how to do it.
(2) men need to be viewed as clients, partners, and agents of change.
(3) gender awareness should be shared to both women and men. Strong cultural and community norms often make women see their role as natural and they are disturbed whenever men want to play a role in it.
(4) male inclusion should be ensured to go hand in hand with the prevention of violence against women instead of providing new privilege to men.
(5) it is important to use positive and non-blame messages, and to be incentivised.

In the activities of the Pulih Foundation, among others with the program MenCare+ and Prevention+, Pulih tried to campaign, advocate, and to develop capacity building to various parties that felt could be a key point for the change.

5.3.4 *Exchange*

Based on the work experience of the Pulih Foundation and its partners, it is recognised that counselling and other specific services to men are less accessible. Voluntary and mandatory interventions to prevent recurring violence, especially in intimate and family relationships with spouses and children were conducted for perpetrators of violence to complement other interventions and prevention. Considering the absence of a special assisted group, we intervened in Jakarta with two programs, MenCare+ and Prevention+. When the MenCare+ and Prevention+ programs were initiated, there was no baseline data on violence against women and children for Jakarta. Therefore, the two programs are a medium to achieve the objective of preventing violence against women and children as well as to collect baseline data. We use a case study to show the increasing number of men who access counselling services due to the campaign, using the number of male clients who accessed the previous Pulih Foundation's service as the baseline.

In order to increase the awareness of the availability of counselling for men, the Pulih Foundation has created several awareness-raising activities. The aim is to increase awareness among males, especially about male violence against women and to show how counselling or seeking help is able to develop positive behaviour towards women and children. The activities include collaborating with several universities and high schools to conduct educational awareness and empowerment activities and training. Psychologists and Counsellors increased capacity for men's counselling, by:

(1) Recruiting and training young male facilitators to become agents of change in their environment as well as the wider environment
(2) Conducting campaigns and public education using social media
(3) Working with print and online magazines that shared concern on the same issues
(4) Incorporating the idea of male inclusion and implementing it using a practical campaign such as resourcing a person on field or for media activities
(5) Creating a discussion forum of fathers recruited through social media
(6) Training health and police personnel in an effort to lead mandatory counselling for male perpetrators
(7) Developing a mobile application related to healthy and gender-equal parenting.

By providing information through various media, it was hoped that awareness about discussing personal issues and seeking help would not be seen as shameful. Rather that it would be seen as being useful to help men's psychological health and their relationships with people around them. As a result, by 2014, the number of male clients in the Pulih Foundation reached 27 men. In 2015 and 2016, this increased to 49 men; and in 2017, the reach increased to 103 men. From these activities, we can assume a minimum increase of 20% per year for male clients. Although the number of clients is still very small considering the scale of the problem and number of violence victims. This positive response has shown that increased awareness of violent behaviour has not only become passive knowledge but some of the men do act by seeking help, which is considered a big leap given the social and cultural context they grew up with, where men are supposed to be strong and tough.

5.3.5 Competition

The program efforts are facing some challenges. First, criticism from women activists that the program is actually providing another privilege for men. Second, there are challenges in changing the social norms and in religious teaching, which still strongly promotes a patriarchal culture, where men are perceived as having leadership role (with all the attributes attached to it) both in the family setting as well as in the public domain. On one hand, the entitlement for men (which creates proneness of violence against women is still strong) on the other hand, seeking help when they feel their violent act is not the right thing to do, becomes a signal of weakness. Considering the challenges, the progress achieved so far can be seen as a promising sign towards gender transformation and although there is still a long way to go, the program can support the prevention of violence against women and children.

To be able to sustain a reduction in violence in the long run, change to the current social norms through maintaining good behaviours and being open to discuss negative behaviours is needed. The discussion at all relationship and gender levels is expected to create a change of mindset in the relationship between men and women. The efforts of this behaviour change program align with situations that already exist and that are acceptable to society.

5.3.6 Segmentation

The program targeted young men aged 15–35 years and included high school kids and young fathers or prospective young fathers as they are considered to have more potential to change. Especially, young fathers because they tend to be eager to engage in parenting, and they appear to be more open to the current state of society. The program is targeting young men because they show interest they question practices and values that intrigue them. Therefore, the right information is expected to encourage behavioural changes in them.

5.3.7 Methods Mix

This programs activity is devoted to the urban community in Jakarta. In terms of availability of services, the place for male counselling as well as the trainer is currently very limited. However, considering, it is a pilot study and there will be continuous efforts, it is expected that in the future, there will be more centres/clinics, which are capable to conduct male counselling, specifically for the case of violence against women.

The campaign was carried out by conducting road shows in high schools, discussions with college students, disseminating information using social and mass media, thematic meetings with men, conducting training to focal parties (NGOs, police, some caring communities), and together with the network, advocating specific law to the government. Given that one of the specificities of urban society is the difficulty of designing activities that take place on a regular basis, specific media chosen and approach for the campaign become important. The influences of celebrities from the global campaign, which promoted engagement in fatherhood, also provided positive influence for this target market. The people, (counsellors that have been trained) also need to be increased so that the process (of counselling) is widely available in order to prevent violence against women. The Social Marketing Methods Mix for this program considers the following principles of marketing:

- **Product**—help-seeking behaviour through Male Counselling Program to prevent violence against women and children
- **Price**—the psychological cost for men, who finally realised that they need help, i.e. they might be seen by their social surrounding as not 'men enough'
- **Place**—Counselling Centre such as the Pulih Clinic and also other counselling clinics, hopefully later will be community health centre
- **Promotion**—through road shows in high schools, discussions with college students, disseminating information using social and mass media, thematic meetings with men, conducting training to focal parties
- **People**—the counsellors that have been trained and conducted male counselling, NGO, police, key influencer in communities
- **Process**—of counselling to prevent violence against women.

5.4 Evaluation and Results

There is an increasing number of clients who consulted at the Pulih Foundation either men or women. In general, they become more aware of issues surrounding violence and are encouraged to attend counselling in order to comprehend what they are experiencing. As a result, men come to the Pulih clinics, initially because of issues related to parenting (custody, understanding children's behaviour), or because they are experiencing violence by their spouses. However, there are also men who receive counselling about personal issues unrelated to their spouses, or in an attempt to understand their relationship with their spouses, or to make changes to their violent behaviour.

In addition, there are regular requests from some universities, especially the faculties of Psychology to discuss specifically about male involvement and male counselling. This is new initiative because previously gender-related information has not been considered as a material that must be given to students. There is also a decree to counsel men who reportedly committed violence by their partners, that is, currently in trial stage in two areas of Depok and East Jakarta. There are mobile applications that need further encouragement since they are related to the development of creative ideas that still compete with the current public interest.

5.5 Conclusion

Men, who want to be involved in care giving, need accessible information about healthy relationships and how to nurture them. This information can be transferred through learning forums, social media, or even by utilising the interactive mobile application. There is a change in the results of behaviour change counselling, where counselling services are still required. In that case, the men's counselling program requires an explanation to the partners/community about how to counsel perpetrators, to make expectations of behaviour change more realistic because not all parties understand the definition of behaviour change counselling. Many people expect behaviour change to occur in a short and permanent time through counselling. In this regard, it is necessary to increase the number of psychologists and counsellors interested in the program. To counsel men, one needs to understand the perspective of violence victims first, and this is a challenge because not all psychologists/counsellors have this perspective.

Campaigning and mobilising communities with positive messages and using public figures as role models are more effective to provide early interest to find out more about this activity. People prefer to ask public figures rather than with psychologists and counsellors so it would be good for psychologist to partner with public figures in public discussion and prepare material together. The selection of media for campaigns needs to be adapted to the conditions of society. As an NGO, the Pulih Foundation must also learn to adapt this condition so it is imperative to change the logic of thinking by also considering changing the community system and not just focusing on the delivery of ideas alone. Delivering the male involvement strategies to the mainstream through an agency will be most effective when it comes to finding key personnel in the agency who have the capacity, power, and commitment. There is a need to document, evaluate, research, and publicise a strategy to obtain strong and scientific evidence as the basis for further development of this activity.

Acknowledgements We collaborated with Riffka Annisa Yogyakarta, PKBI Lampung, PKBI East Java, Alinasi New Men in MenCare+ and we collaborated with Riffka Annisa Yogyakarta, Damar Lampung, Rahima, and the Alliance of New Men in Prevention+ with Rutgers WPF as the lead for both projects. The authors would like to thank Pulih colleagues IGAA Jackie Viemilawati, Jane L. Pietra and Danika Nurkalista for discussions and their extensive contribution to this paper.

References

Anderson AA (2006) The community builder's approach to theory of change. a practical guide to theory development. The Aspen Institute Roundtable on Community Change, New York

Catatan Tahunan (CATAHU) Komnas Perempuan Tahun 2018 Tergerusnya Ruang Aman Perempuan dalam Pusaran Politik Populisme Jakarta, 7 Maret 2018. https://www.komnasperempuan.go.id/file/pdf_file/2018/SIARAN%20PERS%202018/Lembar%20Fakta%20Catahu%207%20Maret%202018.pdf

Kelly D (2012) Applying quality management in healthcare; a systems approach, 3d edn. Reference and Research Book News, vol 27, no 1

McMichael AJ (2002) Population, environment, disease, and survival: past patterns, uncertain futures. Lancet 359:1145–1148

Rutgers (2016–2020) Prevention+. https://www.rutgers.international/programmes/prevention

Sword W (2001) A socio-ecological approach to understanding barriers to prenatal care for women of low income. J Adv Nurs 29(5):1170–1177

Yayasan Pulih (2015) Annual report

Yayasan Pulih. (2018) Prevention+ Annual report

UNESCO (2003) Gender mainstreaming implementation framework, baseline definitions of key concepts and terms

Nirmala Ika Kusumaningrum are clinical psychologist graduated from University of Indonesia and has been working on gender based violence, state based violence, domestic violence, crisis and risk management, conflict resolution, gender inequality, and family crisis issues since 2003. Nirmala has had a long engagement with Pulih Foundation—a not for profit organization working for trauma recovery and psychosocial intervention.

Irma S. Martam is a clinical psychologist graduated from University of Indonesia and has been working on gender based violence, state based violence, domestic violence, crisis and risk management, conflict resolution, gender inequality, and family crisis issues since 2003. Irma has also had a long engagement with Pulih Foundation—a not for profit organization working for trauma recovery and psychosocial intervention.

Chapter 6
Behavioural Change Regarding the Declaration of Intent to Donate Organs: The Case of Share Your Value Project

Yoko Uryuhara

Abstract This study aimed to apply the stages-of-change model to develop effective interventions to increase the declarations of intent towards organ donation in Japan. The 'Share Your Value Project' established at the Uryuhara Lab of Doshisha University, Kyoto, designed and implemented multiple social marketing approaches. The effectiveness of the approaches was measured through a survey exploring stages of change among Japanese individuals. Survey results suggested transforming the meaning of the declaration of intent from something directly connected to donating organs, to a message for the family left behind. The survey also revealed that the most significant barrier was encouraging people who had already made up their minds regarding declaring intent to actually put that intent into action. To target people who had already made up their mind, a three-step intervention was performed. First, to provide accurate knowledge to those misinformed regarding organ donation, second to provide time to think about the significance of declaring one's intent to transform the image of the issue into something favourable, and third to provide a specific method to declare one's intent while the participant's level of involvement was high. Because of the intervention, which included an event for a Guinness World Record™ challenge, 18% of the participants declared their intent to become organ donors. This chapter makes a major contribution to the literature that explores factors affecting the intent to donate organs by applying the stages-of-change model through action research.

Keywords Organ donation · Declaration of intent to donate organs · Stages-of-change model · Share your value project · Japan · Behaviour change

---

Y. Uryuhara (✉)
Faculty of Commerce, Doshisha University, Kyoto, Japan
e-mail: yuryuhar@mail.doshisha.ac.jp

R. Hay et al. (eds.), *Broadening Cultural Horizons in Social Marketing*,
https://doi.org/10.1007/978-981-15-8517-3_6

6.1 Introduction

Opportunities to declare one's intent regarding organ donation exist in people's daily lives in Japan. However, many people are not aware of them, for example driver's licenses, insurance cards and 'my number cards' that are similar to a social security card. According to a public poll conducted in 2017, only 12.7% of 1,911 people over the age of 18 reported they had declared their intent to become an organ donor or not (Cabinet Office 2017). This issue causes three serious problems, affecting individuals, families, and society.

First, individuals' rights are not being respected. Since Japan adopted the Explicit Consent Law that determines organ donation based on an individual's will, it is important to declare intent regarding organ donation during one's lifetime otherwise a person's wishes may not be respected after death. This can particularly be a problem for those who do not wish to donate their organs; therefore, individual declaration of intent plays an important role (Uryuhara 2016).

Second, the psychological burden placed on one's family. If an individual does not declare his or her own intent clearly, the decision is then left to the surviving relatives. If someone in the family has been declared brain dead or, for example, dies suddenly due to cardiac arrest, it is very difficult for surviving relatives to make a decision regarding organ donation within a limited time. The same public poll showed that 87.4% of respondents stated that they would like to respect an individual's written will of donating organs. Actual 90.1% of families agreed to donate a relative's organs after he or she was declared brain dead because they wanted to respect the person's wishes. Therefore, not leaving a declaration of one's intent may impose a psychological burden on the surviving family members.

The third problem—which affects society as a whole—is that organ transplant opportunities are being lost. Nationwide, more than 4,000 people die annually after failing to receive a transplant. Furthermore, there is global ethical criticism regarding Japan's reliance on other countries for organ donation as the lack of local treatment opportunities forces patients to seek transplantation abroad. The percentage of people in Japan who want to donate their organs after death (43.1%) is not smaller than that of other countries. The fact is that donors intent is not being declared, resulting in Japan having the lowest number of organ donors worldwide. Only 2% of those waiting for an organ transplant in Japan will receive one (Japan Organ Transplant Network 2019). The current situation in Japan in which patients are unable to enjoy the benefits of having a high standard of scientific technology, harbours problems related to the United Nations Sustainable Development Goal 3 'Good Health and Well-Being' and Goal 10 'Reduced Inequalities'.

Taking action to clearly declare one's intent regarding organ donation is essential for individuals, families, and society. The rate of declaring intent for organ donation remains low in Japan, despite the largest variety of means in the world to provide this information, including driver's licenses, medical insurance cards, my number cards, organ donation intention cards, and internet registration. Therefore, we need to clarify why this is so and how we can promote the declaration of intent regarding organ donation.

During the past 20 years in Japan, national and local governments have carried out activities to raise awareness about organ donation, but the effectiveness of these efforts has not been clearly demonstrated. Previous research suggests that disseminating information through mass media and public awareness campaigns alone does not lead to individuals taking action regarding the declaration of intent to be an organ donor (Jacob 1996; Thomson 1993; Wolf et al. 1997). In addition, that measures involving active participation, with a focus on younger generations, would be more effective (Callender et al. 1997; Cantarovich 2004; Mandell et al. 2006; Matesanz and Dominguez-Gil 2007).

This study aimed to clarify the mechanisms and interventions necessary for people to be interested in organ donation as well as to promote action regarding the declaration of intent for organ donation.

6.2 Stages-of-Change Model

Previous studies were examined from the viewpoint of values, knowledge, and external interventions. Although all of the reviewed studies provided important viewpoints and suggestions to effectively promote organ donation decision-making, it is still necessary to find out where and how these factors will work during the process as well as their mechanisms. The core is the application of the 'stages-of-change' model (Prochaska and DiClemente 1983).

The stages-of-change model was originally developed from research on smoking cessation (Prochaska and DiClemente 1983), and it posits that change in health-related behaviours progresses through stages: pre-contemplation (no intent to change behaviours), contemplation (considering changing behaviours, but no action taken), preparation (some action taken to change behaviours), action (less than 6 months after the behavioural change), and maintenance (over 6 months after the behavioural change). The theory implies that suitable interventions for behavioural change differ according to the current stage of the target audience. The stages-of-change model (Prochaska and Velicer 1997) was applied to design effective interventions towards organ donation. However, the model is limited when discussing the declaration of intention for organ donation.

Applying the stages-of-change model towards organ donation, suggested five stages of declaring behaviours: Stage 1 (not interested), Stage 2 (interested and considering making a decision), Stage 3 (made a decision regarding organ donation; yes or no, but has not yet declared intent), Stage 4 (declared intent), and Stage 5 (shared the decision with one's family) (Fig. 6.1).

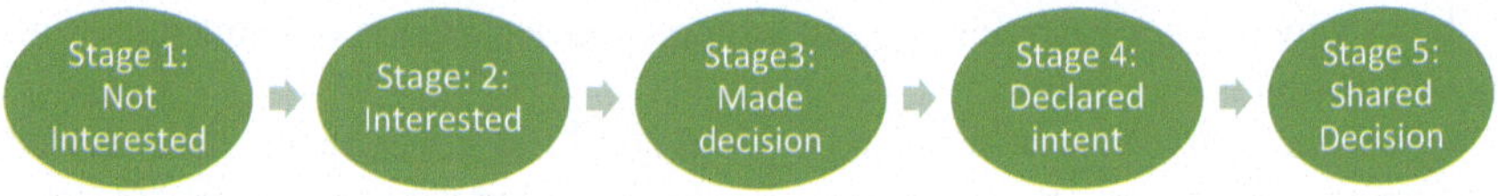

Fig. 6.1 Stages of intention to donate organs

6.3 Overview of the Survey of 10, 000 Japanese People

Preceding the action research, an online survey was conducted to better understand the current state and background of attitudes and behaviours regarding organ donation among Japanese individuals over the age of 20. The questionnaire covered diverse aspects regarding the declaration of intent and other related prosocial behaviours and included outcome variables (level of interest, attitude, behaviours), factors related to transplantation (knowledge, thoughts on transplantation medicine, commitment), personal beliefs (prosocial behaviours, behavioural norms, aid norms, empathy) and impression management, and individual attributes (Table 6.1).

Respondents who chose the same answers for all 10 questions at the end of the survey as well as those with high impression management were excluded from the analysis to minimise bias. The responses were weighted using national Japanese population statistics by prefectures, age, and gender as of 1 October 2013 (Ministry of Internal Affairs and Communications Statistics Bureau 2013a) to match demographics by prefectures in Japan and came up with 10,000 responses for analysis rendering the survey to be sufficiently representative of the state of Japanese citizens.

A factor analysis was conducted on responses regarding transplantation medicine to confirm reliability and validity, using IBM SPSS version 21. For factors affecting outcome variables (e.g. level of interest and behaviours), average values for each item were calculated by groups: interested/not interested in organ donation and declared/not declared one's intent. Two-tailed t-tests were conducted using SPSS (significance level: $p < 0.05$).

6.4 Behavioural Change Stage of Organ Donation Intent Declaration

Using data on levels of interest, attitudes, and declaration behaviours, samples were classified into four stages: Stage 1, Stage 2, Stage 3, and Stage 4.

The results showed that 43.4% of the respondents were in interested in donating their organs (Stage 2) and 36.9% had made the decision to donate (Stage 3). Nineteen per cent of respondents were in Stage 4 and had declared intent. As shown in Fig. 6.2, the findings revealed two barriers in the process of declaration behaviours: from Stage 1 to Stage 2—becoming interested and from Stage 3 to Stage 4 taking action based

Table 6.1 Survey question topics

Dimension 1	Dimension 2	Number of questions	Summary	Number of scales/choices given
Outcome variables	Level of interest	5	Volunteer, donation, blood donation, bone-marrow donation, organ donation	7
	Attitude (behavioural intention)	7	Volunteer, donation, blood donation, bone-marrow donation/registration, organ donation/registration	7
	Behaviours	7	Volunteer, donation, blood donation, bone-marrow donation/registration, organ donation/registration	5
Factors related to transplantation	Knowledge	7	State of transplantation, condition to donate organs	4
	Thoughts on transplantation medicine	22	Anxiety for transplantation, use of physical body, value of transplantation, value of declaration of intent	7
	Commitment	5	Opportunity to share, opportunity to perceive	5
Personal beliefs	Prosocial behaviours	21	Family, friends, others	7
	Behavioural norms	12	Consideration for others, public benefit, conforming with others	7
	Aid norms	11	Self-sacrifice, relief for the vulnerable	7
	Empathy	10	Acquiring perspectives, sympathetic attention	7
Bias	Impression management	7	Social desirability	7
Attributes	Individual attributes	7	Age, gender, profession, academic background, address, family structure	N/A

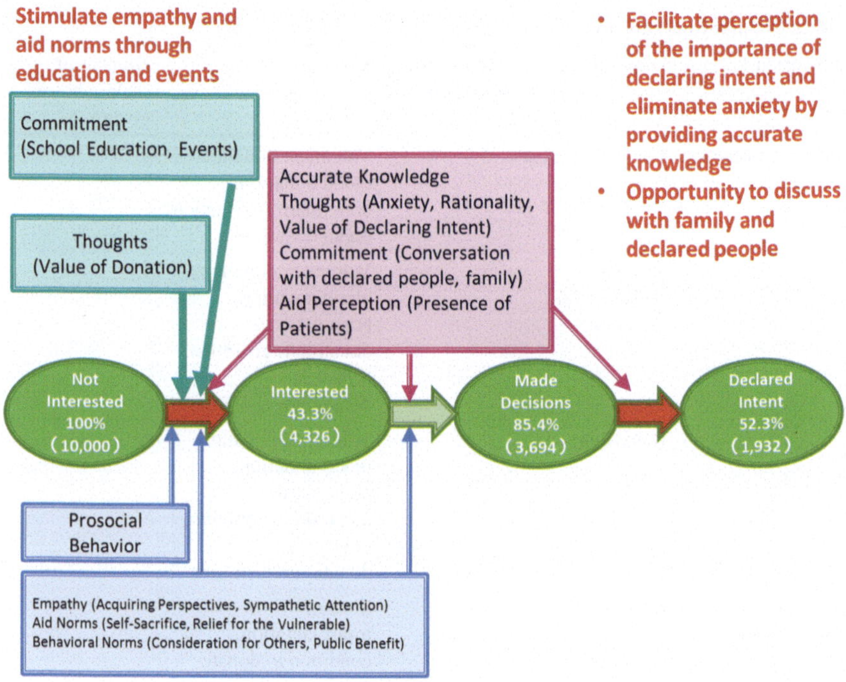

Fig. 6.2 Identified candidates for intervention

on a decision. The study found that providing interventions between these stages is the key to promoting behaviours regarding declaring intent for organ donation.

6.5 Factors Affecting Interest and Declaration of Intent

Respondents were grouped as either interested or not interested in organ donation as well as those with declarations or no declaration of intent regarding organ donation. Two-tailed t-tests examined statistically significant differences (significance level: $p < 0.05$). Results revealed that anxiety towards donating organs was higher among those not interested in organ donation compared with those who were interested. Other factors regarding perception, knowledge, and commitment were lower in the former group. However, when compared with those who had declared their intent to be organ donors, differences in some aspects of commitment were not statistically significant. No significant differences were observed in 'opportunities for school education' or 'event participation'. However, scores for 'presence of someone nearby who has already declared intent' and 'conversations with family' were statistically lower.

These results revealed that necessary interventions to eliminate barriers differ by stages of behavioural change (Fig. 6.2). For example, effective interventions to increase interest towards organ donation (Stage 1 to Stage 2) may provide knowledge on the 'positive value of organ transplantation' in school education and/or at events to enhance empathy and encourage norms. To encourage actual declaration based on one's decision (Stage 3 to Stage 4), creating opportunities to have a conversation with family, friends, and those who had already declared their intent as well as to remove anxiety, appeared effective.

6.6 Share Your Value Project

In April 2015, the 'Share Your Value Project' (SYVP) was established in the Uryuhara Lab of Doshisha University, Kyoto, as a research organisation that implements social marketing. This action research project organization focused on third-year undergraduate students. The name 'Share Your Value Project' indicates the expectation of a society in which each person considers various social issues, proactively thinks and acts, shares and recognises valuable thoughts and actions, and links them together.

Through establishing their independence university students have more opportunities to interact with 'intention-display media', e.g. getting their first health insurance card, or obtaining their first driver's license for the first time, than other potential donors. Stefanone et al. (2012) reported that activities led by highly conscientious students transform the behaviour of college students' more than traditional media campaigns. Therefore, SYVP was considered appropriate for university students to try to implement behavioural change activities.

6.7 Overview of the 'MUSUBU 2016 Campaign'

The intervention targeted university students, primarily those studying social sciences and humanities in the Kansai area of Japan. According to the Ministry of Internal Affairs and Communications Statistics Bureau (2013b), over 90% of university students in Japan are non-medical students, and the most popular discipline is social sciences. As such, we believe our sample is well representative of the demographics of university students in Japan.

Based on the findings from the survey, as well as prior research, we designed and implemented a series of interventions called the 'MUSUBU (connecting people's hearts) Campaign' (Fig. 6.3). The campaign was created to encourage people to declare their intent regarding organ donation in tangible forms. The campaign took participants through the process of gaining interest in declaration of intent for organ donation, acquiring accurate knowledge, and having conversations with family and other important people in their lives. The campaign aimed to bring participants to Stage 4 of the stages-of-change model—declared intent.

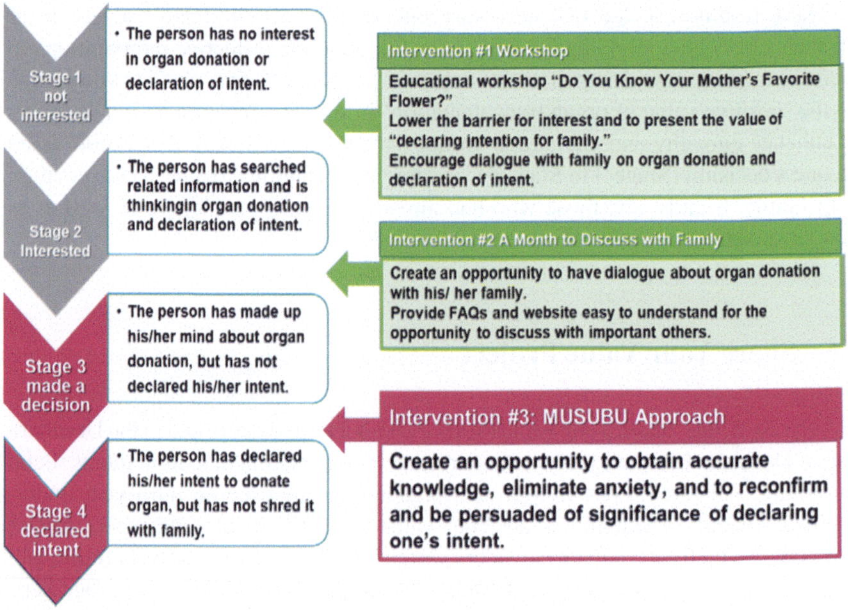

Fig. 6.3 Series of interventions in the MUSUBU campaign

6.8 Interventions to Facilitate Behaviours

Our findings from the aforementioned survey led to the emergence of a theory to facilitate declaring behaviour. The theory focused on moving people from Stage 3 to Stage 4. In this phase, it is important to have eliminated misunderstandings and fear with sufficient knowledge (Shultz et al. 2000). Another effective intervention after having spent much time thinking about organ donation (Carducci and Deuser 1984; Carducci et al. 1989; Skumanich and Kintsfather 1996) is to provide access to a specific way to declare one's intent after interests have increased (Sanner et al. 1995). Thus SYVP designed a series of interventions. First, an intervention must provide accurate knowledge and enough time to think so that people who lack information or have misconceptions regarding organ donation and declaring intent can consider their options. Doing so will enable potential donors to perceive its importance and reduce fear as well as anxiety. Second, providing time to think about the significance of declaring one's intent may transform the image of organ donation and declaring behaviours into something favourable. Finally, providing a specific method to declare one's intent while the participant's level of involvement is high will facilitate declaring behaviour.

Based on the theory, our intervention called the 'MUSUBU approach' (Fig. 6.4) first raised awareness through a Guinness World Record™ challenge (https://guinnessworldrecords.jp/news/2016/10/largest-organ-donation-awareness-lesson#). This intervention aimed to provide accurate knowledge on organ donation and declaration

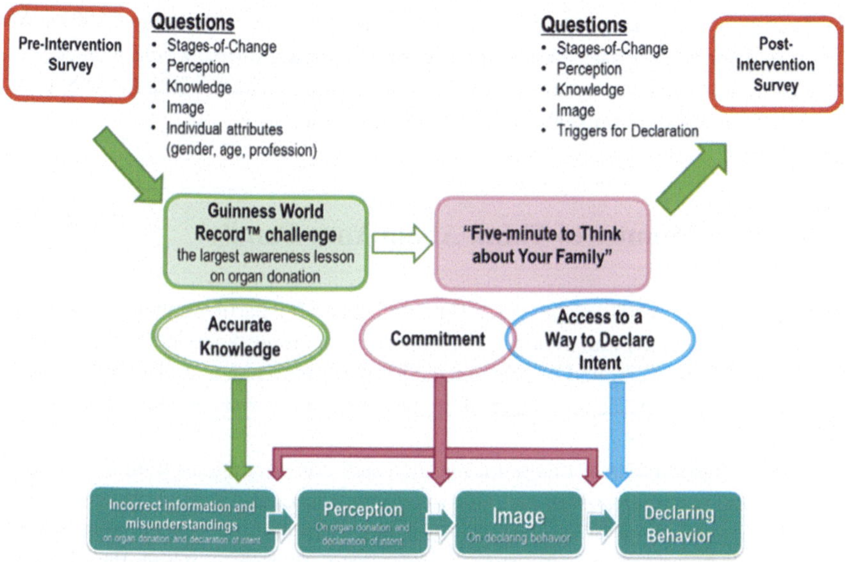

Fig. 6.4 Detail of intervention 3 'MUSUBU approach'

of intent, as well as eliminate participants' anxiety. We then designed an activity we called 'Five Minutes to Think about Your Family' to provide time for individuals to think about the significance of declaring one's intent. At the end, we provided SYVP's original organ donation intention card. Each component of this intervention is described below.

6.9 'Largest Organ Donation Awareness Lesson' as a Guinness World Record™ Challenge

To explore what knowledge should be provided in the lesson, SYVP first conducted an interview with 11 university students studying social sciences and humanities at Doshisha University. The findings revealed the potential effectiveness of knowledge on brain death, organ donation, and declaration of intent.

'Knowledge on brain death' included the fact that brain death is different from a vegetative state, and that human beings will not recover once they are brain dead. 'Knowledge on organ donation' included the actual procedure of organ donation and the state of the human body after organs have been removed for donation. 'Knowledge on declaration of intent' depended on information such as 'when a family member becomes brain dead, 87% of Japanese citizens want to respect his or her will' and 'you can revise your declaration of intent'.

SYVP developed the awareness lesson plan based on these items. We also referred to studies on science, technology, and society that indicated that the general public places emphasis on the trustworthiness of information sources (Hirakawa 2001). We therefore decided that a lesson given by 'experts with authority' would enhance the effectiveness of the intervention, and SYVP invited the president of the Japan Society for Transplantation, to present the lecture for the lesson.

6.10 'Five Minutes to Think About Your Family'

In this part of the event, we provided time and space for participants to think about declaration of intent immediately after acquiring accurate knowledge through the awareness lesson. We aimed to enhance the level of involvement to declare of intent, influence change in perceptions regarding declaring behaviours and have individuals take action on the spot.

Since all participants at this point had invested considerable time in the event—coming to the venue and listening to the opening remarks and awareness lesson—we assumed that their level of involvement was already quite high. A member of SYVP made a speech and asked participants to imagine a situation in which they became brain dead, and their family members must make the final decision regarding whether to donate their organs.

Participants were asked to think about their family and other important individuals in their lives. Furthermore, to deepen the level of involvement among participants, we used the 'foot-in-the-door technique' (Freedman and Fraser 1966). A five-minute speech included seven specific requests (including one minute to declare one's intent). Requests consisted of small requests where a high rate of agreement was expected, large requests including the key action for agreement, and dummy requests with lower necessity to achieve our primary objective. This technique can increase the probability of agreement to declare one's intent with smaller and larger requests, while mitigating the sense of being forced to declare with dummy requests at the end.

The speech included three small requests (close your eyes, think about your family, and imagine a scene where you can no longer declare your intent), one large request (declare one's intent), and three dummy requests (hold the organ donation intention card, share the knowledge and thoughts gained at the event with family after going home, and rewrite one's intent every time it changes). The main request of declaring one's intent was complemented by related requests, such as take out a paper and pen, write down one's intent on the original organ donation intention card, write a message for family on the card, and fold the card, intentionally asking participants to perform the final task. We expected these segmented requests would facilitate smoother access to the media necessary to declare one's intention.

6.11 Providing a 'Letter Card to Your Family'

We distributed an original organ donation intention card to the participants (Fig. 6.5). The aim was to promote declaring behaviours among participants whose level of involvement towards declaration of intent has heightened following the 'Five Minutes to Think about Your Family', and to reduce barriers to actual declaration.

In creating the original card, SYVP referred to the findings from an exploratory study conducted in 2015 targeting people who received organ donation intention cards, but had not yet declared intent. We also conducted interviews with 24 university students to understand their thoughts regarding existing organ donation intention cards. As a result, we found that the new media for declaring one's intent must be

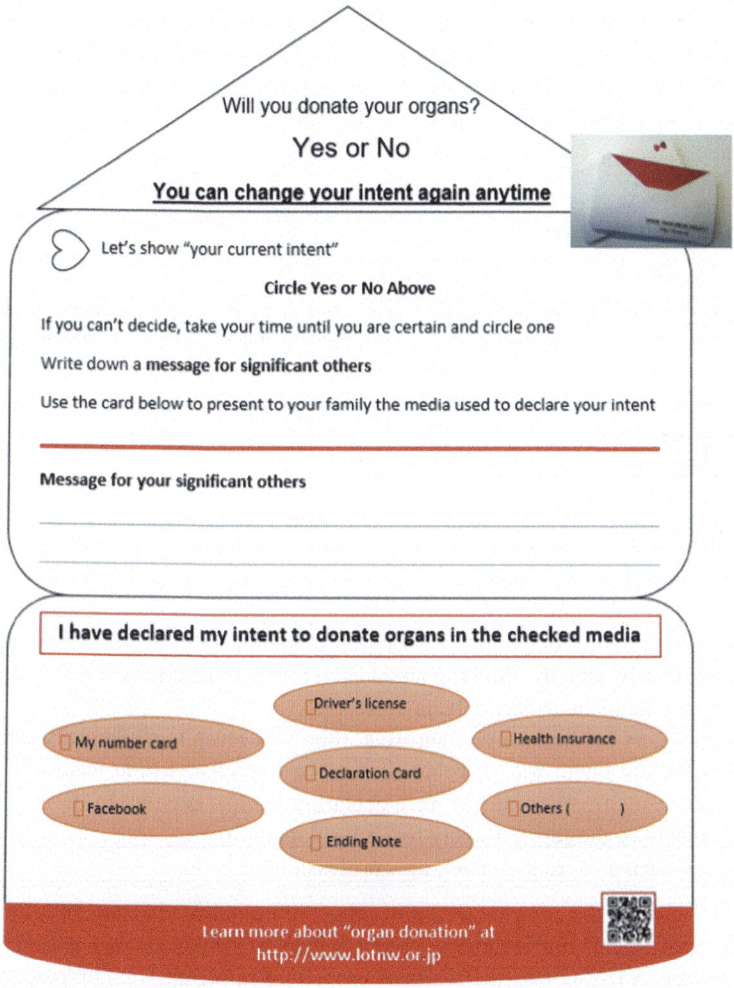

Fig. 6.5 Original organ donation intention cards

something that people would not want to postpone declaring one's intent on and also it must be something they would want to carry around. We also found that descriptions on the card should be simple and written so that anyone could easily understand them.

SYVP came up with the concept of a 'letter card' and worked out the design and content of an original organ donation intention card. The result was an envelope-shaped card with a space to write a message for important individuals in one's life. This space provided an additional function for the card other than only declaring one's intent of YES or NO to become an organ donor and symbolically presented a new value that 'declaration of intent is a message that one should leave for others'. Warm colours, such as orange, were used to give the impression of familiarity and comfort (Hirayu 2002). Corners of the card were also rounded to give the impression of ease.

SYVP then came up with easy-to-understand descriptions to print on the card. We decided to use the phrase 'your current intent' to demonstrate ease and lower the hurdle. The card also included the phrase 'you can change your intent again anytime' to mitigate the mental burden of making a declaration.

6.12 Measurement

A total of 433 participants in the MUSUBU approach were asked to answer a questionnaire on paper twice during the event, before the Guinness World Record™ challenge and after the 'Five Minutes to Think about Your Family'. We considered answers provided in the former as 'pre-intervention' and those provided in the latter as 'post-intervention'.

Items used on the questionnaire were as follows:

'knowledge on brain death, organ donation, and declaration of intent'
'perceptions of organ donation and declaration of intent' (e.g. if the person is brain dead and he/she is dead; I am worried about organ donation; it is important to declare one's intent)
'image of declaring one's intent for organ donation' (e.g. useful, scary, proud, familiar, family, anxiety, thinking about each other, connection)
'stages of change in declaring behaviour'
'direct triggers for declaring intent' (e.g. time was given to declare my intent; a card to declare intent was distributed; those sitting close to me were declaring their intent; many others at site were declaring intent; listening to the lesson by the president of the Japan Society for Transplantation)
'personal attributes' (e.g. gender, age, profession).

When analysing the answers for knowledge on brain death, organ donation, and declaration of intent, one point was given for each correct answer and zero for incorrect ones. A five-point Likert scale was used for rating answers on perceptions of organ donation and declaration of intent, image of declaring one's intent for organ donation, and direct triggers for declaring intent. 'Strongly agree' was given five points, 'agree' four points, 'neutral' three points, 'disagree' two points, and 'strongly disagree' one point. A five-point Likert scale was also used for capturing stages-of-change for declaration of intent. Five points for having shared one's declared intent with family and important others (Stage 5). Four points for having declared one's intent (Stage 4), three points for having decided to declare one's intent, but not taking action (Stage 3), two points for being interested and thinking about declaring one's intent (Stage 2), and one point for not being interested (Stage 1).

Pre- and post-intervention averages were calculated for all items. We conducted two-tailed t-tests using SPSS to examine whether knowledge was acquired, perceptions were changed, and participants moved up in the stages-of-change.

6.13 Evaluation and Results

The response rate for questionnaires was 95%. We excluded respondents who did not answer one or more questions regarding stages-of-change, knowledge, or perceptions. A total of 362 samples qualified for analysis. Nearly 17% of the sample had declared their intent to donate. This was slightly higher than the national average of Japan (12.7%) according to a survey conducted by the Cabinet Office (2017).

6.13.1 *Knowledge*

Average values were calculated for each item.

knowledge on brain death (e.g. human beings will not recover once they are brain dead; the brain will become 'soupy' 1–2 weeks after becoming brain dead; human beings may recover from a vegetative state),
knowledge on organ donation (e.g. the body will be returned in a clean condition; the body will be returned in 2–6 h; there is no cost for donating organs), and
knowledge on declaration of intent (e.g. your family will bear the burden if intent is not declared; if your family member becomes brain dead, about 90% want you to respect his/her will; 'do not donate organs' is an option for declaring intent; you can rewrite your declaration of intent).

Two-tailed t-tests using SPSS revealed that all knowledge was acquired in a statistically significant manner ($p < 0.001$) except for 'your family will bear the burden if intent is not declared' and 'you can rewrite your declaration of intent'.

Table 6.2 Changes in perception towards organ donation

Item	Pre-intervention	Post-intervention
If the person is brain dead, and he/she is dead	3.15	4.15
I am worried about organ donation	3.24	2.77
It is important to declare one's intent	4.23	4.66

Numbers are averages scored on five-point Likert scale Strongly Agree (5)—Strongly Disagree (1)

Table 6.3 Changes in images towards the declaration behaviour

	Pre-intervention	Post-intervention
Useful	4.51	4.78
Proud	3.11	3.80
Familiar	2.48	3.67
Family	3.41	4.24
Thinking about each other	3.65	4.25
Connection	3.68	4.35
Scary	2.96	2.54
Anxiety	3.16	2.70

Numbers are averages scored on five-point Likert scale Strongly Agree (5)—Strongly Disagree (1)

6.13.2 Changes in Perceptions Towards Organ Donation

Average values for items on perceptions towards organ donation and declaration of intent were calculated. Pre- and post-intervention scores are shown in Table 6.2. Two-tailed t-tests revealed that all items on perception changed favourably in a statistically significant manner ($p < 0.001$).

6.13.3 Changes in Images Towards Declaration Behaviour

Average pre- and post-intervention scores were calculated for each item on the perception of images towards declaration behaviours. See results in Table 6.3. Two-tailed t-tests using SPSS revealed that all items on images showed favourable changes post-intervention ($p < 0.001$).

6.13.4 Change in Stages of Declaration Behaviour

The average value for scores of stages-of-change was 2.29 pre-intervention and 2.93 post-intervention. Two-tailed t-tests using SPSS revealed that changes in stages were

Table 6.4 Stages before and after the intervention

Stage 1: not interested Stage 2: interested and considered making a decision Stage 3: made a decision regarding organ donation, but has not yet declared intent Stage 4: declared Intent Stage 5: shared the decision with the family		Stages after intervention					Total
		Stage 1	Stage 2	Stage 3	Stage 4	Stage 5	
Stages before intervention	Stage 1	20	65	10	10	0	105
	Stage 2	1	76	38	24	0	139
	Stage 3	0	5	21	28	3	57
	Stage 4	0	0	1	27	3	31
	Stage 5	0	0	0	0	30	30
Total		21	146	70	89	36	362

promoted in a statistically significant manner ($p < 0.001$). Regarding the percentage of participants whose stages changed post-intervention, 48% experienced no change, 37% moved up one stage, and 13% moved up more than two stages.

Table 6.4 presents the number of participants in each stage pre- and post-intervention. Ten out of 105 Stage 1 participants, 24 out of 139 Stage 2 participants, and 31 out of 57 Stage 3 participants moved up one or more stages and newly declared his or her intent to donate organs. Thus, a total of 65 participants showed direct behavioural change, accounting for 18% of all participants. As a result, the declaration of intent for organ donation among the participants of the MUSUBU approach reached 34.5%.

6.13.5 *Triggers for Declaration of Intent*

Regarding direct triggers to declare one's intent among the 65 participants who did so at the venue, over 90% of them answered 'yes' to 'time was given to declare my intent', 'a card to declare intent was distributed', and 'listening to the lesson by an expert'. 'Those sitting close to me were declaring their intent' and 'many others at site were declaring intent' were not identified as direct triggers.

6.14 Discussion and Conclusion

After the series of interventions, knowledge acquisition was statistically significant, with the exception of only two items. We assume that one of the exceptions—you can rewrite your declaration of intent—was not statistically significant because many participants had already answered this question correctly before the intervention. The other exception—your family will bear the burden if intent is not declared—may have been affected by the fact that no specific case nor data were presented to support this claim.

The research, using a series of interventions matching the stages-of-change led to desirable changes in knowledge, perceptions of organ donation, and declarations of intent, as well as the image of declaring behaviours. Our findings revealed the importance of providing a specific method to declare one's intent, after providing accurate information and facilitating an enhanced level of commitment.

Our study had several limitations. First, we were not able to assess individual effects of providing accurate knowledge, enhancing commitment, and providing access to a specific method to declare one's intent regarding organ donation. We were also not able to examine how each factor directly contributed to behavioural change. Second, the study could not examine in detail how much the atmosphere of an unusual event—a Guinness World Record™ challenge—affected participants' behaviours. Finally, our study only examined behavioural change immediately after the intervention.

We intend to conduct a similar study with a population with diverse attributes—in addition to university students—and accumulate findings to develop evidence for promoting declaration of intent for organ donation. The 'Share Your Value Project' hopes to bring to reality a society that respects the wishes of people who have died and that respects where people can express their intent for organ donation with pride.

Acknowledgements This work was supported by Japan Society for the Promotion of Science (JSPS) KAKENHI Grant Number JP16K45678 'empirical research on diffusion of social value on transplantation medicine to public' (principal investigator: Yoko Uryuhara, 2013–2016) and Yoshida Hideo Memorial Foundation Research Grant 'solving challenges of transplantation medicine through social marketing' (principal investigator: Yoko Uryuhara, 2016–2017). I also appreciate proffesor Aya Okada (Tohoku University) and all members of Share Your Value Project (SYVP) for their joint effort in this research.

References

Cabinet Office, Government of Japan (2017) Survey on organ transplantation (August 2017 Survey). https://survey.gov-online.go.jp/h29/h29-zouki/. Accessed 31 Mar 2019

Callender C, Burston B, Yeager C, Miles P (1997) A national minority transplant program for increasing donation rate. Transpl Proc 29(1–2):1482–1483

Cantarovich F (2004) The role of education in increasing organ donation. Ann Transpl 9(1):39–42

Carducci BJ, Deuser PS (1984) The foot-in the door technique: initial request and organ donation. Basic Appl Soc Psychol 5(1):75–81

Carducci BJ, Deuser PS, Baucer A, Large M, Ramaekers M (1989) An application of the foot in the door technique to organ donation. J Bus Psychol 4(2):245–249

Freedman JL, Fraser SC (1966) Compliance without pressure: the foot-in-the-door technique. J Pers Soc Psychol 4(2):195–202

Hirakawa H (2001) What is "STS"? La Pensee Contingente. https://hideyukihirakawa.com/sts_archive/sts_general/what_is_sts.html. Accessed 24 April 2017 (In Japanese)

Hirayu F (2002) Towards attractive information space: furniture and layout of libraries. J Inf Sci Technol Assoc 52(1):3–8 (In Japanese)

Jacob F (1996) Regional awareness campaign concerning organ sharing. Transpl Proc 28(1):393

Japan Organ Transplant Network (2019). https://www.jotnw.or.jp/. Accessed 31 Mar 2019

Mandell MS, Zamudio S, Seem D, McGaw LJ, Wood G, Liehr P, Ethier A, D'Alessandro AM (2006) National evaluation of healthcare provider attitudes toward organ donation after cardiac death. Crit Care Med 34(12):2952–2958

Matesanz R, Dominguez-Gil B (2007) Strategies to optimize deceased organ donation. Transpl Rev 21(4):177–188

Ministry of Internal Affairs and Communications Statistics Bureau (2013a) Japanese population by prefectures, age, and gender. https://www.stat.go.jp/data/jinsui/2013np/. Accessed 25 Apr 2013

Ministry of Internal Affairs and Communications Statistics Bureau (2013b) Education: number of students by Departments in Graduate Schools, Universities, Junior Colleges, and Higher Professional Schools, Junior Colleges. https://www.stat.go.jp/data/nihon/back13/22.htm. Accessed 1 Feb 2013 (In Japanese)

Prochaska JO, DiClemente CC (1983) Stages and processes of self-change of smoking: toward an integrative model of change. J Consult Clin Psychol 51(3):390–395

Prochaska JO, Velicer WF (1997) The transtheoretical model of health behavior change. Am J Health Promot 12(1):38–48

Sanner MA, Hedman H, Tufveson G (1995) Evaluation of an organ donor card campaign in Sweden. Clin Transpl 9:326–333

Shultz KH, Meier D, Clausen C, Kuhlencordt R, Rogiers X (2000) Predictors of intention to donate organs: an empirical mode. Transpl Proc 32(1):64–65

Skumanich S, Kintsfather D (1996) Promoting the organ donor card: a causal model of persuasion effects. Soc Sci Med 43:401–408

Stefanone M, Anker AE, Evans M, Feeley TH (2012) Click to "Like" organ donation: the use of online media to promote organ donor registration. Prog Transpl 22(2):168–174

Thomson NM (1993) Transplantation—the issues: a cross curriculum programme for secondary schools. Transpl Proc 25(1):1687–1689

Uryuhara Y (2016) Practice of social innovation in higher education and its effectiveness: from the case of "Share Your Value Project" in solving issues on medical transplantation. Doshisha Shogaku (Doshisha Bus Rev). 67(5–6):45–85 (In Japanese)

Wolf J, Servino R, Natahn H (1997) National strategy to develop public acceptance of organ and tissue donation. Transpl Proc 29(1–2):957–963

Yoko Uryuhara is an associate professor at faculty of commerce, Doshisha University. She is also appointed as a board member of Japan Organ Transplant Network, a member of Science Council of Japan. Her research fields are social marketing and organizational behaviour. She worked for global pharmaceutical company for 20 years and has engaged in clinical development, strategic planning, marketing, public relations and health care policies. She aims to creating evidence, which contributes to a solution to a social problem from a diversified view point taking advantage of the education of both social science (marketing, management) and natural science (medicine, pharmacy), and an experience in actual business.

Chapter 7
Mental Health, Alcohol and Other Drugs: After Hours Crisis Marketing Initiative, a Unique and Local Social Marketing Campaign Within a Regional and Rural Australian Setting

Carolyn Loton and Melissa Neal

Abstract Addressing and combatting suicide, mental health, drug and alcohol issues are complex challenges. Seeking help at a time of crisis is associated with improved long-term health outcomes and improved quality of life measures. In parts of the Murrumbidgee area, rates of mental illness, suicide and self-harm are significantly higher than state and national averages. Community organisations in the Murrumbidgee area commissioned a social marketing campaign aimed at increasing awareness of services, reducing stigma and encouraging help-seeking at a time of crisis. The campaign reflected the unique needs of the area and the diversity of the local community, and was part of a wider suite of initiatives. This social marketing campaign was unusual in taking an approach aligned with larger campaigns yet applying local insights and a very local approach to execution. The use of multiple media channels was effective in ensuring a wide reach. Social media, and in particular Facebook and Instagram, played a significant role in the campaign's success. The approach used in this mental health awareness campaign demonstrates the value of using community consultation, co-design and quantitative research, in developing an effective behaviour change marketing campaign. It shows that taking a localised approach to the design and development of social marketing campaigns can lead to effective outcomes.

Keywords Social marketing campaign · Behaviour change marketing · Behaviour change campaign · Co-design · Reducing stigma · Social marketing · Mental health awareness campaign · Mental health campaign · Local social marketing campaign · Marketing for good

---

C. Loton (✉)
Juntos Marketing, Sydney, Australia
e-mail: carolyn@juntosmarketing.com.au

M. Neal
Murrumbidgee Primary Health Network, Wagga Wagga, Australia
e-mail: melissa.neal@mphn.org.au

R. Hay et al. (eds.), *Broadening Cultural Horizons in Social Marketing*,
https://doi.org/10.1007/978-981-15-8517-3_7

7.1 Introduction

If you live in Australia, it is likely that either the person sitting on your left or the person sitting on your right has experienced mental illness in their lifetime. One in five will experience mental illness this year. In the majestic Murrumbidgee, with its unique landscape of vast Australian beauty and its heartland of agriculture, your chances of experiencing a mental health concern are even greater. The incidence of self-harm and alcohol use requiring hospitalisation is significantly higher in many areas of the Murrumbidgee compared to national and state averages. Even more so for those who identify as Lesbian, Gay, Bisexual, Transgender, Intersex or Queer (LGBTIQ), those who are of Aboriginal or Torres Strait Islander background, those from a culturally and linguistically diverse background; and for those living in an isolated rural situation. Yet mental health, drug and alcohol concerns are often taboo. Many people do not seek help because of the associated stigma and when people do look for help, options can be confusing or hard to find.

In 2016, the Murrumbidgee community and the Murrumbidgee Primary Health Network (Murrumbidgee PHN) in partnership with the Murrumbidgee Mental Health Drug and Alcohol Alliance (MMHDAA) commissioned a social marketing and communications campaign to increase awareness of services and improve pathways to information at the time of a mental health crisis, particularly when occurring after hours. More than 20 stakeholder groups informed the campaign and it was an important response alongside a suite of initiatives developed to provide greater mental health support and better mental health outcomes for people living in the area.

The brief was to create a social marketing campaign aimed at reducing stigma, increasing awareness of services and encouraging people to reach out for help at a time of crisis. The campaign was to reflect the unique needs of the local region and to be culturally appropriate for all members of the diverse Murrumbidgee community. The solution needed to be suitable for use well beyond the initial campaign, in order to support longer-term changes in behaviour, attitudes and perceptions (Fig. 7.1).

7.2 Understanding the Social Issue

7.2.1 The State of Mental Health in Australia

One in three Australians experience mental illness in their lifetime and 20% of Australians have experienced a mental illness in the past 12 months—equating to more than 4.8 million people (Australian Bureau of Statistics 2009). An additional 600,000 young Australians aged between four and 17 years of age experience a clinically significant mental health concern (Sawyer et al. 2000). The number of Australians experiencing mental illness is increasing, from 9.6% of the population in 2001, to 11.2% of the population in 2007–08, 13.6% of Australians in 2011–12,

Fig. 7.1 Promotional material featuring the theme of the social marketing campaign. *Source* Murrumbidgee PHN

17.5% of the population in 2014–15 and to 20.1% in 2017–18 (Australian Bureau of Statistics 2008, 2009, 2013, 2015, 2018) (Fig. 7.2).

Of those living with mental illness in Australia, over 80% will experience anxiety or depression. The remaining 20% will experience more severe episodic or persistent illness including (but not limited to) schizophrenia, severe depression, eating disorders and bipolar disorder (National Mental Health Commission 2014).

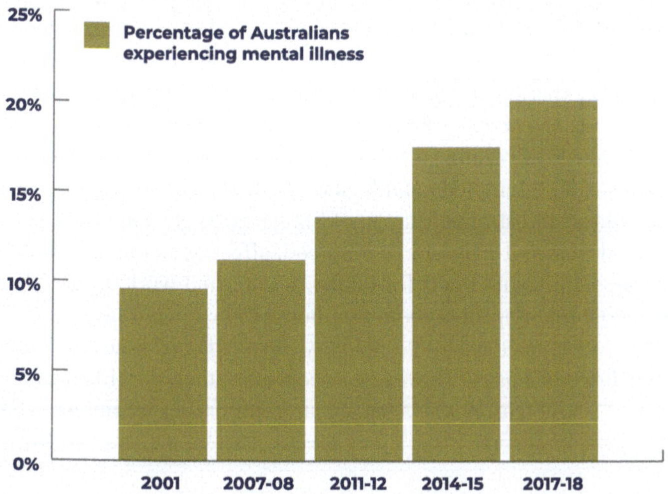

Fig. 7.2 Percentage of Australians experiencing mental illness 2001–2018. *Source* Australian Bureau of Statistics (2013, 2018)

A number of groups have been identified as having a higher risk of experiencing mental illness, including people living in rural and remote areas, Aboriginal and Torres Strait Islander people and young people. Those from Culturally and Linguistically Diverse (CALD) groups, people from LGBTIQ communities, people who use alcohol and/or other drugs, individuals with chronic illness or a disability, and families and carers of people with mental illness, including young carers and people who have experienced trauma (Murrumbidgee Primary Health Network 2016). In addition, '…many people with mental health difficulties face compounding disadvantage…' particularly those from many of the higher risk groups noted above (National Mental Health Commission 2014).

7.2.2 *Suicide and Self-harm in Australia*

It is estimated Across Australia, that there are over 3000 deaths by suicide each year. In 2015, the overall suicide rate was 12.6 per 100,000 in Australia, the highest rate in over 10 years (Australian Bureau of Statistics 2015). In 2009, over 31,500 Australians were admitted to hospital because of self-harm. Rates of attempted suicide in 2015 were estimated to be 65,300 (AIHW: Harrison et al. 2009). It is believed that over half of people who self-harm are not formally known to any mental health support services (Murrumbidgee Primary Health Network 2016). Less than half of those with a clinical level of need will seek help from a mental health professional for their condition (Andrews et al. 2001; Harris et al. 2015).

Across the country rates of suicide, attempted suicide and self-harm are significantly higher in males, and the suicide rate among Aboriginal and Torres Strait Islander people is over double that of non-Indigenous people. Other groups with an increased risk of suicide in Australia are young people aged 15–24, men living in rural and remote areas, those who have previously attempted suicide or engaged in self-harm, people experiencing mental illness, those who identify as LGBTIQ, those bereaved by suicide and people who have experienced an unexpected stressor such as job loss or relationship breakdown (Murrumbidgee Primary Health Network 2016). The rates of suicide, attempted suicide and self-harm are increasing, in line with rising mental illness (Australian Bureau of Statistics 2018). The impacts of suicide, attempted suicide and self-harm are widespread, affecting not just the individual but with long term, significant effects for family, friends and workplaces. There is also a significant economic impact on both families and the community.

This chapter acknowledges the complexity of suicide and notes the multi-faceted issues surrounding self-harm. 'The reasons people take their own life are complex and often there is no single reason why a person attempts or dies by suicide' (Everymind 2019).

7.2.3 Stigma and Mental Health

Stigma has been defined as 'something that detracts from the character or reputation of a person', or a mark of disgrace (Collins English Dictionary 1994). It is common and significant for people with mental health concerns to experience stigma and discrimination. In relation to depression, a recent study found 79% of 1000 adults attending specialist mental health services across 35 countries reported experiencing discrimination (Lasalvia et al. 2013). Stigma and discrimination impede optimal outcomes, can contribute to a worsening of mental illness and can discourage people from seeking help, impacting recovery and quality of life. In a US study of over 1300 people, most respondents '…tended to try to conceal their disorders and worried a great deal that others would find out about their psychiatric status and treat them unfavourably' (Wahl 1999).

In Beyond Blue's 2014 Depression and Anxiety Monitor, which surveyed 2000 Australians from the general population, over half of those experiencing depression and/or anxiety reported they had concealed or hidden their mental health problem from others. Within the monitor, significant numbers of respondents agreed with statements such as, 'people with depression are unpredictable', 'people with anxiety are unstable' and 'I would not employ someone if I knew they had depression' (Beyond Blue 2015).

7.2.4 Understanding the Needs of Those with Lived Experience

In 2014, the Review of Mental Health Services in Australia received over 1800 submissions from Australians with lived experience of mental illness. Analysis of submissions suggested that services and programmes were often not 'designed in consultation with the aim to serve'. The review observed that where services were not based on formal analyses of need there resulted '…high levels of unmet mental health need…highlighted particularly strongly in relation to people living in regional, rural and remote areas of Australia, including farmers and fly-in-fly-out workers (National Mental Health Commission 2014).

The Murrumbidgee PHN responded to this community need when they undertook the 2016 Needs Assessment, and when they utilised the findings of the report as a core basis for planning the provision of services and initiatives that address key priority areas. The co-design approach taken to inform, plan and develop this initiative aimed to have the views and perspectives of those with lived experienced at the core of all aspects of the campaign.

7.3 Background and Social Context

7.3.1 *Mental Health Incidence in the Murrumbidgee Region*

While in some respects, the Murrumbidgee region reflects many other rural and regional areas of Australia, the area has its own unique strengths and characteristics and faces its own unique challenges. Parts of the Murrumbidgee region experience rates of mental illness, alcohol and drug use, suicide and self-harm that are significantly higher than state and national averages. Rates of self-harm (all ages) within the Murrumbidgee PHN area are approximately double the NSW state average. While the rates of self-harm are higher among young people aged 15–24 years across NSW, the rates of self-harm for young people in the Murrumbidgee are among the highest in the state (Healthstats NSW 2019b).

The population falling within the Murrumbidgee PHN region is estimated at approximately 242,840 (Australian Bureau of Statistics 2015). In more than one-third of the Murrumbidgee PHN region, rates of suicide deaths are recorded at between 1.2 and over 2.0 times the national average (Murrumbidgee Primary Health Network 2016). Multiple specific communities within the area have experienced suicide contagion. Alcohol-related hospitalisations in the region are 10% higher than the NSW state average and the region has the second-highest rate of alcohol-attributable deaths in the state (Healthstats NSW 2019a).

Between 2015 and 2017, the levels of self-reported psychological distress in the area doubled. In 2015, the rate of psychological distress was estimated at 10.8%, with similar rates in most of the prior years. In 2017, the incidence of reported psychologic distress rose to 21.5%. This compares to the overall NSW state incidence of 15.1% in 2017 (Centre for Epidemiology and Evidence, NSW Ministry of Health 2017). As noted elsewhere in this chapter, the factors contributing to suicide are multiple and complex and this is the case in the Murrumbidgee area too (Fig. 7.3).

The Murrumbidgee region is strongly agriculturally based and at the time of writing was in the midst of the worst drought in over 100 years. Beyond the direct impact of drought on the incidence of suicide, a lack of rainfall and elevated temperatures have impacted many people's livelihoods resulting in '…geographic and economic dislocation as people move in search of work. This contributes to social and community fragmentation in rural communities… [and]…this impact is particularly severe for Indigenous communities' (Murrumbidgee Primary Health Network 2016).

There is increasing recognition of the major role of social determinants in the prevalence of mental illness and suicide. Social disadvantage is associated with poorer mental health outcomes, particularly for young people. A significant number of suburbs and sub-regions within the Murrumbidgee region have been identified as areas of social and economic disadvantage. The region has a higher proportion of low-income families and higher smoking rates in pregnancy and early parenthood when compared to state and national averages (Murrumbidgee Primary Health Network 2016).

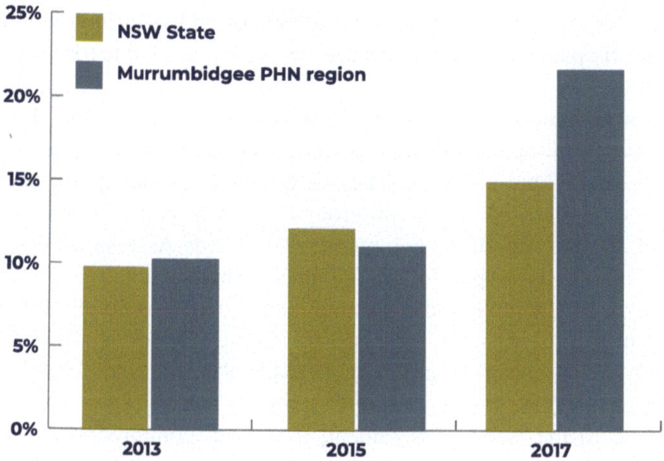

Fig. 7.3 Incidence of self-reported high or very high psychological distress 2013–2017: NSW State vs Murrumbidgee PHN region (People aged 16 years and over). *Source* Centre for Epidemiology and Evidence 2017

7.3.2 Community and Stakeholder Support in the Murrumbidgee Region

In the Murrumbidgee area, over recent years there has been '…a growing commitment in the community to talk about and address mental health, suicide and alcohol and other drug issues, this is a key strength of the region' (Murrumbidgee Primary Health Network 2016).

As for all Primary Health Networks across Australia, the Murrumbidgee PHN is tasked by the Australian Commonwealth Government with seven key target priorities, two of which are addressing mental health and addressing alcohol and other drugs health-related concerns. The other priorities are Aboriginal and Torres Strait Islander health, population health, health workforce, digital health and aged care.

The Murrumbidgee PHN worked closely with the Murrumbidgee Mental Health Drug and Alcohol Alliance (MMHDAA), which brings together key stakeholders from the health, community and social sectors across the region with the joint vision to improve mental health and drug and alcohol outcomes for the Murrumbidgee population. Importantly, the focus of the alliance is on consumer outcomes rather than organisational outcomes, ensuring the consumer is the central focus of all initiatives of the alliance.

Together the groups identified the importance of ensuring consumers and carers have access to information at the point of a mental health, alcohol or other drug crisis, particularly one that arises outside of normal business hours. They identified that there was no strategy or marketing collateral used consistently across MMHDAA organisations, nor in the broader mental health and drug and alcohol sectors in the region. There was recognition of the potential benefits of having appropriate, consistent, strategically planned material 'at hand' and easily accessible for future possible times of need.

During consideration and planning, the stakeholders resolved that the approach should incorporate the aim of reducing stigma associated with seeking help for mental health, drug and alcohol concerns. They determined that the approach needed to appeal to specific high-risk target sub-groups and to be culturally appropriate and respectful of all members of the community. The Needs Assessment demonstrated widespread community support for this approach, in both qualitative and quantitative research stages, for example in the community survey it was noted, 'Education for school age children needs to be increased so that we can potentially prevent more children from suicide and drugs and alcohol… Mental health needs to be more widely discussed' (Respondent, Murrumbidgee Community Survey).

In response to the Needs Assessment, the group commissioned a marketing and communications campaign, with the objectives of encouraging help-seeking behaviour and improving access to information at the time of a mental health, alcohol or other drug crisis. Juntos Marketing was awarded the contract to undertake this work in mid-2017, and the high-profile, integrated social marketing campaign was developed and launched later in 2017.

7.4 Behavioural and Social Change Goals

The overarching objective of this project was to encourage greater help-seeking behaviour among people experiencing a mental health, alcohol or other drug crisis within the Murrumbidgee region of NSW, Australia, as well as their families and carers, particularly after hours. Given that many crises occur out of usual business and service hours, the stakeholders identified the importance of encouraging help-seeking behaviour during evenings, weekends and public holidays, directing those in crisis to identified pathways, specific to the local area.

In addition to these out-of-hours periods, the Christmas period was identified as a peak time of need including the run-up to and the post-Christmas timeframe. Return to study at the beginning of the Australian academic year was also identified as an important secondary peak.

The specific behavioural objectives of the project were:

- To increase awareness of, and calls to the *Murrumbidgee AccessLine*—a local triage service catering specifically to mental health, drug and alcohol concerns this telephone helpline has the capability to direct callers to appropriate identified, location-specific pathways
- To increase awareness of other emergency or mental health support services
- To increase calls or online visits to other mental health support services at a time of crisis, in particular the government mental health support website, 'www.headtohealth.com.au', *Lifeline* and/or Triple 000

- To reduce the stigma around mental health among the general community in the Murrumbidgee area.
- To increase community engagement and awareness regarding the need to reach out and seek support at a time of crisis.

From a marketing and communications perspective, the goals were to develop an effective digitally focused communication campaign, which would work well across a range of channels and media, which was founded in co-design and community consultation, and which would continue to be relevant and contemporary over future months and years.

The specific marketing and communications campaign objectives were:

- To maximise the reach of the campaign across the region using a range of media
- To achieve frequency of message on social media of at least 6 views spread over the 12-week campaign period, and a frequency of individual social media message to be seen not more than twice per person for any one sponsored post
- To skew the planned timing of messages and communications to after usual business hours, to holiday periods and to expected peak times of mental health distress and concern

7.4.1 Citizen Orientation

In more traditional product and service marketing, segmentation, target audience definition and customer orientation are paramount to successful outcomes. At the core of any strong traditional marketing initiative is the concept of ensuring marketing messages and objectives are relevant to the customer, in order to maximise interest and uptake or purchase. So too with social marketing and behaviour change marketing, where building campaigns based on a deep understanding of the target audience/s provides one of the vital foundations for strong outcomes. Yet with social marketing and behaviour change marketing campaigns, the 'target audience' can be very broad. A citizen is defined as someone who is legally accepted as belonging to that country' or alternatively people who live in a town or city (Collins English Dictionary 1994).

Considered in the context of this initiative, we understand 'citizen orientation' to mean seeking, understanding and incorporating the views, perceptions and preferences of those with lived experience of mental health, drug or alcohol concerns, as well as the wider community, in the planning and execution of the campaign. The wider community could potentially equate to every person living within the Murrumbidgee PHN-defined geographical area. In considering citizen orientation for this campaign, it was, therefore, seen as extremely important that the initiative should target, resonate and appeal to high-risk groups as well as the wider Murrumbidgee community.

7.5 Research and Insight

7.5.1 Methodology

To inform the Murrumbidgee suicide, mental health, alcohol and other drugs needs assessment, community consultations were held with key stakeholder groups across the region, and this was supplemented by an online survey available to all community members in the region (n = 799). The online survey was distributed to partner organisations, the local community, carers and consumers with the aim of maximising broad representation of the wider community (Murrumbidgee Primary Health Network 2016).

A structured co-design stage was undertaken prior to commencing work on the communications campaign. Qualitative research included a facilitated strategic stakeholder workshop, a series of four focus groups and fourteen in-depth telephone interviews. The purpose of these was to confirm and agree all foundational aspects of the campaign, including the objective, main campaign messages, style, look of creative execution, terminology, language and tone. The initial stakeholder workshop included representatives from over 20 MMHDAA member organisations. This workshop provided a framework for the overall project.

Focus groups were held in a range of locations across the Murrumbidgee region, with participants from a diverse range of key identified target groups including high-risk sub-groups, differing age groups, demographic and community profiles. Examples included a younger persons group with heterosexual and LGBTIQ members, an Aboriginal and Torres Strait Islander group and a group conducted in one of the regional towns.

In-depth interviews were conducted with health professionals, mental health experts and a wide variety of community members and stakeholders, including those with lived experience.

Focus groups and in-depth interviews were used to explore and confirm the most relevant mechanisms for communication. Input from these citizens determined the direction of the campaign and provided invaluable insights into aspects such as language and tone for key messages.

A small number of participants from the focus groups and in-depth interviews also joined a community advisory panel, providing ongoing input and feedback during the campaign development. The panel was consulted at regular points to further qualify specific aspects of the programme.

7.5.2 Key Co-design Insights and Research Themes

Key themes and insights identified in the research and co-design stages were:

7.5.2.1 There Are Significant Issues Relating to Mental Health, Alcohol and Use of Other Drugs in the Murrumbidgee Areas

As noted in detail in Sect. 7.3.1, in many areas within the Murrumbidgee region, levels of self-harm and alcohol use requiring hospitalisations are significantly higher than the national and NSW state averages (Murrumbidgee Primary Health Network 2016).

The reasons behind this trend are many and complex. Drought and climatic events have a significant impact, as does dislocation and fragmentation of community when people need to search for work. Parts of the area have higher than average social disadvantage, a further risk factor as noted earlier in this chapter. Geographical barriers to accessing services and reduced availability of trained mental health professionals also contribute.

7.5.2.2 At a Time of Crisis, People in the Murrumbidgee Area Can Find It Difficult to Know Where and How to Seek Support, Especially in the After-Hours Period

People with lived experience, their families, carers and communities in the Murrumbidgee area consistently reported that information about service options is lacking, confusing or not readily available (Murrumbidgee Primary Health Network 2016).

> Awareness and service availability is key. There are still …such limited services for people with mental health conditions and alcohol and other drug issues. Furthermore, suicide is such an issue in rural communities—in particular I believe there needs to be more support for our farming communities, more targeted education for high school aged people, more people trained with mental health first aid and awareness for the LGBTIQ community (Community Member, Murrumbidgee PHN area).

The ability of people in the region to access subsidised mental health-related services is lower than state and national averages. Service gaps partly result from the rural and regional nature of the area, for example, the notable distance required to travel for treatment for those living in a number of regional towns and surrounds (Murrumbidgee Primary Health Network 2016). A further factor affecting access to appropriate services in the region is '…the size of the mental health workforce, where the number of allied mental health clinicians is approximately half that of the NSW average' (Murrumbidgee Primary Health Network 2016). During the qualitative research stage, community members also indicated that during a time of imminent crisis, it is often difficult for the person themselves to reach out, therefore, carers and loved ones were an important target consideration.

7.5.2.3 Stigma is a Barrier to Seeking Support

As is the case in the rest of Australia, in the Murrumbidgee, stigma relating to mental illness, drug and alcohol use creates barriers to seeking support and negative impacts (Murrumbidgee Primary Health Network 2016; The Australian Senate 2010). Key community concerns for those living in the Murrumbidgee region included the need for education aimed at reducing stigma and the need to increase support via both awareness of services and accessibility. In community stakeholder interviews, local mental health professionals indicated that 'the level of shame experienced by the individual often leads to self-loathing; very secretive behaviour and reluctance to seek help' (Murrumbidgee Primary Health Network 2016), supporting comments include:

> …there is still such a huge stigma creating barriers within our community… (Community Member, Murrumbidgee PHN area).
>
> [We need]…education for [the] public [to] remove stigma associated with mental illness (Community Member, Murrumbidgee PHN area).
>
> Education for school age children needs to be increased so that we can potentially prevent more children from suicide and drugs and alcohol. Mental health needs to be more widely discussed (Community Member, Murrumbidgee PHN area).

7.5.2.4 Any Campaigns and Initiatives Need to Combine Consideration of Identified High-Risk Groups with the Community at Large

Within the area, the high-risk groups are known and the need to address their risk is acknowledged. Yet 54% of people with mental illness do not access any treatment. In some populations, up to three quarters of all people who end their lives are not in contact with mental health services (Australian Institute of Health and Welfare 2014; Hewlett 2016).

In the Murrumbidgee area, it is believed that over half of people in the region who self-harm are not formally known to any mental health support services (Murrumbidgee Primary Health Network 2016). In other words, a mental health, alcohol or drug crisis can happen to anyone.

7.5.3 Campaign-Related Insights

The co-design stage of the research provided specific direction and guidance for the development of all aspects of the campaign, including the creative concept, the look and feel, tone, messaging, the choice of language and channels for communication. Careful consideration of high-risk groups as well as the wider population was central to the development of messages, language, images and the overall campaign. To address and combat the stigma associated with seeking help, the campaign aimed to normalise mental health, drug, and alcohol concerns and encourage help-seeking behaviours across all members of the community.

7.6 Foundational Campaign Planning

Based on the research, co-design and insights development stage, detailed foundational planning was undertaken to develop key messages, campaign style and positioning. Choice of language and care with the visual execution were essential to ensure non-stigmatisation.

7.6.1 Key Messages and Language

Initial draft versions of the campaign messages were tested and refined in the co-design stage with a range of consumers and health professionals. Findings from the research and community consultation then informed the language and terminology, ensuring non-stigmatisation and maximising message cut-through. The following key messages were agreed on:

- Life doesn't always go according to plan

This message was considered engaging, non-judgmental, inclusive and broad reaching.

- Tough times can happen to anyone

This message was seen as more appealing and broad reaching than specifically referring to mental illness, drugs or alcohol. It was considered inclusive and relevant.

- If you need support, reach out

Referring to 'reaching out' resonated, more than 'seeking help' although the latter was also used in some secondary messaging. An overarching message of the campaign was that mental health concerns can happen to anyone.

7.6.2 Creative Direction Including Imagery

A range of visual styles and creative looks were presented to participants within the focus groups and in-depth interviews. Feedback and guided discussion led to key insights including the importance of the characters having their faces showing. Faceless images were seen as stigmatising. Participants favoured incorporating graphics and images that conveyed the unique Murrumbidgee location, including ensuring the river was seen within the imagery and the use of animals such as cockatoos.

7.6.3 *Media*

Consumers in the co-design stage saw the planned use of television as an appropriate core media in the context of the area. The choice of radio was highlighted as a very important channel, particularly to reach those located outside the main regional town centres, as many rural populations may not have reliable cellular access at all times.

Social media has widespread use and appeal within the overall community and was, therefore, seen as well suited to help normalise help-seeking behaviour. Participants reinforced the planned choice of Facebook as the most likely effective social media channel, and confirmed Instagram as a favoured platform for many younger community members.

7.7 Segmentation and Targeting

The programme was tailored to meet the needs of the identified high-risk target groups, within the context of the wider Murrumbidgee community. The specific high-risk groups were identified as primary target groups. As noted previously, these high-risk groups included young people, those from an Indigenous background, those from CALD backgrounds, those from the LGBTIQ community and rurally isolated males (Murrumbidgee Primary Health Network 2016). Given the prevalence of suicide, self-harm and mental health crises among people who are previously unknown to services, the initiative also needed to have broad community appeal and thus the wider Murrumbidgee community were the secondary target market. The development of the campaign materials and media strategy, therefore, considered both specific high-risk groups and the broad community. This was achieved by using a creative execution that was inclusive of a wide range of groups and people, yet also allowed us to target specific groups.

7.8 Value

Addressing and combatting suicide, mental health, drug and alcohol issues are a complex challenge. Seeking help at a time of crisis is associated with improved long-term health outcomes and improved quality of life measures (Department of Health and Ageing 2013). Given appropriate mental health care and services, most people experiencing a crisis are able to recover well and return to a productive and satisfying life. Therefore, by increasing the number of people living in the Murrumbidgee area who seek help at a time of crisis, it is expected that there will be a long-term community and economic benefit.

7.9 Competition

In the context of social marketing, 'competition' can be seen as what the target audience may do or prefer to do, in place of the desired behaviour (Kotler and Lee 2008). The 'competition' for the behaviour change targeted by this programme could be seen as:

- Complacency—doing nothing (and potentially living with high levels of stress, long-term untreated mental health concerns and/or substance use or abuse, reduced quality of life and reduced productivity), or
- Taking drastic action—including self-harm or attempting suicide.

7.10 Theory

The transtheoretical stages of change model suggests the following important stages of behaviour change: pre-contemplation, contemplation, preparation, action and maintenance (Prochaska and DiClemente 1986) (Fig. 7.4).

Applying the transtheoretical model of change, this behaviour change marketing campaign aligned with the stages of contemplation, preparation and action. Flay (1989) suggests that social marketing campaigns are generally most effective in initially creating awareness, modifying and influencing perceptions by providing motivations to change attitudes. The authors suggest that as attitudes change, propensity to make behaviour change increases (Flay 1989). For this reason, when evaluating the campaign effectiveness, the following were identified as relevant campaign aims:

- Awareness of the Murrumbidgee AccessLine
- Awareness of other sources of support, specifically *LifeLine*, 000 in an emergency, www.headtohealth.com.au and local general practitioners
- Propensity to seek help (for oneself or a loved one)
- Increases in intended help-seeking behaviour

As a further point of reference, when seen in the context of Andreasen's (2002) social marketing benchmarks, this initiative meets the suggested criteria to be considered a social marketing campaign. Examples of alignment with Andreasen's criteria include the use of behaviour change (seeking help), which was integral to the message, the execution and all aspects of the intervention, the use of the marketing mix and the use of market research to understand the target market and to monitor the implementation and outcomes of the campaign (Andreasen 2002).

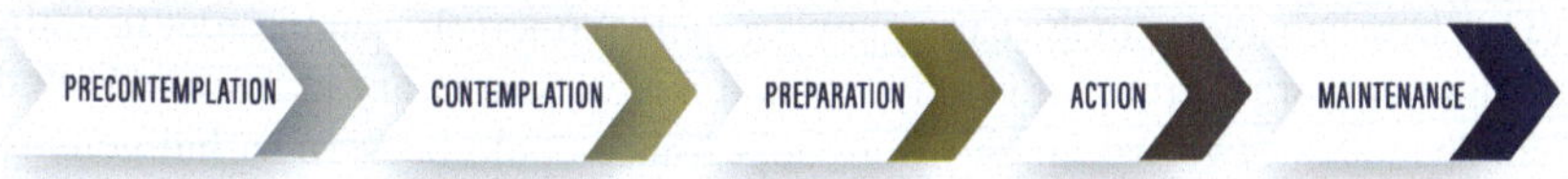

Fig. 7.4 The transtheoretical stages of change. *Source* Prochaska and DiClemente (1986)

7.11 Social Marketing Intervention Mix

With the campaign objectives defined, the citizen orientation and segmentation completed, and detailed recommendations from the co-design and qualitative research stages determined, direction for the campaign's creative and media became clearer.

7.12 Creative Style, Look and Feel

The creative style and execution needed to cater to identified high-risk groups including Aboriginal and Torres Strait Islander background, those from CALD groups, rurally isolated males and those identifying as LGBTIQ, yet the style, look and feel also needed to have broad appeal, given the need to communicate with individuals not known to service providers. Given the sensitive nature of this campaign, the visuals and all aspects of the creative development had to be inclusive and work towards reducing stigma.

At the core of the campaign was the development of an animated motion style video, in the form of a 30-s Television Commercial (TVC). The animated motion style was chosen for multiple reasons. This animated creative style is frequently used in mental health executions, as it facilitates the depiction of diversity yet lends itself to a non-judgmental tone. Examples of other mental health awareness campaigns where the style has been successfully used include those of The Black Dog Institute and Headspace. Using an animated motion style execution is generally seen as contemporary, and, therefore, likely to appeal to a wide range of age groups and life stages. It is a cost-effective way to depict a range of different looking people. Additionally, this approach can be very adaptable to different media formats, including both print and digital.

7.13 Campaign Characters

The characters and scenery were specifically designed to reflect the unique Murrumbidgee area and community. The creative concept included diversity within the animated characters in order to represent the community at large as well as specific high-risk groups, with depictions designed to reflect community diversity, convey inclusiveness and to underpin one of the key messages: mental health, drug and alcohol concerns can happen to anyone.

The 'hero' character in the campaign is a male farmer, chosen in order to address the important target group of rurally isolated males. Secondary characters included a female student, two workmen whose appearance could be seen as Indigenous or representing those from a CALD background, two young women suggestive (but not directive) of a same sex couple and a father and baby. Other characters included older people, someone in wheelchair, more professional-looking people, a heterosexual couple and younger people.

The body shape of characters needed to reflect typical community members, and for this reason, there are a range of weights, heights and body shapes within the broader range of characters. The faces of all the characters are depicted throughout all materials. It was seen as important that people were not 'faceless' as this could be construed as representing shame or stigma. Particular care was taken with the facial expressions of the hero male farmer, as he needed to capture and guide the viewer to take action (and reach out).

7.14 Unique Location

Communicating 'place' and in particular the unique attributes associated with the Murrumbidgee area were crucial to the creative development. The majestic Murrumbidgee River is Australia's second-largest river with its source in the Snowy Mountains of New South Wales. It is the chief tributary feeding in to the Murray River. While much of the area is flat, low-lying land, the backdrop of the Snowy Mountains is iconic to the area. It is a major food-producing region of Australia.

The background scenery needed to accurately convey that this was a Murrumbidgee campaign at a glance. The Snowy Mountain ranges were used in the background, complemented by a depiction of the Murrumbidgee River itself in a number of the video frames, and unique flora and fauna such as the cockatoos, sheep and gum trees throughout the clip.

7.15 Marketing Mix

The 30-s video was used on local television stations as well as on social media (with the addition of captions) (Fig. 7.5).

The script for the TVC was adapted and recorded for radio. A series of 14 social media tiles and two web tiles were developed, with visuals adapted to feature the same creative design and characters as in the TVC. Within the social media tile set, there were general tiles encouraging people to reach out at a time of crisis, as well as occasion-specific tiles for Christmas and for return to study. Several depictions utilised main characters from the high-risk groups, such as a same sex couple and the hero, a rurally isolated male (Fig. 7.6).

Small geographical pockets within the Murrumbidgee PHN area, in particular the Hay and Deniliquin areas are serviced by a Victorian-based regional television station. Due to restrictions with the *Murrumbidgee Accessline*, it was not possible to incorporate these towns in the television schedule, and we, therefore, used print advertising to reach these populations.

Fig. 7.5 Frame from the mental health campaign television commercial. *Source* Murrumbidgee PHN

Fig. 7.6 Selection of social media tiles used in the mental health campaign. *Source* Murrumbidgee PHN

A3 posters and postcard flyers were designed for distribution through community locations across the entire Murrumbidgee area, in order to extend reach and frequency, and to reinforce the call to action (Fig. 7.7).

The creative campaign was complemented by a locally based public relations launch and campaign.

7.15.1 Media Strategy

The media strategy used a combination of television, radio, social media, and print to build awareness and to encourage help-seeking behaviour among the target groups.

Television and radio were chosen as core media for the campaign, due to their ability to reach a mass market. The media strategy included:

- 12-week TV, radio and print campaign across the Murrumbidgee region
- Heavy bursts across all media targeting known peak periods of crisis, particularly Christmas, New Year, Australia Day and return to study in early February

Fig. 7.7 Postcard flyer used in the Murrumbidgee mental health awareness campaign. *Source* Murrumbidgee PHN

- TV and radio scheduling skewed towards after hours, due to this being a time of higher need

Social media was the third core pillar of the media strategy, with a focus on Facebook and Instagram. Facebook was seen as providing the greatest digital reach opportunity to the widest demographic range, due to its wide audience and geo-targeting capabilities. The use of Instagram allowed greater access to a younger skew of social media users. Twitter was used to share key campaign messages, with a focus on other health professionals, Murrumbidgee PHN existing followers and partner organisations.

The social media strategy included:

- Regular weekly timeline posts on all social media
- Sponsored Facebook and Instagram posts scheduled to achieve maximum reach. Posts were geo-targeted and time specific with an after-hours focus
- Social media target for overall campaign frequency of 5–6 and targeted weekly post frequency of 1–2

As mentioned previously, print media was used in selected locations where regional constraints meant that television coverage was not achievable.

The campaign was supported in a wider context by a range of other initiatives undertaken by the Murrumbidgee PHN and the MMHDAA. These included a stepped-care approach for access to mental health services; drug and alcohol services tailored to the most vulnerable groups within the community and primary and secondary suicide prevention strategies.

Policy change, advocacy around mental health service provision and other mental health initiatives continue in the wider state and federal context.

7.16 Partnerships

The project and campaign were undertaken under the auspices of the Murrumbidgee PHN. The success of the initiative also lay with the involvement of a wide range of partnerships, particularly the MMHDAA and their individual members, as well as many community groups.

7.17 Evaluation and Results

The following measures were used to evaluate the programme:

- Pre- and post-campaign awareness across the Murrumbidgee region
- Achievement of social media, television and radio campaign objectives
- Reported impact by MMHDDA and community members (social impact)

Table 7.1 summarises the quantitative results from the pre- and post-campaign evaluation, demonstrating uplifts in all key awareness measures. While indicative, the small sample size should be noted.

The planned schedule for television and radio was exceeded across all media activity, as outlined in Table 7.2.

Social media was particularly effective in this campaign, far exceeding previous Murrumbidgee PHN campaign results. Overall, the campaign achieved over 1.1 million impressions. Almost 95,000 people saw the most successful social media post, which equates to nearly 40% of the Murrumbidgee population.

Detailed scheduling for time and day of post was undertaken with the aim of ensuring the posts coincided with target out-of-hours timeslots. This resulted in very strong reach, video views, campaign impressions and social media actions, as outlined in Table 7.3. Post scheduling was carefully managed to ensure frequency objectives were achieved.

Table 7.1 Results from pre- and post-campaign online survey

	Pre-campaign 9–13th Dec 2017 N = 76 (%)	Post-campaign 6–11th Feb 2018 N = 75 (%)
Awareness of phone services available in the Murrumbidgee (for support with drug, alcohol or mental health concerns)	22.4	37.3
Unprompted awareness of *AccessLine*	3.9	13.3
Prompted awareness *AccessLine*	35.5	45.3
Unprompted awareness of any other relevant services	18.4	29.3
Unprompted awareness of *LifeLine*	79.0	81.0

Source iLink online survey, conducted across Murrumbidgee postcodes

Table 7.2 Planned versus delivered television, radio and press

	Planned	Delivered	Delivered as % of planned media (%)
Total TV spots	1110	2059	185
Total TV TARPS	2485	2834	114
Total radio spots	353	571	162
Press ad insertions	5	8	160

Source Murrumbidgee PHN

Table 7.3 Social media campaign outcomes

Measure	Result
Total campaign impressions	1,144,741
Reach per sponsored post	40,180–94,925
Individual post frequency achieved per person per week	1–2
Campaign post frequency achieved per person	5–6
Video views—full	12,930
Video views—partial	52,800
Social media actions—likes, comments, shares	54,944

Source Facebook, Instagram, Twitter analytics

Almost all comments incorporated individuals tagging other individuals, indicating people were sharing information with those who they deemed in need of support or information.

A key call to action of the campaign was to visit the federal government's Head to Health mental health support website. Figure 7.8 shows Head to Health website visits from web users located in major towns in the Murrumbidgee area. The peak seen in February 2018 corresponds with the timing of the 'return to study' social media post promotions, which included a specific call-to-action to visit the Head to Health website.

In addition to the above outcomes, there was a notable increase in the number of calls received by the *Murrumbidgee Accessline* during the period of the campaign, although specific data was not available for inclusion in this chapter. Reported impact, as seen in community and MMHDAA feedback, included the following:

> Clear message, communicated in a non-judgmental way.
>
> Radio worked well. Heard it many times.
>
> I liked the simplicity of the message. Low stigma vibe….
>
> The campaign worked really well & good that it ran at targeted times.

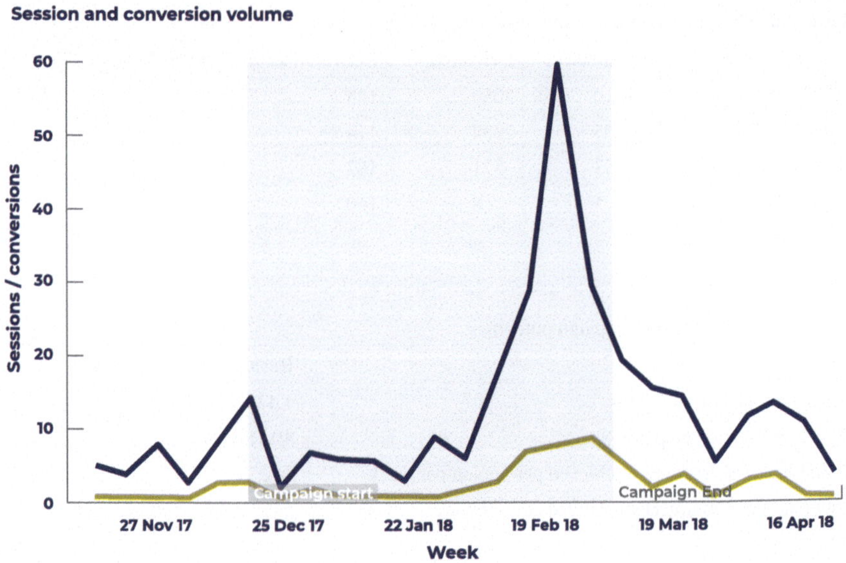

Fig. 7.8 Visits to head to health website from users in major towns in the Murrumbidgee region. *Source* Head to health website

7.18 Longer Term Impacts

Since its initial launch in 2017–2018, the campaign has been extended within the Murrumbidgee in a number of ways. At the conclusion of the initial infield campaign, all MMHDAA members were provided with the full suite of digital and printed materials, enabling them to continue promoting the key messages and visuals through their own channels.

7.18.1 Extending Original Campaign Over Time

During the following year, a second wave of the campaign was run on television, radio and social media, for twelve weeks between October 2018 and February 2019. The most significant results of this second major wave were a 10% point increase in awareness of phone services available in the Murrumbidgee for support with drug, alcohol or mental health concerns when compared to the previous year (37% awareness in February 2018 (n = 76) compared with 47% awareness in February/March 2019 (n − 91)), and increases in unprompted awareness of other relevant services, (29% in February 2018 (n = 76) to 36% awareness in February/March 2019 (n = 91)) (Source: iLink online survey, conducted in the Murrumbidgee.) (Fig. 7.9).

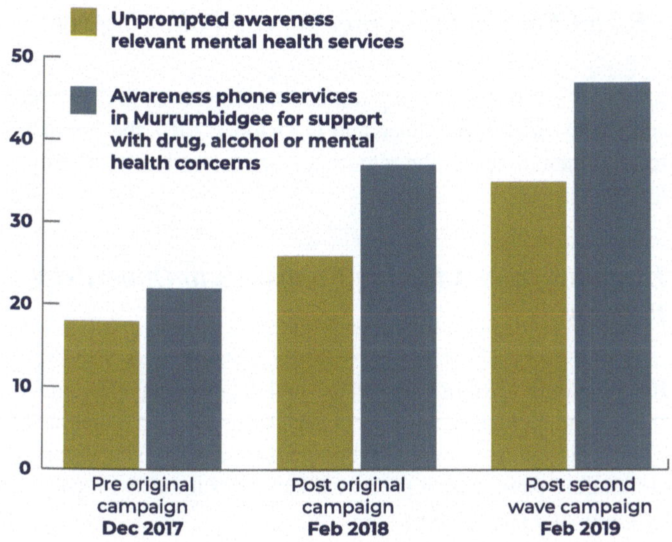

Fig. 7.9 Changes in prompted and unprompted awareness of pathways to support, Dec 2017–Feb 2019 (N = 76, N = 75, N = 91, respectively). *Source* iLink online survey conducted across all demographics and a range of postcodes across Murrumbidgee region

7.18.2 Extending Original Social Media Materials to Feature Drought-Themed Visuals

Campaign materials have been extended to specifically address the current drought circumstances. The messages remained as in the original campaign, with a focus on 'if you need support, reach out' and 'tough times can happen to anyone'. The visuals in some of the materials were adapted to subtly portray the current serious drought conditions. The Murrumbidgee steering committee report that this adaptation was well received in the community.

7.18.3 Extending Campaign Theme to Mental Health Diaries in Response to Drought Emergency

During 2019, 30,000 mental health diaries were produced incorporating the campaign key messages and hero image. As part of the local drought response, the diaries have been distributed across the region via thirteen government and corporate partner organisations. Over 9000 campaign postcards have also been distributed across the region as part of mental health resources packs.

7.18.4 PR and Editorial Coverage for the Following year

In 2019, Farmers NSW featured and promoted the campaign in their regular publication together with an editorial piece and in October 2019, the original TVC was aired in selected cinemas across the region during Mental Health Month.

7.18.5 Extending and Adapting Campaign to Other PHN Regions

In early 2019, the South Eastern NSW PHN, which covers an adjacent regional and rural area of NSW, adapted the campaign materials for use across mass media in their region. The call to action was altered to direct consumers to a specifically developed local website landing page, which promoted key ways to access local assistance as well as a digital mental health app.

7.18.6 Australian Marketing Institute National Award: Best Not-for-Profit Campaign 2018

Finally, in late 2018 the Australian Marketing Institute's National awards recognised the campaign as the Best Not-for-Profit campaign of the year.

7.19 Discussion

In Australia, many mental health campaigns, behaviour change campaigns and social marketing initiatives are developed either on a large scale by state governments, federal governments and large national not-for-profit organisations, or alternatively on a much more modest scale by smaller grassroots groups.

This campaign was unusual in taking an approach aligned to larger campaigns yet applying local insights and a very local approach to execution, to create a local social marketing campaign. Engaging community and stakeholders regularly throughout programme planning and execution ensured widespread support and maximised impact.

The authors suggest that insights elicited during the co-design component were essential to the development of strong, effective messaging and to ensuring the campaign's local relevance. The strategic development of the campaign built on these messages, combining them with a highly professional creative concept that reflected the unique locale and encouraged engagement.

The campaign was specifically tailored to local needs and input, from the messaging and creative development to the choice of media and the specific media scheduling strategy. The timing of the campaign to coincide with a known peak time of need was also of significance.

This overall approach, which builds on co-design and community-generated insights, is highly transferable for use in other places and settings, to encourage help-seeking behaviour. The experience of this initiative suggests that taking a localised approach to the design and development of social marketing and behaviour change marketing campaigns can lead to effective outcomes.

7.20 Conclusion

Mental health, drug, and alcohol concerns are widespread and the incidence is increasing. In parts of the Murrumbidgee area, rates of mental illness, suicide and self-harm are higher than state and national averages. Addressing and combatting suicide, mental health, drug and alcohol issues are complex challenges. Seeking help at a time of crisis is associated with improved long-term health outcomes and improved quality of life measures (Department of Health and Ageing 2013).

Based on the 2016 Needs Assessment, the Murrumbidgee PHN and over 20 key stakeholder groups belonging to the MMHDAA identified the need to increase awareness of services and improve pathways to information at the time of a mental health crisis, particularly when occurring after hours. By increasing the number of people living in the area who seek help at such times, it is expected that there will be a long-term community and economic benefit.

A local social marketing campaign was commissioned with the aim of reducing stigma, increasing awareness of services and encouraging people to reach out for help at a time of crisis. The campaign reflected the unique needs of the area and the diversity of the local community. Multiple media channels were effective in ensuring as wide a reach as possible. The television and radio campaign ran for 12 weeks from late 2017 to early 2018. Optimal frequency targets were achieved and planned media reach was exceeded. Social media, in particular Facebook and Instagram, played a significant role in the campaign's success. Social media was highly effective in supporting and extending the awareness and messaging achieved using television and radio.

The social media components of the campaign achieved strong reach, frequency and reactions, including over 1.1 m impressions. The most successful social media post reached nearly 40% of the Murrumbidgee population.

The campaign played a role in:

- Increasing community awareness about specific support services and pathways
- Assisting in reducing stigma and normalising the conversation around mental health, drug and alcohol issues, so that people felt more comfortable to seek help

- Increasing help-seeking behaviour, as seen in additional calls to the *Murrumbidgee Accessline* and in a spike in visits to the Head to Health website from major towns in the area
- Contributing to the positive reputation of the Murrumbidgee PHN and the MMHDAA Alliance within the Murrumbidgee community, in taking proactive steps to address recognised local health concerns

The authors acknowledge that issues around suicide, self-harm, mental health, and drug and alcohol concerns are complex and that a single social marketing campaign is just one approach that can contribute to better outcomes within a suite of comprehensive, strategic services, systems and processes at multiple touchpoints.

The approach used in this social marketing campaign demonstrates the value of using community consultation, co-design and quantitative research, in developing effective social marketing campaigns to reduce stigma and encourage help-seeking behaviours.

The campaigns' continued uptake in the region suggests that the messages and creative approach continue to be relevant within the community.

References

AIHW: Harrison JE, Pointer S, Elnour AA (2009) A review of suicide statistics in Australia. Australian Institute of Health and Welfare, Sydney

Andreasen A (2002) Marketing social marketing in the social change marketplace. J Public Policy Market 3–14

Andrews G, Henderson S, Hall W (2001) Prevalence, comorbidity, disability and service utilization: overview of the Australian National Mental Health Survey. Br J Psychiatry 145–153

Australian Bureau of Statistics (2008) National survey of mental health and wellbeing 2007: summary of results. Cat. no. 4326.0. ABS, Canberra

Australian Bureau of Statistics (2009) National survey of mental health and wellbeing: summary of results, 4326.0, 2007. ABS, Canberra

Australian Bureau of Statistics (2013) 4338.0—profiles of health, Australia, 2011–12. ABS, Canberra

Australian Bureau of Statistics (2015) 3303.0—causes of death, Australia. Australian Bureau of Statistics

Australian Bureau of Statistics (2018) Catalogue 3303.0 cause of death Australia 2017. Australian Bureau of Statistics

Australian Institute of Health and Welfare (2014) Australia's health 2014. AIHW, Canberra

Beyond Blue (2015) Information paper—stigma and discrimination associated with depression and anxiety. Beyond Blue, Melbourne

Centre for Epidemiology and Evidence, NSW Ministry of Health (2017) NSW population health survey (SAPHaRI). NSW Ministry of Health, Sydney

Collins English dictionary (1994) HarperCollins Publishers, Glasgow

Department of Health and Ageing (2013) National mental health report 2013: tracking progress of mental health reform in Australia 1993–2011. Commonwealth of Australia, Canberra

Everymind (2019) suicide-facts-and-stats, September 22. www.lifeinmindaustralia.com.au: https://www.lifeinmindaustralia.com.au/about-suicide/suicide-data/suicide-facts-and-stats

Flay BR (1989) Strategies for evaluating mass media prevention campaigns. In: Rice R (ed) Mass media and health. Sage

Harris MG, Hobbs MJ, Burgess PM, Pirkis JE, Diminic S, Siskind DJ, Andrews G, Whiteford HA (2015) Frequency and quality of mental health treatment for affective and anxiety disorders among Australian adults. Med J Aust 185–189

Healthstats NSW (2019a) beh_alcafhos_phn_trend, September 22. www.healthstats.nsw.gov.au: https://www.healthstats.nsw.gov.au/Indicator/beh_alcafhos/beh_alcafhos_phn_trend

Healthstats NSW (2019b) men_suihos_phn_trend, September 22. www.healthstats.nsw.gov.au: https://www.healthstats.nsw.gov.au/Indicator/men_suihos/men_suihos_phn_trend

Hewlett E (2016) Mental health analysis profiles: OECD working paper no. 81, January 19. www.oecd.org: https://www.oecd.org/officialdocuments/publicdisplaydocumentpdf/?cote=DELSA/HEA/WD/HWP(2015)4&docLanguage=En

Kotler P, Lee N (2008) Social marketing. Influencing behaviors for good. Sage Publications Inc, Los Angeles.

Lasalvia A, Zoppei S, Van Bortel T, Bonetto C, Cristofalo D, Wahlbeck K, Bacle SV, Van Audenhove C, Van Weeghel J, Reneses B, Germanavicius A (2013) Global pattern of experienced and anticipated discrimination reported by people with major depressive disorder: a cross-sectional survey. The Lancet 55–62

Murrumbidgee Primary Health Network (2016) Moving ahead: Murrumbidgee mental health, suicide prevention, alcohol and other drugs needs assessment. FirstHealth T/A Murrumbidgee Primary Health Network, Wagga Wagga

National Mental Health Commission (2014) The national review of mental health programmes and services. National Mental Health Commission, Sydney

Prochaska JO, DiClemente CC (1986) Toward a comprehensive model of change. In Miller WR, Heather N (eds) Treating addictive behaviors. Applied clinical psychology, vol 13. Springer, Boston, MA

Sawyer MG, Arney FM, Baghurst PA, Clark JJ, Graetz BW, Kosky RJ, Nurcombe B, Patton GC, Prior MR, Raphael B, Rey J (2000) The mental health of young people in Australia. Commonwealth Department of Health and Aged Care, Canberra

The Australian Senate (2010) The hidden toll: suicide in Australia report of the Senate Community Affairs References Committee. Commonwealth of Australia, Canberra

Wahl O (1999) Mental health consumers' experience of stigma. Schizophrenia Bulletin 467–478

Carolyn Loton is an Occupational Therapist who has worked as a marketing specialist for over 20 years. In 2006, she founded Juntos Marketing, a full service marketing agency whose focus is to use marketing for good, working across health, health promotion, aged care, professional services and industry. The business now also incorporates Leapfrog Research. Carolyn has a passion for communicating positive health and ageing messages more effectively. She strongly believes in the power of strategy, insights, creativity and professionalism as the foundations of marketing that deliver long term value. She holds a Master of Business Administration from Edinburgh University, as well as a Diploma in Marketing and Bachelor of Applied Science (Occupational Therapy).

Melissa Neal has spent most of her life working and living in regional and rural New South Wales where she has developed an understanding of the challenges faced by these communities. She possesses a commitment to developing local partnerships and working in organisations that improve outcomes for people in regional and rural areas. Prior to joining the Murrumbidgee Primary Health Network in 2015, Melissa worked at senior levels in the not for profit sector across the industries of community services and primary health. She has also served as a board director in disability services and is currently a director of a family support organisation. During her 20-year career, Melissa has utilised her experience in the areas of governance, leadership, strategic management, human resources and industrial relations to manage and lead organisations. Melissa holds a Master of Human Resources (Employment Relations), is a professional member of the Australian Human Resources Institute, and is currently completing a Master in Business Administration.

Chapter 8
MAKE IT COUNT 2018: Industry Case Study

Kathy Knox, Joy Parkinson, Ali Ahani, and Filippa Neilsen

Abstract This chapter is presented in five sections. In the first section, we present a summary of the background and analysis of the organ donation context, highlighting international issues and current challenges in post-mortem organ donation. In the second section, we present a case study style summary and critical explanation of the MAKE IT COUNT 2018 campaign from Australia, which aimed to raise awareness and increase Australian Organ Donor registrations at the 2018 Brisbane Ekka in Queensland. We conducted targeted awareness and registration activities before and during the Ekka festival, featuring a team of roving recruiters, a concurrent social media campaign, and a unique smart device application, the DonateLife App. Developed in consultation with young adults and the culturally and linguisti cally diverse (CALD) community, the App delivers facts about organ donation, links users to the Australian Organ Donor Register and gives practical support on talking with family about donation decisions. In the third section, campaign strategies and tactics are mapped against social marketing principles. The fourth section presents the results and examples of the campaigns impact. The fifth section concludes by considering the campaign in the broader context, it discusses limitations and gives future recommendations for promoting pro-donation behaviours.

Keywords Post-mortem organ donation · Social marketing · Campaign · Principles · Evaluation

8.1 Background and Context Analysis

The first successful transplantation of a human organ from a post-mortem donor was conducted in 1954 (Adenwalla and Bhattacharya 2012). Thanks to rapid dramatic improvements in transplantation science and medicine, recent advances

K. Knox (✉) · J. Parkinson
Social Marketing@Griffith, Griffith Business School, Nathan, QLD, Australia
e-mail: k.knox@griffith.edu.au

A. Ahani · F. Neilsen
Griffith Business School, Griffith, QLD, Australia

R. Hay et al. (eds.), *Broadening Cultural Horizons in Social Marketing*,
https://doi.org/10.1007/978-981-15-8517-3_8

have resulted in international firsts including the first baby from a post-mortem transplanted uterus (Ejzenberg et al. 2018). Transplantation survival rates are high and successful outcomes for recipients abound. The large body of academic literature about post-mortem organ donation behaviours demonstrates that stated attitudes towards consenting as a potential donor are (almost) unanimously positive. However, revealed preferences represented through international and national registration rates, and the number of Donors Per Million Population (DPMP) indicates widespread hesitancy and conservatism. Although transplantation is the only therapeutic option for terminal organ failure, the number of organ transplants reaches less than 10% of world need annually, resulting in significant organ shortage (Citerio et al. 2016).

The discrepancy between public sentiment of support and private pro-donation behaviours such as deciding whether one wants to pledge as a potential donor, communicating with one's family that decision and making an enduring registration or record of the decision suggests a social desirability effect. The discrepancy is necessarily influenced by many systemic and structural factors. Continuing efforts in awareness raising, communication and general information to the public, education of professionals, and organisational measures at hospital level have had so far no or little effect on the number of transplantable organs (Gevers et al. 2004), resulting in closer attention to the consent model, registration processes, ethics and legislative changes to organ procurement processes.

8.2 Legal and Ethical Factors

In terms of legislation, national policy varies according to consent models (opt-in, opt-out). To address organ shortage, consent model approaches can be considered. Presumed consent model (opt-out) as opposed to opt-in consent model. Some studies have demonstrated economic advantage of opt-out model (see Silva et al. 2007). For example, Gevers et al. (2004) conducted a comparative analysis of the consent systems in ten European countries. Variations in DPMP were not associated with opt-in or opt-out consent policy. There are countries currently with an opt-out consent model where the DPMP rate is lower than countries with an opt-in consent model. However, there are multiple factors that contribute to DPMP rates. Authors concluded that systemic change to the prevailing consent model has limited capacity to alone effect change in a country's DPMP rates. In terms of administrative processes and systems, there are a multiplicity of ways to register as a donor, including the need to go online to register, to have identity and medical details to hand and to pass through multiple layers of security and authorisation. Registration systems vary across countries.

Legislation and ethical practices around the definition of death and the conditions of donation are further factors in DPMP rates. Since inception, organ donation and transplantation has been guided by the 'Dead Donor' principle. The overarching ethical requirement is that a patient must be declared dead before the removal of organs for transplantation. However, Donation after Circulatory Death (DCD) presents an adaptation to this definition and rule. Legislation on and current international practices surrounding DCD differ significantly across nations (Guzik-Makaruk et al. 2019). Grey areas remain. For example, Truog and Miller (2008) suggest the DCD pathway challenges the dead donor rule and is problematic, creating an logical paradox: 'the hearts of patients who have been declared dead on the basis of the irreversible loss of cardiac function have in fact been transplanted and have successfully functioned in the chest of another' (Truog and Miller 2008). They argue reliance on the dead donor rule has potential to undermine trust in transplantation, has questionable ethical cover that does not withstand scrutiny. Respect for valid consent may be more important than concerns about whether a patient is already dead (Truog and Miller 2008).

In terms of ethics and practices, there are regional variations in the clinical evaluation practices for determining death and post-mortem organ donation. Reportedly, the persisting shortfall of available organs for transplantation has prompted countries to introduce DCD schemes for kidney retrieval and other organs including liver, pancreas and lungs (Manara et al. 2012). For example, practices around DCD vary across European NeuroTrauma Effectiveness Research in Traumatic Brain Injury study centres (van Veen et al. 2018).

Against the present background of international legislative changes to consent models, registration processes and enduring questions of the ethics of organ procurement processes, some contemporary observations emerge (Fraser 2016). The absence of unified guidelines and regulations present challenges for clinical, research, ethical, legal and transplant communities worldwide. For example, the prevalence of trafficked organs and illegal practices creates difficult consequences for clinical and medical transplantation science and research conducted with, on, or about those organs (Rogers 2017). Portugal recently developed the first protocol for trafficked organs (Silva 2019). There is an international absence of guidelines contained in current ethical statements about detecting, reporting and preventing trafficked organs and research stemming from potentially trafficked organs.

8.3 Psychological and Cultural Factors

The discrepancy between stated pro-donation opinions and actual pro-donation behaviours is further impacted by the psychologically and temporally distant, hypothetical, morbid and low probability nature of post-mortem organ donation. Post-mortem organ donation differs in this regard from other forms of altruistic donation such as blood or blood-product donation behaviour (Holdershaw et al. 2011), reproductive tissue donation and living organ donation: contexts where the donor is alive and healthy, and gains positive reinforcement encouraging repeat donation. As a high involvement decision context, contemplating post-mortem donation is characterised by fear, anxiety, doubt, and is subject to myth and misconception. Accordingly, research reveals a range of conscious attitudinal misconceptions that predict organ donation registration (Shepherd and O'Carroll 2014). Including medical mistrust

(Williamson et al. 2018; Williamson et al. 2017) and death anxiety (Wu 2008; Morse et al. 2009; Wu et al. 2013) as well as unconscious visceral responses that act as barriers including the ick factor reflecting disgust (O'carroll et al. 2011) and jinx or superstitious beliefs (Quick et al. 2016). Inviting people to engage in recommended behaviour like discussing organ donation with family members, or registering intent as an organ donor is thus extremely complex issue that differentiates promotion of organ donation from other health-related promotions (Kamin et al. 2017). Practitioners and advocates who promote registration behaviours must consider target-group relevant barriers and misconceptions in the context of the unique nature of post-mortem organ donation.

Furthermore, pro-donation behaviour is culturally bound: that is, influenced by a wide variety of social, religious and moral norms (Alden and Cheung 2000; Hyde and White 2009; Kamin et al. 2017; Morgan et al. 2003; Morse et al. 2009; Phillipson et al. 2013; Sadic et al. 2016). For example, Phillipson and colleagues explored knowledge, attitudes and beliefs in Greek, Serbian and Macedonian Orthodox communities in Australia, showing unique barriers such as deficits of knowledge and taboos surrounding discussion of death (Phillipson et al. 2015). Within specific cultural groups, individual differences in social, psychological and demographic factors interact in understanding willingness to donate or to discuss donation [Kamin & Knox, under revision]. For example, in their study of 298 Chinese young adults, Wu (2008) found that 88% reported reluctance to discuss organ donation. After taking into account variance related to intentions to discuss, individual differences in subjective norms and death anxiety explained variance in discussion behaviour (Wu 2008). These studies highlight the need for active behaviour change strategies that consider cultural and individual differences in precipitating factors. Culturally based values and attitudes interact with communication effects from mass media, TV and movies, which further contribute, to myths and misbeliefs about donation, donors and recipients (Richardson 2017; Long et al. 2012; Morgan et al. 2005; Quick et al. 2014). Promoting pro-donation decisions and behaviours is therefore a complex and challenging context for governments, health practitioners and social researchers internationally.

Given the importance of increasing the supply of organs to meet demand, this industry case study describes the development, delivery and evaluation of a community awareness and pro-donation behaviour promotion campaign in terms of social marketing principles and practices. This case study focuses on a campaign delivered in Australia to illustrate the application of social marketing principles to this complex and impactful topic.

8.4 Pro-Donation Behaviours in Australia

In 2018, 1,782 people received a life-changing organ or tissue transplant from 554 deceased organ donors and their families in Australia (DonateLife 2019). Despite growth in national donation rates, around 1,400 Australians are waitlisted for a transplant (DonateLife 2019). Organ shortage presents a substantial economic and social burden for those who need an organ, their family, community and the national health-care system. To boost donation rates requires that donation is encouraged as an optimistic social value and requires normative changes to make family discussion about organ and tissue donation an accepted and valued experience. Lack of awareness is a critical factor in behaviours of registration and family discussion about donation (Hyde and White 2013). Research indicates that the general public has insufficient knowledge about organ donation and this lack of awareness is an obstacle to developing target behaviours.

Only 8% of young Australians aged 18–24 are registered on the national Australian Organ Donor Register (Opdam 2016). Talking about organ donation is an important behaviour because under Australia's opt-in policy, a potential organ donor's family will be asked to confirm donation and will have the final say. In general, social marketing would need to influence three interrelated organ donation behaviours: discovery, decision and discussion [Kamin & Knox, paper in review]. Therefore, talking with family about a decision to donate organs is an important factor in addressing organ shortage. However, relevant research on effective methods for promoting communication about organ donation among young adults is currently lacking. Hence, there is a research gap concerning young adults' barriers to organ donation discussion.

Set against a complex background, the purpose of the MAKE IT COUNT 2018 campaign was to dissolve known barriers to discussing organ donation with family among young Australian adults. Developed with social marketing principles at its core, the campaign informs future researchers and practitioners, and has broad social impact in terms of increased awareness and registrations among young adults; and reduction of the percentage of young adults who are undecided by increasing family discussion about organ donation. In Sect. 8.2, the setting for the case study is described, and the aims, outcomes, and campaign activities are presented.

8.5 Case Study: MAKE IT COUNT 2018

Public education campaigns and large-scale community events can improve donor recruitment behaviour (Feeley and Moon 2009; Rady and Verheijde 2016; Tumin et al. 2014) as they provide real opportunities for transferring thoughts among locals and visitors and help as an educational and teaching platform (Dwyer et al. 2000). Evidence shows that short-term interventions that (i) address commonly held concerns about organ donation and (ii) provide an immediate registration opportunity, boost efficiency of community-based interventions (Golding and Cropley 2017). Therefore, large-scale community events provide a unique opportunity to increase organ donation awareness, discussion and registration behaviours among target demographic groups. The 2018 Royal Brisbane Show (known locally as 'The Ekka') was the setting for MAKE IT COUNT 2018; a community campaign supported by an OTA Community Awareness Grant, to improve organ donation awareness, discussion and registration behaviour. The Ekka is an annual event drawing crowds of 400,000 across five days. The Ekka festival represents a significant opportunity to promote and generate interest in organ donation and to prompt people to 'Make it Count' by discussing and registering their donation decision on the Australian Organ Donor Register.

The existing evidence in the academic literature and current statistics demonstrates the clear need for an innovative and customised approach to engaging members of the target demographic of young Australians. Specifically, there is currently no mobile App available to facilitate the target demographic groups to engage with the topic of post-mortem organ donation, to discover culturally appropriate facts, to break down barriers and myths surrounding organ donation, and to motivate and enable individuals to decide, register, and communicate that decision to their families. A clearly identified need exists to engage young Australians using the convenience and ease of an App platform. Mobile apps provide a unique and powerful platform to reach the target audience. Smartphones are owned by 15 million Australians and are the most used digital device, averaging more than one hour a day [source: IAB Australia and Nielsen Mobile Ratings Report, 2015].

To facilitate registrations and prompt opportunities to discover, decide and discuss facts about organ donation, the DonateLife App was previously developed by the authors with funding from the Australian Government Organ and Tissue Authority. App design was informed by pilot exploratory research and a series of focus groups with the target group. Content generated by the intended end user was integrated into App functionality, including why it is important to discuss with family, and prompts for starting a discussion with family about organ donation (Fig. 8.1). The DonateLife App is a unique smart device application designed to increase Australians' awareness of donation and directly facilitate online donor registrations. The DonateLife App engages guides the user through discovery of facts, providing multimedia resources and links to video clips, animation and multicultural and faith resources. We integrated the app in the current campaign as part of the strategy. Features and functions of the app are described throughout.

8.6 Aims

MAKE IT COUNT 2018 aimed to increase registrations on the Australian Organ Donor Register (AODR) for priority demographic segments that are underrepresented including young adults, males and people from multicultural communities, using a targeted awareness campaign before and during the Ekka week in 2018. Specific aims were to raise awareness of the campaign before the Ekka period commenced, to generate social media engagement and discussion behaviours among the diverse community of visitors and population attending The Ekka festival, and ultimately to increase registrations on the AODR (Fig. 8.2).

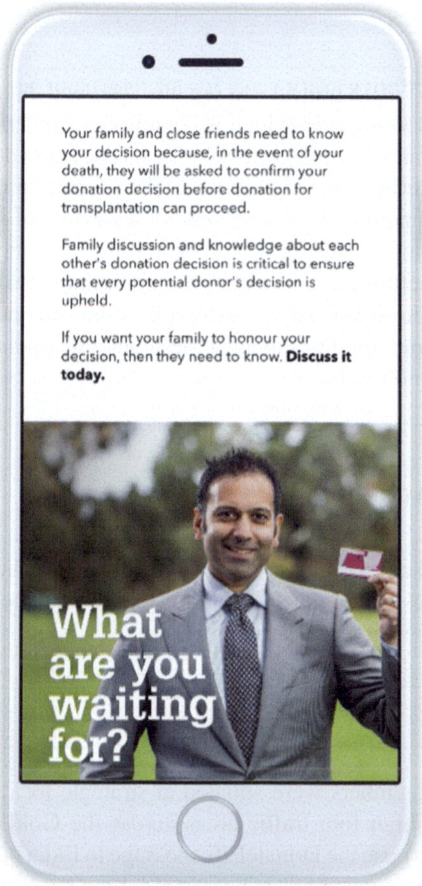

Fig. 8.1 DonateLife App: The importance of family discussion

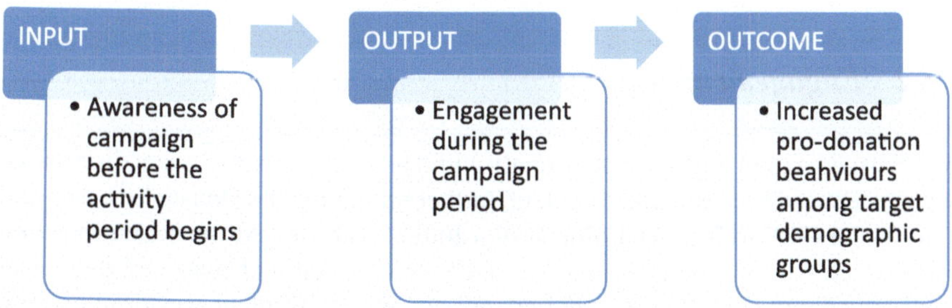

Fig. 8.2 Theory of change for the MAKE IT COUNT 2018 strategy

8.7 Key Outcomes

The MAKE IT COUNT 2018 project was organised around three measurable targets. The first outcome was awareness, which we enumerated in terms of the quantity of information, flyers, marketing materials and DonateLife app downloads distributed into the community during campaign period. The second outcome was discussion of organ and tissue donation, operationalised by the extent of social media engagement and activity during campaign period. The third outcome was registration on AODR around the timeframe of The Ekka. Although individual registrations cannot be matched to a specific source, aggregate numbers indicate the fluctuation in registration activity over time. Findings against these outcome indicators are reported in Sect. 8.3, evaluation and results.

8.8 Campaign Activities

A social marketing-based project was delivered to implement the campaign activities. Key campaign activities were defined in relation to key outcomes. Activities included

8.8.1 Exposure

Trained roving team members were situated at strategic locations including transport connections and high foot traffic areas across the Gold Coast, Brisbane and surrounding areas to expose the DonateLife message to Ekka patrons and the general public. Team members wore and distributed OTA branded t-shirts, hats, wristbands and logos. Team members were conversant in OTA messages and facts.

8.8.2 Engagement

A social media campaign ran concurrently with The Ekka festival to foster discussion, engagement and user-generated content, further promoting the Donate Life App and links to the OTA website and official registration pathway. An internship position was created for the Social Media Manager who created and managed the social media campaign content and analytics reported herein. Content was developed in consultation with OTA and DL QLD and recognition of the role of the funder was incorporated. The online social media presence extended before and during the Ekka 2018 event with daily prize draws.

8.8.3 Recruitment

Team members were trained in the registration process, distributed registration flyers and promoted the Donate Life App to collect registrations. During transit or while waiting to board transport presented a convenient opportunity to disseminate information, answer questions, talk about and register on the AODR. Team members were from CALD backgrounds and most were fluent in language other than English. Team members engaged with the public and demonstrating how to register online, download the App and fill out a registration form.

In Sect. 8.3, we define and specify social marketing principles applied in the MAKE IT COUNT 2018 campaign. Qualitative and quantitative evaluation data were gathered to evaluate the project against stated outcomes. Section 8.4 presents a summary of evaluation data and outcomes achieved.

8.9 Social Marketing Principles and MAKE IT COUNT 2018

Social Marketing is a praxis framework that draws knowledge from many disciplines to understand how to influence peoples' behaviour and achieve social change (French and Gordon 2015). The MAKE IT COUNT 2018 case study described here draws on social marketing benchmark criteria in conjunction with context-specific organ donation constructs drawn from empirical research (Schulz et al. 2000). To differentiate this campaign from pure communications or awareness raising, MAKE IT COUNT 2018 employed social marketing principles to stimulate behaviour change. The benchmark criteria include behaviour change goals, customer orientation and insight, exchange, competition, segmentation, marketing mix and theory. Implementation of the benchmark criteria of social marketing is described in turn.

8.10 Behavioural Change Goals

Pro-donation behaviours fall squarely within the realm of social marketing, in the sense that forming a decision about post-mortem donation, registration of consent as a donor and discussing that decision with family are voluntary behaviours under an individual's free will, and each behaviour has a social implication for common benefit. The decision to consent to donating one's own organs (and tissues) can also be thought of as a specific aspect of consumer disposition behaviour in that 'one may by will, or contract, supervise the disposition of her own remains' (Silver 1988). Following the social marketing consensus definition benchmark criteria, specific behavioural targets were allocated for the MAKE IT COUNT 2018 campaign.

Behavioural goals were based on national baseline data for organ and tissue registrations and national goals for increasing registration among target demographic groups:

(1) Increase awareness of organ donation and the AODR among young adults during the Ekka festival period,
(2) Generate social engagement and discussion among to the community during the Ekka festival period and
(3) Increase number of registrations on the AODR for priority demographic segments (young adults, males, CALD community) during the Ekka festival period.

To facilitate registrations and prompt opportunities to discover, decide and discuss facts about organ donation, a range of tools and strategies were implemented as described further in the following sections.

8.11 Customer Orientation and Insight

Development of the 'Make It Count' campaign draws on evidence and consumer insights from two previous funded formative research projects: i) a systematic review of previous studies (Knox et al. 2017) to understand target-group sentiments, and ii) a programme of qualitative (focus group) and quantitative (survey) formative research with target-group members (unpublished). Together, previous formative research yielded the following actionable insights relevant to young adults, males and CALD communities in southeast QLD:

- Despite positive sentiment towards organ donation, perceived lack of opportunity and low levels of awareness prevent supportive attitudes being converted to pro-donation behaviours;
- Positive reinforcement from social reference groups could facilitate conversion of pro-donation attitudes to pro-donation behaviours for young adults;
- Simple prompts, nudges and reminders could be sufficient to initiate discussion and decisions about donation in the target group.

In response, the MAKE IT COUNT 2018 campaign was designed to be customer oriented, using social media, and peer-to-peer communication, and to provide convenient opportunities to register online, and to find out more about organ donation via the Donate Life app. A diverse team of young adults were trained as recruiters to provide verbal and written information in vivo. Recruiters provided a physical point of contact to 'Make it Count', i.e. by facilitating conversation, distributing information and registration materials, by communicating the Donate Life message, promoting the App and answering questions and supporting actual online registrations.

8.12 Exchange

According to French and Blair-Stevens (2006), in the context of social marketing exchange describes something that a person has to give up in order to get the proposed benefit of the intervention. In their view, the exchange was treated as 'direct exchange', meaning that there is something tangible or intangible that needs to be given up simultaneously to get a direct benefit (French and Blair-Stevens 2006). As discussed earlier, thinking and talking about organ donation carries with it a considerable psychological burden. Contemplation or discussions of mortality and the psychologically distant hypothetical event of becoming and organ donor create anxiety and negative affect. Exchange in a pro-donation context can be conceptualised as giving up comfort: experiencing negative emotion, in return for the benefit of positive emotion, self-esteem and reinforcement from social reference groups. The MAKE IT COUNT 2018 campaign activities were intended to activate the underlying positive public attitudes to behaviours, accentuate positive social reinforcement and encouragement from the social reference group, and emphasise the value, sense of altruism, and positive affect that is associated with pro-donation behaviours.

8.13 Compctition

Andreasen's (2002) social marketing benchmark criteria require recognising and addressing the competition of the behaviour targeted by an intervention. The social marketer must understand what other behaviours are competing for the chosen target audience's time and attention in order to develop strategies that minimise the impact of the competition (Andreasen 2002). According to Hastings (2003), social marketing needs to offer '*unique and meaningful benefits*' (p. 307) that provide better value than the competing behaviours. MAKE IT COUNT 2018 sought to diffuse commonly held misconceptions and myths surrounding organ donation, overcoming acknowledged barriers to engaging with the topic to minimize costs and maximise the benefits by creating convenient opportunities to

(1) *Discover*: Teams of trained campaigners were present at venues and transport hubs during The Ekka with flyers, merchandise and information to facilitate behaviour in vivo;
(2) *Decide*: Present facts and dispel myths via content delivered in multiple modalities (in person, in print media, via animations and images presented in the DonateLife app);
(3) *Discuss*: Formative research showed that discussing donation does not have to be serious, dark or difficult (Knox et al. 2017).

By providing a prompt (i.e. talking points generated by peers), an opportunity (in the form of a registration flyer) and various reminders or nudges (in the form of merchandise, social media content, and reminders scheduled through the Donate

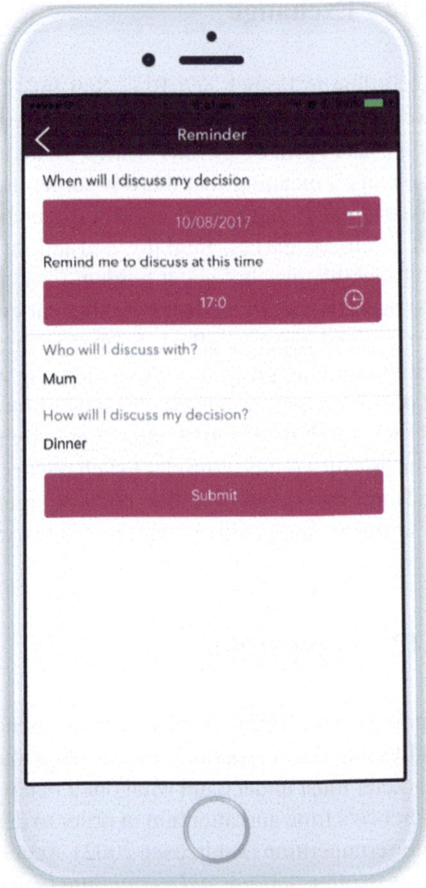

Fig. 8.3 Scheduling a reminder to discuss with the DonateLife app

Life app), the MAKE IT COUNT 2018 campaign addressed some pertinent elements of competition for pro-donation behaviour. The DonateLife App prompts the user via push notification to discuss with family and supports individuals to overcome barriers and set a goal and schedule a reminder to initiate a discussion (Fig. 8.3).

8.14 Segmentation

Andreasen (2002) stated that careful segmentation of target audiences is necessary in order to '*ensure maximum efficiency and effectiveness in the use of scarce resources*' (p. 104). Segmentation is the process of dividing a total market (population) into groups with relatively similar needs to design a social marketing intervention that addresses needs by each group; uses a developed segmentation approach, going beyond simple targeting approaches, and avoids the use of generalisation (Donovan and Henley 2010; Rundle-Thiele et al. 2015). The more specifically the target audience can be defined, the greater the potential for an intervention to change behaviour. Segmenting the target group provides insights and can identify the group that is most likely to change their behaviour therefore ensuring that return on investment is maximised.

It is important to note that segmentation and targeting are two different concepts. Segmentation can be based on one or more of demographic, psychographic, geographic, behavioural and epidemiological factors. Segmentation is based on the principle understanding that populations are typically heterogeneous (Evers et al. 2013) and that groups with similar needs and wants can be identified. Following the identification of groups, social marketers can chose different strategies to reach the market.

The MAKE IT COUNT 2018 campaign employed a *differentiated or multi-segment marketing strategy* in which we provided tailored offerings to each different market segment by means of the trained recruiters and the App. Thus, segmentation enabled distinct target audiences to be identified, each of which may have unique needs and wants (Evers et al. 2013). The content within the DonateLife App was tailored to segments in multiple ways:

(1) *According to individual stages of change* (i.e. readiness to decide or discuss): acknowledging that target-group segments at different points along a continuum from pre-contemplation to action require tailored information, strategies of engagement and have diverse needs. To achieve segmentation in the application, a staging algorithm classifies individuals according to 'stage of readiness' from pre-contemplation to action for deciding and communicating their organ donation intentions (Prochaska and DiClemente 1982).
(2) *According to language*: The DonateLife App presented content in six languages (simplified Chinese, Punjabi, Hindi, Arabic, Vietnamese and English) which the user could designate. Team members from diverse cultural backgrounds and fluent in languages other than English conducted campaign activities at the Ekka.
(3) *According to cultural beliefs*: The users' interaction within the DonateLife App contains content relevant to their individual cultural beliefs or religion.

8.15 Marketing Mix

The Marketing Mix benchmark criterion recommends that social marketing campaigns involve the use of multiple strategies, including the 4Ps of the traditional marketing mix: product, price, place and promotion (Evers et al. 2013). MAKE IT COUNT 2018 utilised multiple marketing tactics to enable behaviour change, including awareness raising to increase knowledge and visibility of the issue (promotion), and practical hands-on support in vivo (place). Design elements of the campaign considered previous findings concerning common barriers and obstacles to achieving the target behaviours (price). Using a personalised communication approach (Arora et al. 2008), the campaign was innovative in that the content presented was entirely determined by and customised to suit the individual user (product). Each component of the marketing mix is explained in turn.

- *Product*: In the context of this intervention, product refers to the benefits derived from an exchange by the target group. The product of this campaign is a service offering, built around three layers of product, the core product of feel-good experience, good vibes and positive self-esteem at thought of helping others, the actual product being the service offerings including the App and registration flyer. The augmented product includes the positive, easy, quick method of registering, the intangible reward, social interaction, positive reinforcement, incentive (tangible reward) for the social media campaign posts.
- *Place/distribution*: Place refers to the distribution channel where and when the target audience enter an exchange. For this campaign, the distribution channel allows the target group to access the product through the App and the personal selling strategy. Using a push strategy, we took the campaign to the people, rather than trying to bring the people to the campaign. Train stations and high foot traffic areas across the gold coast to Brisbane corridor and around the central business district were the physical distribution points. Social media platforms were the virtual distribution points.
- *Price*: Price represents the cost or sacrifice exchanged for the product that audience will pay for the promised benefits, which can be psychological, emotional, cultural, practical, financial or physical. For this campaign, the product benefits came at a price including psychologically complex choice, anxiety, fear, uncertainty, morbidity, so we aimed to minimise the non-monetary price by implementing the app to make it easy, fast, available and relevant to the consumer.
- *Promotion*: According to Andreasen (2002), the promotion tools are used to communicate desired outcomes and inform, remind and persuade the target audience about the value of the product and behaviour change. The social marketing campaign in this study used a variety of promotional tools to raise awareness of the campaign, reinforce the message and to promote social marketing activities. The strategy aimed to increase awareness and generate social media engagement among the diverse community of visitors and population attending the Ekka 2018, in the weeks leading up to the Ekka, during the campaign period, and after the campaign. Thus, the marketing mix included face-to-face peer-to-peer promotion and the social media strategy (Table 8.1).

8.16 Theory

Where social marketing interventions are theoretically based, the Theory of Planned Behaviour (Ajzen 1991) typically dominates, despite a known gap between intentions and behaviour (Judith et al. 2011). MAKE IT COUNT 2018 campaign draws on the Trans-theoretical Model (Prochaska and DiClemente 1982) and traditional exchange theory.

Table 8.1 Promotional strategy for the MAKE IT COUNT 2018 campaign

Before	During	After
Scheduled social media posts build positive awareness of recruiters and their role at the Ekka	Scheduled social media stories to drive engagement among target demographic groups between 9th–18th August	Recognition and thanks to recruiters
Social and traditional media demonstrate the purpose of the campaign, the App and the recruiters' role	Face to face distribution of AODR flyers and merchandise, promotion of the App among target demographic groups between 9th–18th August	Continued messages to drive visitors to the OTA website and reminders to register a decision about organ donation on the AODR
Images and videos show that recruiters will be supportive and helpful	Photo competition to encourage and reward user-generated content and engagement during the CAG activity	Announcement of winners and reward

Traditional economic exchange theory stipulates that for an exchange to take place, the target audience must perceive the benefits as equal to or greater than perceived costs: otherwise known as a 'cost–benefit analysis'. In the context of the MAKE IT COUNT 2018 campaign, a social marketing mix of strategies was developed to emphasise and promote the benefits, while reducing the barriers and perceived physical and psychological costs.

Based on the Trans-theoretical Model (Prochaska 2013; Prochaska et al. 1992), people cycle through modifiable 'stages of readiness' for behaviour change (pre contemplation; contemplation; preparation; action; maintenance; relapse). Trans-theoretical model has been widely applied to various health behaviours, and research supports the applicability of stages of change to organ donation decision-making (Robbins et al. 2001). The DonateLife app ascertains the users' stage of change with measures drawn from published studies (Morgan and Miller 2002; Smith et al. 2004), tailoring the content presented according to responses. Hence, this project is innovative both in the application of a social marketing approach to organ donation, and in that few social marketing interventions are theoretically based.

8.17 Evaluation and Results

The MAKE IT COUNT 2018 project effectiveness and impact were evaluated against the three outcomes:

(1) Awareness—A proxy for awareness was measured in terms of the quantity of information, merchandise, flyers and marketing materials distributed. Downloads of the DonateLife App during the campaign period provide an additional index of awareness.

Table 8.2 Campaign outcomes and performance indicators

Outcome measure	Summary of campaign performance
(1) Awareness indicators	• 4,500 AODR registration brochures • 1,850 DonateLife pens • 1,650 DonateLife wristbands • 165 DonateLife water bottles • 50 DonateLife Hats • 50 + DonateLife App downloads Total distribution 8,265 branded items
(2) Discussion behaviour—Social and traditional media reach	• Facebook—Organic reach of 12,394 with $0 budget, 18.7% increase in followers to the host page, average total reach 513 per post • Twitter—89 total interactions, 25 retweets • LinkedIn—950 additional views internationally, • Photography competition • Traditional media—Griffith University research news article 268 views and 12 shares Total reach 13,650 viewers
(3) Registrations behaviour	• 45,000 new registrations on AODR in QLD for 2018 • 1,000 registrations on AODR around The Ekka Festival

(2) Discussion behaviour—Discussion of organ and tissue donation was operationalised in terms of social and traditional media engagement and activity during campaign period. Total campaign reach was estimated along with metrics to measure social media engagement and reach across multiple platforms during the campaign period.
(3) Registration behaviour—Registrations on AODR around the timeframe of The Ekka (aggregate numbers). In addition, reported state-wide annual registration activity was tracked over time to support the overall impact of promotional efforts across Queensland.

Performance against outcome measures is summarised in Table 8.2 and described further below.

8.18 Awareness

In this project, we promoted awareness of the DonateLife brand and message by taking full advantage of the supporter kits, printed material and the range of promotional materials and merchandise available from the OTA. Team members were trained at a dedicated training day using the supporter kit material and information contained therein, specifically, approaching the public, answering questions and discussing facts, how to fill in the registration form, how to download and use the DonateLife app. Roving recruiters utilised the existing resources supplied by the OTA including multicultural flyers, posters, signage and handouts. Team members received training to become conversant in the key OTA messages and facts surrounding organ donation. Team members wearing OTA branded t-shirts, hats, wristbands and logos were rostered to present at key locations and transport routes and distributed 8,265 items of branded merchandise across the campaign (Table 8.2). Recruiters promoted the custom-built DonateLife App. The App was originally uploaded on 29 Oct 2017 and updated 11 March 2018 for this campaign. The DonateLife App was downloaded 50+ times from Google Play at campaign close.

8.19 Discussion Behaviour

The social media campaign ran before and during the Ekka festival 2018. The content appeared from 7 August through to 24 August 2019, thus, measures were collected over an 18-day active campaign period. Equivalent measures for the 18 days prior to the active period give a relative comparison of empirical data during the campaign. We selected two social media platforms for the campaign namely Facebook and Twitter. A programme of 18 posts were developed in consultation with funding organisation.

Statistics collected from the host Facebook and Twitter accounts indicate the campaign was successful in generating awareness by creating media coverage via press and social media prior to the CAG activity and creating user-generated content and engagement during the CAG activity. Facebook was the primary social media platform for the campaign. In general, the statistics from the campaign show that the content contributed to raised awareness amongst the target group and engaged the community. Facebook engagement statistics (page followers, reach and views) at baseline were tracked on 6 August. The number of page followers increased from 187 at baseline to 222 on the day after the active campaign ended (Table 8.2). The campaign therefore resulted in 18.7% increase in followers to the host page.

In the 18 days ahead of the campaign, the organic reach was measured as 2,428. During the 18-day campaign period, the total organic reach was 12,394, reflecting an increase of 9,966 reach with zero budget. The total number of page views in the 18 days ahead of the campaign was 46 where two of these views are post views. The total number of page views during the 18-day campaign period was 46 of which 26 views relate to post views (Table 8.2).

The best performing post 'Meet Kate Rootsy' featured a high-profile local transplant recipient (Fig. 8.4). The total reach for this post was 3,300, ranking as the second top-performing post on the host page in 2018. The post generated 202 clicks and 123 likes/shares/comments during the campaign period.

Activity and engagement on the Twitter platform were modest, with 77 likes/comments/shares in total across the campaign. The post that reached the highest engagement related to myth-busting (Did you know this about organ donation?), and

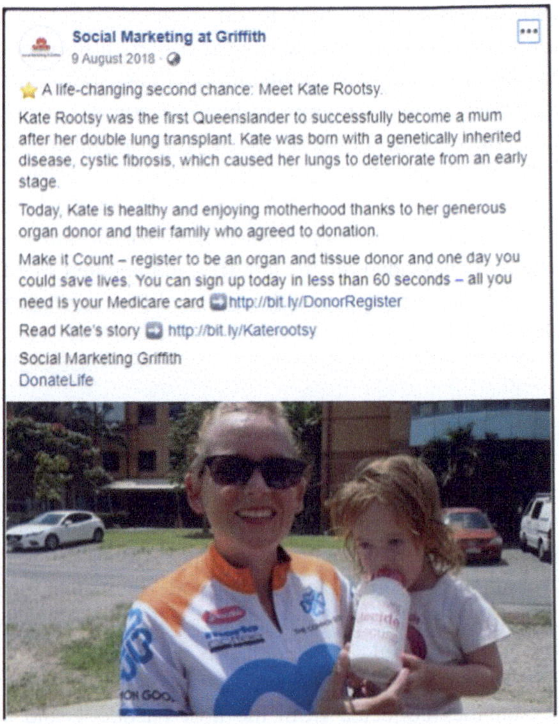

Fig. 8.4 Highest performance social media post

received 8 retweets and 22 likes, making the total interaction count of 30. Five of the content posts were shared to an authors' personal LinkedIn account, which contributed to further raising reach with 950 additional views. Individual posts were viewed between 83 and 333 times. The LinkedIn audience generated diverse international contact with viewers ranging from academia to industry professionals located around Australia, USA, United Kingdom, Europe, India and Malaysia.

The photography competition created user-generated content and engagement amongst the target community of young adults, males and the CALD demographic. The outcome of the competition was better than expected (Fig. 8.5). Winners were selected, and prizes were allocated. A media release article titled 'Make It Count during the Brisbane Ekka to boost organ and tissue donor registrations' was published on 8 Aug 2018 (https://blogs.griffith.edu.au/business-school-research/department/griffith-marketing-department/2018/08/08/make-it-count-during-the-brisbane-ekka-to-boost-organ-and-tissue-donor-registrations/). According to Google Analytics, the article was viewed 268 times and was shared 12 times on Facebook at close of the campaign.

Fig. 8.5 Example photography competition entries

8.20 Registration Behaviour

According to DonateLife Queensland, around the time of the Ekka festival a spike of approximately 1,000 new registrations on the AODR were observed. Although it is not possible to identify the source of individual registrations, the pattern is consistent with previous years and time-bound to the Ekka festival period.

Across the state, there were approximately 45,000 new registrations on the AODR in QLD in 2018, increasing from 36,000 in 2017. The growth rate in new AODR registrations for Queensland is therefore comparable to the neighbouring state of NSW, which has a population of 2 million higher than QLD. The increasing rate of growth in registration behaviours suggests campaigns across the state including community champions such as Kate Rootsy and local promotional activities such as MAKE IT COUNT 2018 are effective in contributing to registration behaviour. Despite the associated increases, the overall proportion of people registered on the AODR in the state of QLD remains lower than desired, indicating the need for continued activity to increase registration behaviours.

8.21 Discussion

There is clearly an identified need to increase pro-donation decision, discussion and registration behaviours among target demographic groups in Queensland. Consistent with the National Community Awareness and Education Program (Australian Government 2009), this project addressed the identified need to increase registrations on the AODR by targeting activities to those that are currently underrepresented, specifically young adults age 18–29 years, males across all age groups and people from the CALD community. This chapter presented the MAKE IT COUNT 2018-campaign aims, outcomes and activities including exegesis of how the programme was developed in accordance with social marketing principles and practices. Evaluation results against outcome measures showed that leveraging a significant annual event, The Brisbane Ekka, represented a well-suited opportunity to raise awareness, generate discussion behaviours and to prompt people to register their donation decision on the Australian Organ Donor Register. Given the global significance of organ shortage, and the whole-of-community engagement focus of the Ekka, this campaign addressed the identified need to increase registrations by targeting activities to under-represented market segments: young adults, males, and culturally and linguistically diverse communities. Having culturally diverse team members enabled engagement with a range of cultures, which aligned to the goals of the programme.

However, the prevailing discrepancy between public support for the concept of post-mortem donation and realised behaviour such as registration on the AODR was reflected in our campaign results. Public support and engagement with the campaign were estimated at 13,650 views, in comparison to actual behaviour which was estimated at 1,000 registrations. As discussed in Sect. 8.1, there are a range of legislative and ethical, psychological and cultural factors that add weight to pro-donation behaviours. While application of social marketing principles and practices as described herein contributed to the development of a user-focused interactive strategy approach, many constraining elements remained outside the scope of the campaign.

8.22 Limitations

Program design considered some barriers and competing forces pertinent to the target groups, based on previous formative research, by way of presenting a convenient opportunity to talk and think about organ donation with a peer, obtain a registration method, or download the app to refer to later. However, the campaign was not able to address all competing forces and focused primarily on bringing the message to the target groups.

In terms of engagement and discussion behaviours, there are limitations to the campaign activities. It is important to note that throughout the campaign duration, the host account posted other unrelated content on their social media platforms. The consequence is that some of the collected data to indicate discussion and engagement, such as number of followers and page views, could be impacted by other content and is therefore not uniquely related to the MAKE IT COUNT 2018 campaign. In addition, none of the posts were paid or promoted to target segments. Therefore, we were unable to report on reach to priority demographic groups (e.g. males) specifically.

There are limitations in terms of tracking registration behaviour. While we can estimate the overall local fluctuations in registration numbers for a specific time period, we were unable to identify actual registrations on the AODR during or following the campaign due to privacy restrictions and the roaming nature of campaigners. The estimates provided are approximate and may reflect concurrent promotion efforts and the effects of factors outside the control of this campaign. Additionally, according to the Trans-theoretical model, awareness and receipt of information would not convert to pro-donation behaviour in a direct or linear manner, therefore registration behaviours might not emerge for some time depending on a person's stage of readiness to change along with other unobserved factors. The overall impact of this campaign should be considered with these caveats in mind.

8.23 Recommendations

Given current knowledge of the complex legislative, ethical, psychological and cultural landscape which surrounds donation of organs and tissue, along with other Substances Of Human Origin (SOHO), a range of recommendations can be drawn. Based on the Theory of Change underlying the MAKE IT COUNT 2018 campaign (Fig. 8.1), awareness precedes engagement and discussion behaviours, which lead to registration behaviours. To complement community-based awareness and registration activities such as the MAKE IT COUNT 2018 campaign, a dedicated host site identity and social media presence would be ideal. To support awareness future campaigns could promote deliberately to target segments using social and traditional media.

Following the Trans-theoretical model, target segments of the community represent various stages of change in readiness to discuss, decide or register. Accordingly, pro-donation campaigners would be well advised to continue to form novel ways to reach target groups at various stages of change, for specific segments. Campaign messages and tactics could be tailored to draw on values to reduce competition and enhance exchange for segments at each stage.

In the international area, changes to consent systems are forecast to create flow-on effects for Australians. When the UK transitions to an opt-out model in 2020 (Davis 2020), pro-donation campaigners, and researchers should anticipate questions, discussion and associated opportunities for awareness and promotion in the ripple effects. Locally, in Australia social marketers might work 'up stream' to influence policy innovations to make decision, discussion and registration behaviours more readily accessible for target segments. Looking 'downstream', social marketers are advised to consider novel and engaging ways to reach the target groups especially males, leveraging other large-scale events, sporting and community festivals, and corporate endorsement for example with workplace-based promotional sign-on days. Researchers are encouraged to consider the wider SOHO context for tactics and strategies applicable to pro-donation promotion.

8.24 Conclusion

The MAKE IT COUNT 2018 campaign was an evidence based, peer-to-peer community awareness campaign featuring a concurrent digital promotion on multiple social media platforms, executed during a large local festival in Brisbane, Queensland. In the short campaign timeframe, the project achieved success in terms of defined outcomes. Designed with social marketing best practice principles, the campaign contributed broadly to increased issue awareness, engagement and discussion, and contributed broadly to growth in the rate of registrations on the AODR in Queensland.

Acknowledgements The Social Marketing @ Griffith research centre at Griffith Business School, Griffith University, supported this work. This project was sponsored by the Organ and Tissue Authority through a Community Awareness Grant No. G1617-013-01 awarded in 2017. The authors extend thanks to DonateLife QLD community education trainers Alana Cresswell, Shelley-Lee Waller and Belinda Chilcott, all members of the recruitment workforce and Victoria Aldred for contributions to the project.

References

Adenwalla HS, Bhattacharya S (2012) Dr. Joseph e. Murray. Indian Journal of Plastic Surgery: Official Publication of the Association of Plastic Surgeons of India

Ajzen I (1991) The theory of planned behavior. Organizational behavior and human decision processes. https://doi.org/10.1016/0749-5978(91)90020-T

Alden DL, Cheung AH (2000) Organ donation and culture: a comparison of asian american and european american beliefs, attitudes, and behaviors. J Appl Soc Psychol

Andreasen A (2002) Marketing social marketing in the social change marketplace. J Pub Policy Market. https://doi.org/10.1509/jppm.21.1.3.17602

Arora N, Dreze X, Ghose A, Hess JD, Iyengar R, Jing B, Neslin S (2008) Putting one-to-one marketing to work: personalization, customization, and choice. Mark Lett

Citerio G, Cypel M, Dobb GJ, Dominguez-Gil B, Frontera JA, Greer DM, Wijdicks EFM (2016) Organ donation in adults: a critical care perspective. Intensive Care Medicine. https://doi.org/10.1007/s00134-015-4191-5

Davis N (2020) All adults in England to be deemed organ donors in 'opt-out' system. The Guardian. Accessed from https://www.theguardian.com/society/2020/may/19/deceased-uk-adults-to-be-deemed-organ-donors-in-opt-out-system

DonateLife (2019) Retrieved from https://donatelife.gov.au/about-donation/get-facts/facts-and-statistics

Donovan R, Henley N (2010) Principles and practice of social marketing: an international perspective. Cambridge University Press

Dwyer L, Mellor R, Mistilis N, Mules T (2000) A framework for assessing "tangible" and "intangible" impacts of events and conventions. Event Management

Ejzenberg D, Andraus W, Baratelli Carelli Mendes LR, Ducatti L, Song A, Tanigawa R, Chada Baracat E (2018) Livebirth after uterus transplantation from a deceased donor in a recipient with uterine infertility. The Lancet. https://doi.org/10.1016/S0140-6736(18)31766-5

Evers U, Jones SC, Caputi P, Iverson D (2013) Promoting asthma awareness to older adults: formative research for a social marketing campaign. J Asthma Allergy Educators

Feeley TH, Moon S-I (2009) A meta-analytic review of communication campaigns to promote organ donation. Communication Reports

Fraser C (2016) An analysis of the emerging role of social media in human trafficking

French J, Blair-Stevens C (2006) Social marketing national benchmark criteria. UK National Social Marketing Centre, UK

French J, Gordon R (2015) Strategic social marketing. Sage

Gevers S, Janssen A, Friele R (2004) Consent systems for post mortem organ donation in europe. Eur J Health Law

Golding SE, Cropley M (2017) A systematic narrative review of effects of community-based intervention on rates of organ donor registration. Progress in Transplantation

Guzik-Makaruk EM, Olesiuk-Okomska M, Matuszkiewicz-Rowińska J, Małyszko J (2019) Selected legal aspects of donation after circulatory death in poland. Annals of Transplantation

Hastings G (2003) Relational paradigms in social marketing. J Macromarketing

Holdershaw J, Gendall P, Wright M (2011) Predicting blood donation behaviour: further application of the theory of planned behaviour. J Soc Mark

Hyde MK, White KM (2009) Communication prompts donation: exploring the beliefs underlying registration and discussion of the organ donation decision. British J Health Psychol

Hyde MK, White KM (2013) A test of three interventions to promote people's communication of their consent for organ donation. Psychology & Health. https://doi.org/10.1080/08870446.2012.731060

Kamin T, Berzelak J, Ule M (2017) Mind the gap: social aspects of willingness for post-mortem organ donation in slovenia. Teorija in Praksa

Knox, K, Parkinson J, Pang B, Fujihira H, David P, Rundle-Thiele S (2017) A systematic literature review and research agenda for organ donation decision communication. Prog Transpl. https://doi.org/10.1177/1526924817715459

Long SDP, Morgan SE, Harrison TP, Afifi WAP, Stephenson, MP, Reichert TP, Morse CP (2012) When families talk: applying interpretative phenomenological analysis to african american families discussing their awareness, commitment, and knowledge of organ donation. J Natl Med Assoc

Manara, A. R., Murphy, P. G., & O'Callaghan, G (2012) Donation after circulatory death. BJA: British J Anaesthesia. https://doi.org/10.1093/bja/aer357

Morgan SE, Miller J (2002) Beyond the organ donor card: the effect of knowledge, attitudes, and values on willingness to communicate about organ donation to family members. Health Commun 14(1):121–134

Morgan SE, Harrison TR, Long SD, Afifi WA, Stephenson MT, Reichert T (2005) Family discussions about organ donation: how the media influences opinions about donation decisions. Clin Transplant. https://doi.org/10.1111/j.1399-0012.2005.00407.x

Morgan SE, Miller J, Arasaratnam LA (2003) Similarities and differences between african americans' and european americans' attitudes, knowledge, and willingness to communicate about organ donation1. J Appl Soc Psychol. https://doi.org/10.1111/j.1559-1816.2003.tb01920.x

Morse, C. R., Afifi, W. A., Morgan, S. E., Stephenson, M. T., Reichert, T., Harrison, T. R., & Long, S. D (2009) Religiosity, anxiety, and discussions about organ donation: understanding a complex system of associations. Health commun. https://doi.org/10.1080/10410230802676755

O'carroll RE, Foster C, McGeechan G, Sandford K, Ferguson E (2011) The "ick" factor, anticipated regret, and willingness to become an organ donor. Health Psychol
Phillipson L, Larsen-Truong K, Jones SC, Pitts L, Nonu M, Filippi M, Trkulja Z (2013) Addressing stigma and engaging orthodox religious communities around organ and tissue donation. Paper presented at the The 12th Congress of the International Society for Organ Donation and Procurement, Sydney, Australia
Phillipson L, Larsen-Truong K, Pitts L, Nonu M (2015) Knowledge of, beliefs about, and perceived barriers to organ and tissue donation in serbian, macedonian, and greek orthodox communities in australia. Progress in Transplant. https://doi.org/10.7182/pit2015550
Prochaska JO (2013) Transtheoretical model of behavior change Encyclopedia of behavioral medicine. Springer, pp 1997–2000
Prochaska JO, DiClemente CC (1982) Transtheoretical therapy: toward a more integrative model of change. Psychotherapy: Theory, Research & Practice
Prochaska JO, DiClemente CC Norcross JC (1992) In search of how people change. Applications to addictive behaviors. Am Psychol
Quick BL, Anker AE, Feeley TH, Morgan SE (2016) An examination of three theoretical models to explain the organ donation attitude–registration discrepancy among mature adults. Health commun
Quick BL, Morgan SE, LaVoie NR, Bosch D (2014) Grey's anatomy viewing and organ donation attitude formation: examining mediators bridging this relationship among african americans, caucasians, and latinos. Communication Research
Rady MY, Verheijde JL (2016) Campaigning for organ donation at mosques. Paper presented at the Hec Forum
Richardson KL (2017) Depiction of organ donation on television negatively affects viewers' decisions to become donors
Robbins, ML, Levesque DA, Redding CA, Johnson JL, Prochaska JO, Rohr MS, Peters TG (2001) Assessing family members' motivational readiness and decision making for consenting to cadaveric organ donation. J Health Psychol 6(5):523–535
Rogers, W (2017) China says it has stopped harvesting organs, but evidence belies its claim. The Conversation. Retrieved from The Conversation website: http://theconversation.com/china-says-it-has-stopped-harvesting-organs-but-evidence-belies-its-claim-72841
Rundle-Thiele S, Kubacki K, Tkaczynski A. Parkinson J (2015) Using two-step cluster analysis to identify homogeneous physical activity groups. Mark Intell Plann
Sadic S, Sadic J, Krupic R, Fatahi N, Krupic F (2016) The influence of information and religion on organ donation, as seen by school teachers in bosnia and herzegovina. Materia socio-medica
Schulz, KH, Meier D, Clausen C, Kuhlencordt R, Rogiers X (2000) Predictors of the intention to donate organs: an empirical model. Paper presented at the Transplantation proceedings
Shepherd L, O'Carroll RE (2014) Do affective attitudes predict organ donor registration? A prospective study. J health psychol
Silva AP (2019, 7 June 2019) Implementation of antitrafficking convention. Paper presented at the Ethical Challenges in Organ Transplantation, University Medical Centre Ljubljana
Silva ENd, Campelo AK, Neto GB (2007) The impact of presumed consent law on organ donation: an empirical analysis from quantile regression for longitudinal data. Paper presented at the Berkeley Program in Law & Economics, University of California, Berkeley
Silver T (1988) The case for a post-mortem organ draft and a proposed model organ draft act. BUL Rev
Smith S, Kopfman JE, Massi Lindsey LL, Yoo J, Morrison K (2004) Encouraging family discussion on the decision to donate organs: the role of the willingness to communicate scale. Health Communication 16(3):333–346. https://donatelife.gov.au/about-us/who-we-are/national-program
Truog RD, Miller FG (2008) The dead donor rule and organ transplantation. New England J Med
Tumin M Noh A Jajri I Chong CS, Manikam R, Abdullah N (2014) Factors that hinder organ donation: religio-cultural or lack of information and trust

van Veen E, van der Jagt M, Cnossen MC, Maas AIR, de Beaufort ID, Menon DK, Participants (2018) Brain death and postmortem organ donation: report of a questionnaire from the center-tbi study. Critical Care. https://doi.org/10.1186/s13054-018-2241-4
Williamson LD, Bigman CA, Quick BL (2018) A qualitative examination of african americans' organ donation-related medical mistrust beliefs. Howard Journal of Communications. https://doi.org/10.1080/10646175.2018.1512064
Williamson LD, Reynolds-Tylus T, Quick BL, Shuck M (2017) African-americans' perceptions of organ donation: 'Simply boils down to mistrust!'. J Applied Commun Res. https://doi.org/10.1080/00909882.2017.1288293
Wu AMS (2008) Discussion of posthumous organ donation in chinese families. Psychol health med. https://doi.org/10.1080/13548500701351992
Wu AMS, Tang CS, Yogo M (2013) Death anxiety, altruism, self-efficacy, and organ donation intention among japanese college students: a moderated mediation analysis. Australian J Psychol

Kathy Knox is a research psychologist with expertise in human development, attitude and health behaviour change across a range of settings. Kathy is a member of the Menzies Health Institute Queensland and Research Fellow with Social Marketing @ Griffith leading a rapidly developing program of empirical research on social behaviour change. Kathy's research focus is on communities and sustainability behaviours including health education and communication, reducing food waste and excess water use, increasing recycling and organ donation. Kathy has direct experience in designing and conducting community-based interventions and has evaluated local and national interventions to change awareness, knowledge and cultures of behaviour.

Joy Parkinson is Deputy Director of Social Marketing @ Griffith at Griffith University. Joy is a consumer behaviour expert. She has worked on the design, implementation, and evaluation of a number of health behaviour change programs. Joy's research includes transformative service design for behaviour change in healthy weight management and particularly for minority and vulnerable consumers. As such, she focuses on systems social marketing, systems evaluation, transformative services research, service thinking in social marketing, and social impact.

Ali Ahani is a sessional lecturer and business analytics researcher at Griffith Business School. His research interests include Business Informatics, Health Informatics, Tourism Marketing and Health Marketing. His contributions have been published in prestigious peer-reviewed journals including the International Journal of Hospitality Management and Scientific Reports-Nature.

Filippa Neilsen is an experienced marketing professional with Masters in Advanced Marketing, with love for all things digital. Filippa develops and leads marketing and communication strategics for challenging and creative environments.

Part III
Social

The final category in this edited volume is devoted to behavioural change in the societal context. Several case studies show innovative campaigns, especially run by the government, resulting in desired behavioural change in targeted communities. The *Relevance of Social Marketing in the Global South Family Planning Programmes: A case from Zambia* chapter discusses a sub-Sahara setting in Zambia, with one of the highest fertility rates in the world, high unmet need for modern contraceptives, high rate of teenage pregnancy, high HIV/AIDS prevalence rate, high occurrence of early marriages and a predominantly young population. Hence, the call for more investment in family planning programmes and research in other developing countries.

Along the same lines, *Declare or dispose'—New Zealand's Border Compliance Social Marketing Programme* case present Ministry for Primary Industries, New Zealand's Border Compliance Social Marketing programme with the aim to 'reduce risk to New Zealand posed by travellers bringing in and failing to declare or dispose of biosecurity risk items'. The programme aims to protect New Zealand's horticultural and agricultural industries, as well as its environment, by ensuring overseas visitors do not bring in items that may contain harmful pests and diseases.

Few case studies evaluated the effectiveness of social marketing campaigns. For instance, the *Mobile Health Apps for Senior Citizens* case shows that several design aspects of health apps in the marketplace require modification to produce robust, reliable and usable technologies. The research uses the lens of social marketing, and draws on the technology acceptance model, to explain the factors that predict the successful adoption of health apps amongst senior members of the population.

The *Effectiveness of public advertisements to influence perceptions towards public transport among young adults in Singapore* case study aims to explore the effectiveness of public information advertisements by the Land Transport Authority (LTA) Singapore to influence perceptions of young adults towards the use of public transport. The LTA is encouraging the general public to use public transport by introducing advertisements near bus stops and train stations that highlight initiatives taken to create more park connectors, cycling paths and covered walkways to make use of public transport more convenient to access. A set of recommendations and actions are proposed to enhance communication reach and to improve outcomes of LTA's social marketing efforts.

The final case study *Marketing Social Change: Improving Bush Internet in Rural Regional and Remote Australia* show how a group of Australian cattle producer women used social marketing and advocacy to improve internet access and educational data allowances in rural, regional and remote Australia. A content analysis of Facebook posts and comments from the Better Internet for Rural, Regional and Remote Australia (BIRRR) shows evidence of a connection between advocacy and the principles of marketing. The results of which led to a change in policy giving rural, regional and remote school children a dedicated education portal to complete their studies and which highlighted deficits in access to internet connectivity.

Chapter 9
Relevance of Social Marketing in the Global South's Family Planning Programmes: A Case of Zambia

Lucy Nyundo, Lynne Eagle, and Maxine Whittaker

Abstract This chapter illustrates the challenges of health-related behaviour change programmes in countries where there are strong cultural influences on behaviours coupled with a strong religious influence; either or both which may be in conflict with 'western' scientific approaches. These issues are common across many countries in the Global South. The term Global South refers to low income, often politically or culturally marginalised countries in Latin America, Asia, Africa and the Oceania region (Dados and Connell in Contexts 11(1):12–13, 2012). The chapter focuses on fertility behaviour in a country within the Sub-Saharan region. Data from Zambia relating to cultural norms, religion and fertility behaviour is presented. Firstly, the potential influence of Pre-Marriage Counselling (Pre-MC) on Family Planning (FP) and contraceptive choice decisions is described. Then the influence of religion and fertility behaviour such as the claimed use of cannabis seed as a contraceptive is presented. This chapter displays a case study that has implications for countries beyond Sub-Saharan Africa and a research agenda for cross-cultural social marketing-related research concludes the chapter.

Keywords Social marketing · Family planning · Contraception · Cultural norms · Pre-marriage counselling · Religion · Fertility behaviour

9.1 Introduction

The Sustainable Development Goals (SDGs) 2030 adopted by the United Nations in 2015 include specific goals relating to sexual and reproductive health, noting that access to contraception enables planned and spaced pregnancies, thereby lowering maternal mortality and reducing economic strain (International Federation of Gynecology and Obstetrics (FIGO) 2018). This explicit attention in SDGs on FP and

L. Nyundo (✉)
National Institute of Public Administration, Lusaka, Zambia
e-mail: lucy.nyundo@jcu.edu.au

L. Eagle · M. Whittaker
James Cook University, Townsville, Australia

R. Hay et al. (eds.), *Broadening Cultural Horizons in Social Marketing*,
https://doi.org/10.1007/978-981-15-8517-3_9

reproductive health may be explained by the numerous studies that show the link between population growth, fertility rate, maternal and child health and development (Abel et al. 2016; May 2016; Starbird et al. 2016). This evidence has led to the development of the Family Planning-Sustainable Development Goals (FP-SDGs) model. The FP-SDGs model is an evidence-based advocacy tool that is used to lobby for investment in FP programmes especially in high fertility regions in the Global South.

Sub-Saharan Africa is characterised by a high fertility rate (Bongaarts and Casterline 2019; May 2016; May and Turbat 2017) and many experts argue that it may not go below 4 births per woman (Casterline and Agyei-Mensah 2017; Schoumaker 2019; Bongaarts 2008). Furthermore, this region's population is largely young, hence, the discourse on the potential of a population dividend in the region. Population dividend is the transformation of a population into a productive and working age population (Fang 2018). Therefore, the call for governments to prioritise reproductive health in their development agenda is timely. In line with this, Zambia's 2017–2021 National Health Strategic Plan (NHSP) is linked to the country's 2030 vision of becoming a prosperous middle-income country. Zambia's NHSP (2012–2021) adopts a multi-sectorial approach in order to manage the sociocultural attributes, family and community pressures that negatively impact reproductive health.

The majority of people in Sub-Saharan countries like Zambia do not oppose birth spacing but may have reservations about limiting births and some types of FP methods. FP methods can be categorised as traditional or folk, natural and modern. Traditional or folk methods include the use of herbs, ornaments and evoking of spirits, etc. Natural methods are based on awareness and observations about a woman's body and menstrual cycle, thus, they are also referred to as fertility awareness methods and do not require the use of medications or physical devices. These include safe days, withdrawal, abstinence and lactational amenorrhoea—a temporary family planning method based on the natural effect of breastfeeding provided the child is less than 6 months old and the mother has no menses, i.e. is not menstruating (World Health Organisation 2018). Modern methods include pills, condoms, implants, injections, sterilisation, etc.

The adoption levels of these FP methods differ and documentation of prevalence levels (adoption) is biased towards modern contraceptives because they are more effective and safe (Ajayi et al. 2018). According to the 2018 Zambia Demographic Health Survey (DHS), 50% of married women use some form of FP method of which 48% use modern contraceptives, and 2% use other methods. However, the Contraceptive Prevalence Rate (CPR) has marginally improved from 49% in 2014 to 50% in 2018. Fertility has only reduced by 1.8 births in 26 years from 6.5 to 4.7 births per woman (Zambia DHS 2018). Given a literacy rate of 68% among Zambian women of reproductive age and a projected 67% intent to use modern contraceptives (Central Statistical Office C, Ministry of Health M, ICF International 2014), it is clear that there are more complex barriers to the uptake of modern contraceptives beyond the frequently cited barriers of constrained access to health centres, frequent stock-outs of preferred contraceptives, cost, lack of information about FP methods (United States Agency for International Development 2014).

The Zambian case is an example that may qualify the assertions by May (2016, p. 308) that the high fertility rate in Sub-Saharan Africa is rooted in cultural and traditional reproductive regimes. Hence, the need for socially and culturally sensitive FP programmes. Therefore, this chapter illustrates how social marketing strategies can be used to overcome sociocultural barriers in the uptake of modern contraceptives.

9.2 Behaviour Change Foundations

Social marketing is a multi-disciplinary approach to solving social problems ranging from dangerous driving, pollution to public health issues. Human behaviour is at the centre of social marketing and behaviour change. A wide variety of behaviour change theories exist- over 80 theories, but many social marketing programmes are not informed by theory (Davis et al. 2014). Furthermore, there is no consensus on what theory or model is suitable for what type of social problem and in what context (Lefebvre 2000). Nonetheless, social marketing scholars and practitioners have expressed the desire to document what worked and did not, about a given programme. This is because it helps to organise knowledge and augment evidence of the role that social marketing plays in eradicating social problems for the good of the public (Donovan 2011; Firestone et al. 2017). For example, even though social marketing programmes in high fertility regions such as Sub-Sahara Africa have been useful in overcoming some sociocultural barriers to the uptake of modern contraceptives, evaluation of these programmes is problematic because many are not grounded in theory (Kiene et al. 2014; Lopez et al. 2016; Truong 2014).

Thus, the need to develop appropriate social marketing theories and models to inform behaviour change programmes remains urgent. As a starting point, Rundle-Thiele et al. (2019) offer guidance on how to develop and use theory in designing behaviour change programmes.

9.3 Implicit Assumptions

Many past behaviour change programmes have been based on implicit assumptions that are inherently flawed. The first assumption is that the provision of products and services will drive uptake. The second is that information provision alone is sufficient to change behaviours. Several studies concede that behaviour change is rarely achieved through information provision alone due to the complexity of decision-making in areas such as health (Simis et al. 2016). In the past, there has been an over-reliance on the information deficit model, i.e. the assumption that the reason people do not take the actions recommended is that they lack knowledge of the issue. This simplistic stance has been criticised for ignoring attitudes and values (Kraft et al. 2015; Simis et al. 2016). Further, it has been observed, particularly in often contentious areas such as vaccinations that '*being right is not enough*' (Camargo and Grant 2015, p. 2) and that there is a need for communication skills within the scientific community (Seiler et al. 2013), and a move from one-way transmission of scientific information to engaging in dialogue with sectors of the public (Lee and VanDyke 2015).

In this chapter, we discuss the complexities of factors influencing reproductive health-related decisions, specifically, fertility choices, highlighting the challenges of traditional beliefs and widespread customs and social expectations. Fertility behaviour and decisions in Sub-Saharan Africa is complex and unique because of the influence of social norms and third parties like in-laws. Generally, demand for FP services and modern contraceptives in Sub-Saharan Africa remains below 50%. Even when there is easily accessible and available FP services, good method mix (i.e. contraceptives) and accurate information, demand is not guaranteed. This may be explained by factors such as spousal/partner communication, power between husband and wife, varying fertility preferences between spouse/partners and social and religious expectations (Ashraf et al. 2014; Bankole and Audam 2011; Hollerbach 1980; Oladeji 2008). These factors constitute the fertility decision-making context thereby rendering the service/product availability and information assumptions in behaviour change programmes flawed.

9.4 Theory

Theory is essential in any given field of study. A theory that is widely used in behaviour change programmes is the Theory of Planned Behaviour (Armitage and Conner 2001; McEachan et al. 2011). This theory reflects an ongoing development of a group of sociocognitive theories, beginning with the Theory of Reasoned Action (TRA) (Ajzen and Fishbein 1980; Fishbein and Middlestadt 1987) and its successor the Theory of Planned Behaviour (TPB) (Ajzen 1991). Neither initially recognised cultural influences, a weakness rectified in the more complex Integrative Model of Behaviour Prediction and Change is shown in Fig. 9.1. The left-hand column shows the additions to the earlier iterations of the theory. In their more recent texts (Ajzen and Hornik 2007; Fishbein and Ajzen 2010), the theory is once again referred to as the Theory of Planned Behaviour.

The updated TPB specifically recognises the influence of a myriad of background factors, including, importantly, the role of cultural factors, past programme activity and media exposure.

A key contribution of research underpinning the effective use of this theory is that different population segments may be driven more strongly by attitudinal factors, normative influences or perceived self-efficacy, i.e. ability to change behaviour and sustain the change (Fishbein 2008). This indicates that different programme strategies may be needed for different population segments (Fishbein and Yzer 2003).

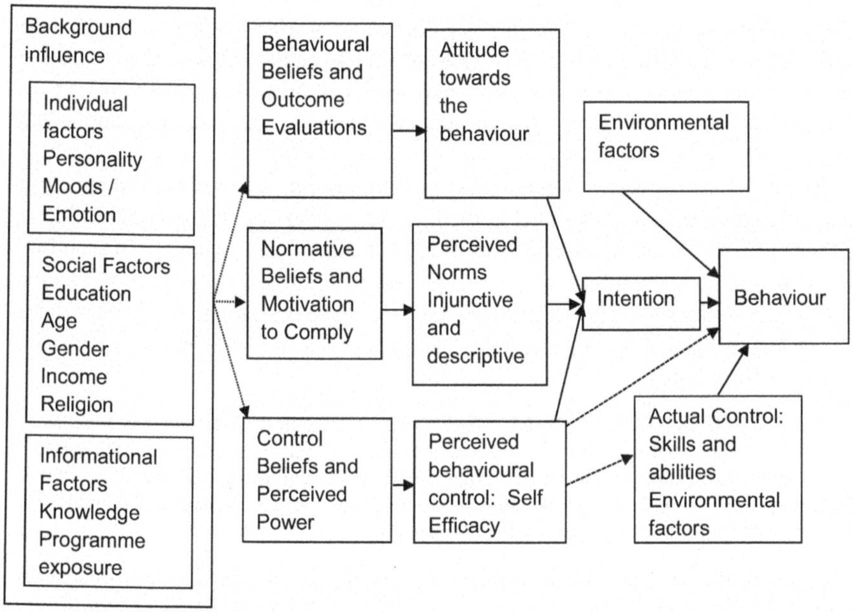

Fig. 9.1 Integrative model of behaviour and change (Ajzen and Hornik 2007; Fishbein and Ajzen 2010)

Further considerations illustrated by this model are the relative importance of attitude, perceived norms and self-efficacy:

> "The relative importance of these psychosocial variables as determinants of intention will depend upon both the behaviour and the population being considered".

and

> "One behaviour may be primarily determined by attitudinal considerations, whereas another may be primarily influenced by self-efficacy. Similarly, a behaviour that is attitudinally driven in one population or culture may be normatively driven in another" (Fishbein and Cappella 2006)

A common misconception regarding models such as the Theory of Planned Behaviour is that these models assume that 'planned' means that all behaviour to be strictly rational. The theories' primary developers state that:

> "The processes… whereby people arrive at their intentions represent a 'reasoned approach to the explanation and prediction of social behaviour only in the sense that people's behavioural intentions are assumed to follow in a reasonable, consistent, and often automatic fashion from their beliefs about performing the behaviour. This does not mean that people are assumed to be always logical and rational. The beliefs they hold need not be veridical; they may be inaccurate, biased or even irrational" (Fishbein and Ajzen 2010, p. 24).

A further misconception is that norms are a single concept, although there is a growing body of research that distinguishes between injunctive norms (portrayal of what people ought to do) and descriptive norms (what people actually do) (Cialdini 2007)—this is reflected in Fig. 9.1. Cialdini (2007) cautions that conflict between perceptions of the two subsets of norms may lead to a failure to engage with an issue in situations where behaviour change may be seen as contrary to prevailing social norms.

Figure 9.1 incorporates amendments and fine-tuning of the background influences shown on the left side panel in the most recent version of the updated Theory of Planned Behaviour (Ajzen and Hornik 2007; Fishbein and Ajzen 2010). The Integrative Model of Behaviour Change recognises different sets of background factors (individual, social and information). These factors have been found to be key in determining health behaviour. For example, when applied to fertility decisions (Fig. 9.2) it was concluded that

> "The TPB can usefully be employed to further our understanding of fertility decisions. By examining behavioural, normative, and control beliefs about having a child we can identify important considerations that influence this decision. The information obtained can also guide adoption of policies or programmes designed to encourage (or discourage) couples to have more children" (Ajzen and Klobas 2003, p. 203).

Although the TPB may explain the *why* and *what* of fertility decision-making by pointing out the various factors (e.g. accessibility of health facilities, method mix) it does not recognise the actors, e.g. spouse/partner and close relatives at play in the fertility decision-making process neither does it explain *how* the fertility decision is determined. Certainly, various factors and actors are at play in fertility decision-making but the TPB does not show how individuals arrive at the decision to either have a child or not, using a method of their choice. Recognising these shortfalls

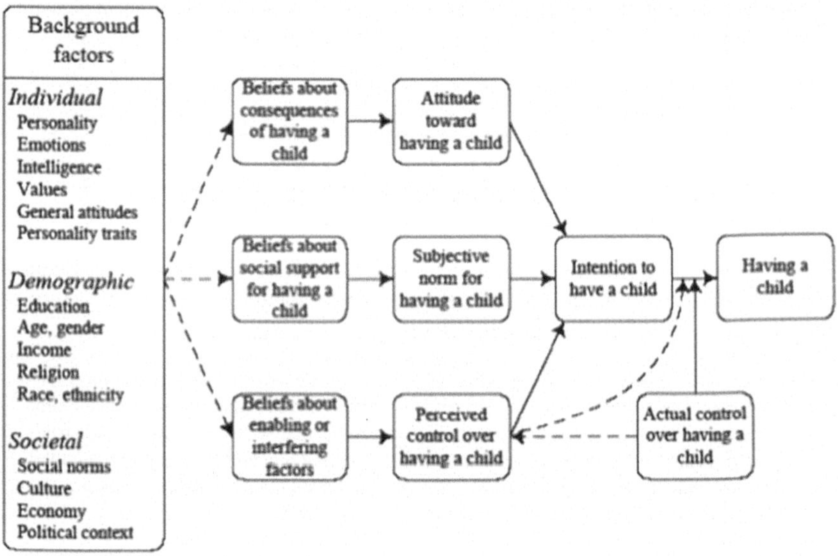

Fig. 9.2 Theory of planned behaviour applied to fertility decisions (Ajzen and Klobas 2003)

of the theory when applied to fertility decisions, Ajzen and Klobas (2003) welcome modifications to the model. Thus, Morgan and Bachrach (2011) propose a metatheory namely, the Theory of Conjuntural Action (TCA) urging that the TPB does not reflect the opportunities and constraints that emerge in the course of one's reproductive life. The TCA (Fig. 9.3) combines social theory and dual processes in the brain.

Unlike the TPB, the TCA takes a long term perspective to fertility decision-making. The TCA assumes that fertility decisions are not rigid, they may change over the life cycle, due to experience or to circumstances. Recognising the possible pitfalls of the TCA and appreciating the validity of other models including the TPB which may explain short-term fertility decisions, Morgan and Bachrach (2011) conclude that '*an appropriate model or theory for human fertility is one that recognises the complexity of the decision and should be flexible enough to allow for life's zig and zags*' (p. 13).

Apart from the TPB, another model that can be considered in social change is the Diffusion of Innovation model (Fig. 9.4). Like other social change theories, this

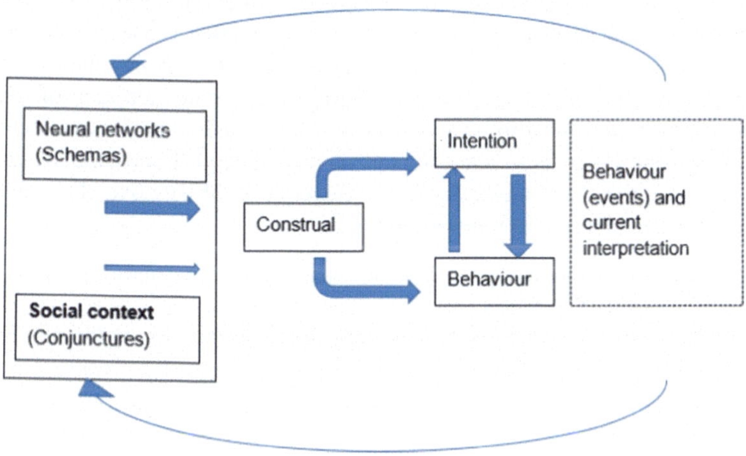

Fig. 9.3 Theory of conjectural action (Morgan and Bachrach 2011, p. 14)

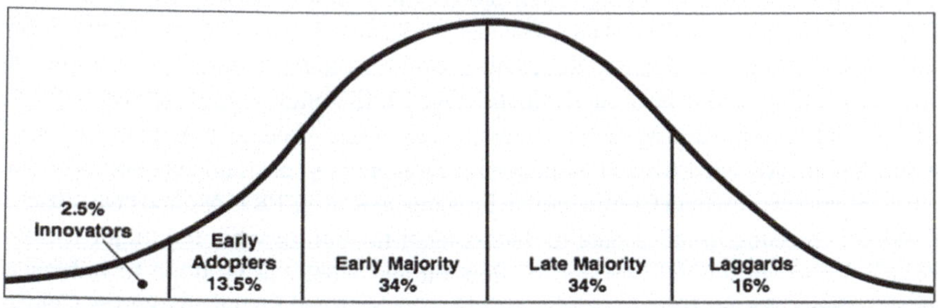

Fig. 9.4 Diffusion of innovation model (Rogers 1962)

model assumes that the population will progressively adopt innovations. Rogers' (1962) Diffusion of Innovation Model assumes progression from adoption by a small number of 'innovators' followed by 'early adopters' through to 'laggards', until 100% diffusion of innovation is completed.

In this model, there is an assumption that innovations will progressively be completely adopted. However, there is the chance that they may not be adopted or may be trialled and then abandoned. For example, diffusion of modern contraceptive (e.g. pills, injections, implants, etc.) has not followed the path assumed by this theory. Many of the early majority may stop use—whether for conception, change in method or another reason. Discontinuation of use of modern contraceptives (in this case the innovation) is becoming a concern because of failure or method-related reasons (e.g. side effects, convenience of use). Ali et al. (2012, p. 5) confirm that on average 38% of women in Africa, Asia, Eastern Europe and Latin America abandon reversible contraceptives after 12 months and 64% by the 36th month.

Moore (2002), in a revised Diffusion of Innovation model has positioned abandonment as a 'chasm' between the early adopters and the remaining adopters, describing a split between an early market and the mainstream market. The early market is made up of innovators and early adopters only. Thus, adoption may cease at a point where only a small percentage of the population has adopted it, with the majority of the population unwilling to do so. This has implications for the adoption of modern contraceptives in high fertility regions because hormonal contraceptive technologies have adverse side effects hence the raising push towards improving natural and herbal contraceptives (old methods). Thus, Rogers (1962) lists five characteristics necessary for the adoption of innovations:

(1) " where the innovation must give relative advantage, that is, it must supersede or perform better than its predecessor.
(2) It must be compatible with existing values, past experiences and the needs of the adopter.
(3) The innovation must be relatively easy to understand, if not then adoption will not occur.
(4) Adoption will occur more readily if the adopter has an opportunity to trial the innovation and
(5) the results of the innovation must be observable by others, for if they can see the results, they will more readily adopt the innovation".

The application of the Theory of Diffusion of Innovation to complex social issues such as fertility transition is not easy because the issue is private or personal but operates in a social ecosystem that includes the community and the state. For example, in the case of China, it may be argued that the 'One Child Policy' forced universal diffusion of modern contraceptives which if left to individuals could not have been the case. In Sub-Sahara Africa, diffusion of modern contraceptives varies and is generally low at 1.92% annual growth in Contraceptive Prevalence Rate (CPR) (Ahmed et al. 2019).

What is missing in the debate to date is identification of what the major factors are that act as substantial barriers to adoption and what can be done to influence greater uptake. As has been long advocated by the National Social Marketing Centre (2005), this requires an in-depth understanding of the lives, behaviour and the issue on which behaviour change is focussed, drawing on a mix of data sources and research methods, including '*A deep understanding of what moves and motivates the target audience, including who and what influence the targeted behaviour*' (Benchmark 4).

Therefore, both the TPB and DOI fail to explain how persons arrive at the point of deciding to adopt or not, a modern contraceptive. This failure leaves a theoretical gap on the fertility decision-making process. In reviewing the various theories, frameworks, models and concepts that can be used to explain fertility behaviour, Bruijn (2006a, b, p. 563) concludes that the respective theories do not provide a complete picture, but each advances propositions that contribute to understanding of fertility behaviour. Building on this, Huinink et al. (2015) suggest the development of an integrative approach in developing a comprehensive model.

Of the many behavioural theories applied in behaviour change research and programmes (Davis et al. 2015), a rarely used model is the Consumer Decision-Making model originally developed by Engel, Kollat and Blackwell in 1968. Like other models, the Engel-Kollat-Blackwell model has gone through numerous revisions (Bray 2008). Figure 9.5 shows the latest version of the model now known as the Consumer Decision-Making model or *Engel*-Blackwell-Miniard model.

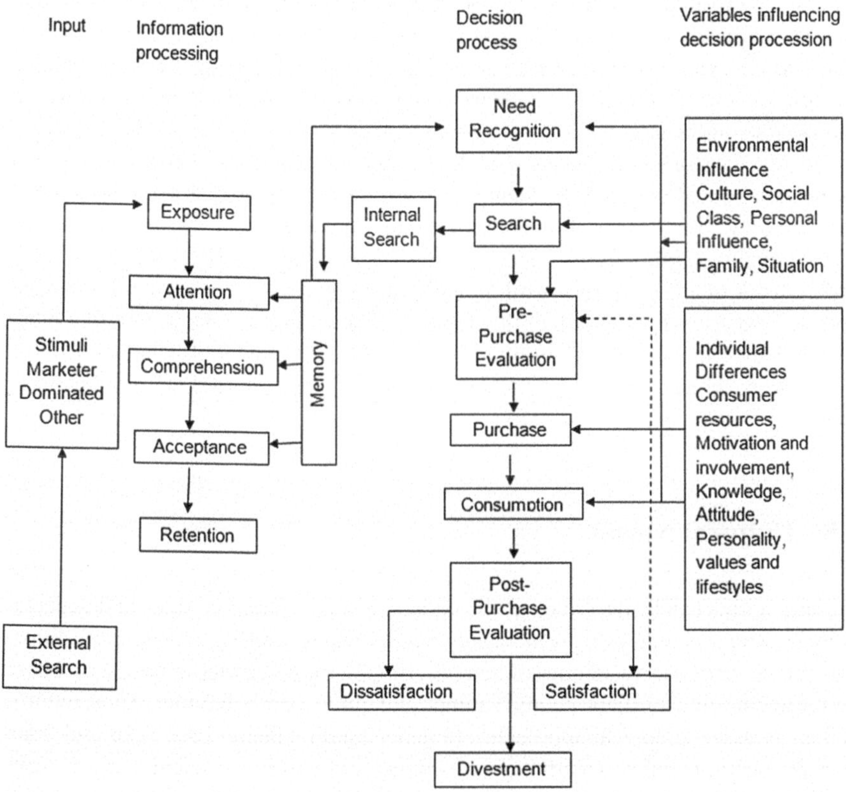

Fig. 9.5 Engel-Blackwell-Miniard model (Bray 2008)

The premise of this model is that consumers recognise needs as a result of input stimuli and the influence of environmental (e.g. culture, social class, situation) and personal (e.g. personality, attitude, etc.) factors. Upon recognition of need, the consumer—depending on the need—consciously or unconsciously engages in the search for alternatives to satisfy the need. This leads them to search for information internally or externally depending on the need or problem. Then the consumer progresses to pre-purchase evaluation, selection (purchase) and post-purchase. The choice can be either satisfactory or unsatisfactory. Satisfactory outcomes lead to repeat purchase (continued use) while the unsatisfactory outcome leads to abandonment (discontinuation). This revised model acknowledges that the purchased product is likely to be disposed of at some point post-consumption, and this is the final stage in the process (divestment).

In spite of this model having key constructs that reflect the decision-making process, it has been seldom applied in fertility research and health behaviour in general. However, this model is dominant in mainstream marketing from which social marketing stems. Therefore, it is worthwhile to consider this model in health behaviour. This is because the model offers an easily adaptable decision-making process. For example, information or awareness is essential in health behaviour and the recent version of this model (Fig. 9.5) reflects the different sources of information (internal and external). However, awareness alone does not warrant behaviour change because of other factors at play, which are recognised by the model (environmental and personal influence).

Clearly, the Consumer Decision-Making model highlights the constructs required to understand and map out the health decision-making process. This is because of the various visible (e.g. health centres, commodities), invisible (e.g. provider bias, service quality, cultural norms, etc.), personal factors (age, gender, education, etc.) and sources of information (institutional and community generated information-personal experiences) that affect the decision-making process. However, like other social change models, the Consumer Decision-Making model also makes a linear assumption, which may not be the case in fertility decision-making and health in general.

9.5 Cultural Norms

Cultural norms can be described as the undocumented standards, rules and expectations by which people in a given society live by (Stanford Encyclopedia of Philosophy 2018; National Academies of Science, Engineering and Medicine 2018). These norms unconsciously shape people's minds and guide their behaviour. Thus, anthropological and sociological approaches to understanding health behaviour and practices are instrumental in unearthing various sociocultural barriers and in some cases facilitators of healthy behaviour.

In the case of fertility behaviour, scholars and practitioners agree that fertility decline in Africa will be different from other regions due to Africa's historical

Table 9.1 Data collection points

Data collection point	Catchment area	No. of respondents	Type
Chilenje Hospital	Urban	37 (excluding pilot)	Public
Bauleni Hospital	Peri-urban	37	Public
UNHCR	Peri-urban	38	Public
Medcross Hospital	Urban	38	Private

distinctiveness, social and cultural norms (Casterline 2017). For example, in Zambian society, pre-marriage counselling (traditional or religious) is a mandatory undertaking for persons who want to get married through a socially acceptable and culturally right marriage procedure. Though, sparsely documented in extant fertility research in Zambia, Kapambwe et al. (2013) confirm that Marriage Counselling (MC) is an alternative platform for effectively disseminating reproductive health information.

Motivated by the need to validate the social marketing opportunity presented by this social-cultural platform, Nyundo et al. (2020) sought to explore the influence of pre-marriage counselling on personal fertility choices. The study was subjected to a lengthy ethical clearance process in Australia and Zambia. Respondents ($n = 149$) were randomly recruited and screened (15 yrs. +, married/engaged/once married/widowed) before the pre-tested questionnaire could be administered. The data was collected from four purposively selected hospitals in Lusaka district as shown in Table 9.1. Epi Info 7 was used to design the questionnaire, enter and store the data. Excel and Epi Info were used to analyse the closed questions while thematic analysis was used to analyse the open questions.

The study involved both female (79%) and male (21%) respondents (Table 9.2). All the respondents were Christian. However, there are various church denominations in Zambia. The majority of respondents (34%) attend Pentecostal churches (a new generation of protestant churches), are aged 30–39 years, reside in low-income areas and the highest level of education attended is secondary level. Lusaka district is within Lusaka the capital city of Zambia where there is a mix of tribes as reflected in the results of this study as majority (52%) of the respondents belong to other tribes other than the major four tribes (Bemba, Tonga, Chewa and Ngoni).

The majority (94%) of the respondents in the study underwent or were currently undergoing Pre Marriage Counselling (Pre-MC). The most common reason among the respondents who did not undergo pre-MC was getting pregnant before marriage (outside wedlock). This is considered to be against the prevailing social or cultural norm, which can be inferred from the explanations that respondents gave such as

> "Because my parents chased me to go to the person who got me pregnant and they refused to look for someone to teach me". **R20**

> "I got pregnant outside wedlock but we are planning to do it though we are constrained because of money". **R140**

Table 9.2 Demographic profile of respondents

Gender	Frequency	Percentage
Female	118	79%
Male	31	21%
Total	149	100%
Residential area		
High income	15	10%
Middle Income	29	19%
Low Income	70	47%
Mixed Income	35	23%
Total	149	100%
Education		
Never been to school	3	2%
Primary	41	28%
Secondary	65	44%
College	20	14%
University	19	13%
Total	148	100%
Age		
15–20	4	3%
21–29	45	30%
30–39	48	32%
40–49	30	20%
50–59	13	9%
60+	9	6%
Total	149	100%
Tribe		
Bemba	27	18%
Tonga	16	11%
Chewa	16	11%
Ngoni	12	8%
Other	78	52%
Total	149	100%
Church		
Catholic	28	19%
Jehovah's witness	9	6%
New apostolic	8	5%
Pentecostal	51	34%
Seventh day adventist	20	13%

(continued)

Table 9.2 (continued)

Gender	Frequency	Percentage
United Church of Zambia	15	10%
Other	18	12%
Total	149	100%

Other reasons included:

Going against other social and cultural norms; a lack of resources e.g. time or money; or not having close family such as parents to make the necessary arrangements for pre-MC.

As one respondent noted:

"I did not have people to organize someone to teach me because my parents are dead, my grandmother who raised me is also dead". **R25**

Undergoing pre-MC is a common practice irrespective of the type of marriage (see Table 9.3). However, it is not mandatory to undertake both the religious and traditional pre-MC.

For couples that wish to wed in church they are required by Church regulation to undertake religious pre-MC. Generally, it is recommended to undertake both types of pre-MC for purposes of appreciating both the religious and traditional values of marriage as explained by some of the respondents.

"Because I wanted to know how to keep my husband using both traditional and church values" (Female, Age 50–59).

"I am told that the religious teachings are not as detailed as the traditional ones and my friends have convinced me to do both" (Male, Age: 21–29).

"To learn how to behave in a Christian marriage but also to appreciate the teaching of tradition about marriage" (Male, Age: 40–49).

Table 9.3 Reported relationship status of respondents and their undertaking of pre-MC

Relationship status	Undergo pre-MC			
	YES		NO	
	Frequency	Percent	Frequency	Percent
Civil marriage	15	11%	0	0%
Customary marriage	69	49%	6	67%
Divorced	5	4%	3	33%
Engaged	11	8%	0	0%
Prefer not to say	1	1%	0	0%
Religious marriage	23	16%	0	0%
Widowed	16	11%	0	0%
Total	140	100%	9	100%

> "We did the traditional one but our counsellors mixed the traditional and some church teachings. We are also planning to go for the church counselling now that we go to the same church" (Female, Age: 21–29).

Church marriage counsellors deliver religious pre-MC and the emphasis is on Christian principles or values about marriage. While traditional counsellors conduct the traditional pre-MC known as '*alangizi*' (Nyanja language common in Lusaka) or '*ifimbusa*' (Bemba tribe of the Northern region). The emphasis is on traditional values and principles about marriage. When asked about the topics of the pre-MC the respondents, predominantly female, stated that there were communication, respect, conflict resolution, budgeting and caring for extended family, husband/spouse (Fig. 9.6).

Using a mix of word cloud and thematic analysis techniques, six topics were identified to be the topics covered during pre-MC (Table 9.4). Furthermore, two

Fig. 9.6 Keywords in pre-MC teachings as reported by respondents

Table 9.4 Topics covered during pre-MC

Theme (code)	Keywords/phrases
Marriage pillars	Patience, respect, love, communication, care for each other, joint decision-making, finance, budgeting and conflict resolution
Bedroom matters	Please a partner in bed/sexually, sex styles, dance, body hygiene, tidiness in the bedroom
Duties	Cooking, caring for a home, providing for a home, welcoming visitors
Social conduct	Respect everyone, do not socialise with persons who are not married, mature behaviour, be generous
Social-cultural pillars	Humility, submission, leadership and authority—'man is the head of the house'—keep secrets/confidentiality
Extended family	How to relate with in-laws, support extended family

topics, namely generic pillars of marriage and managing extended family, were found to be common in both traditional and religious pre-MC. However, Traditional pre-MC focuses on bedroom conduct (e.g. please partner sexually, tidiness in the bedroom, etc.) and social-cultural pillars (e.g. submission, humility, leadership and authority) of marriage while Religious pre-MC focuses on Marriage pillars (e.g. respect, love, communication, etc.).

Basic statistical analysis (frequency and cross-tabulation) shows that most of the respondents agree that pre-MC (religious or traditional) influences a couple's sexual relationship and their understanding of communication, roles and responsibilities, leadership and authority in marriage. However, they disagree that pre-MC influences a couple's family size decision (i.e. number of children to have or not) or contraceptive choice (i.e. family planning method) (Table 9.5).

Furthermore, this influence is constant irrespective of the demographic variables (e.g. age, church, education, household income, etc.) except for tribe. The influence of pre-MC on authority and leadership, and family size, differs across different ethnic groups (see Table 9.6). This is logical because of possible differences among the different ethnic groups—Zambia has 72 tribes or ethnic groups. Traditional pre-MC often reflects tribal individual ethnic group's values or principles.

Africa is said to be heterogeneous in terms of culture and this can be explained by the variety of tribes and ethnic groups across the continent. This has implications because population segments within a country maybe organised according to ethnic groups which may have different social norms resulting in a myriad of social-cultural barriers or facilitators of desired health behaviour and practices. In this case, it is evident that pre-MC may have spillover influence on fertility choices because pre-MC (a social practice in Zambian society) affects key aspects of the husband-wife relationship (sexual relationship, communication, roles and responsibilities, authority and leadership) which are fundamental in fertility decisions (see Table 9.7).

The study by Nyundo et al. (2020) on the potential influence of pre-MC (religious and traditional) on fertility choices (family size and contraceptive choice) shows that this social-cultural practice is generally appreciated by people irrespective of their demographic profile. This confirms the conclusion by Kapambwe et al. (2013) that '*although the social context in Zambia is changing, marriage counsellors are still considered relevant by most communities and are potentially an important avenue for health information*'. From this data (Nyundo et al. 2020), it is evident that pre-MC counsellors are perceived to be a credible source of information about key marital issues but not sources of accurate information about FP as indicated by some of the responses.

> I am not sure if she can give me the right information when it comes to this issue **R19**, **Age 30–39**

> I don't think they have much knowledge about this **R23**, **Age 40–49**

However, pre-MC counsellors are potential socially acceptable community health workers who can be incorporated in social marketing programmes to increase the

Table 9.5 Influence of pre-MC

	Strongly agree		Agree		Neither agree/disagree		Disagree		Strongly disagree		Total		Chi-square	*df*	Probability
Sexual relationship															
Female	45	30%	59	40%	7	5%	3	2%	4	3%	118	79%	9.69	4	0.05
Male	14	9%	8	5%	6	4%	2	1%	1	1%	31	21%			
Total	59	40%	67	45%	13	9%	5	3%	5	3%	149	100%			
Communication															
Female	46	31%	56	38%	9	6%	4	3%	3	2%	118	79%	5.25	4	0.26
Male	12	8%	10	7%	6	4%	2	1%	1	1%	31	21%			
Total	58	39%	66	44%	15	10%	6	4%	4	3%	149	100%			
Roles and responsibilities															
Female	31	21%	64	43%	12	8%	9	6%	2	1%	118	79%	9.22	4	0.06
Male	14	9%	11	7%	6	4%	0	0%	0	0%	31	21%			
Total	45	30%	75	50%	18	12%	9	6%	2	1%	149	100%			
Leadership and authority															
Female	31	21%	66	44%	12	8%	5	3%	4	3%	118	79%	7.97	4	0.09
Male	11	7%	10	7%	8	5%	1	1%	1	1%	31	21%			
Total	42	28%	76	51%	20	13%	6	4%	5	3%	149	100%			
Family size															
Female	2	1%	11	7%	13	9%	61	41%	31	21%	118	79%	7.66	4	0.1
Male	1	1%	5	3%	7	5%	8	5%	10	7%	31	21%			

(continued)

Table 9.5 (continued)

	Strongly agree		Agree		Neither agree/disagree		Disagree		Strongly disagree		Total		Chi-square	*df*	Probability
Total	3	2%	16	11%	20	13%	69	46%	41	28%	149	100%			
FP method															
Female	2	1%	40	27%	12	8%	39	26%	25	17%	118	79%	6.27	4	0.18
Male	1	1%	5	3%	7	5%	9	6%	9	6%	31	21%			
Total	3	2%	45	30%	19	13%	48	32%	34	23%	149	100%			

Table 9.6 Chi-square Test results

	Sexual relationship	Communication	Roles and responsibility	Leadership and authority	FP method	Family size
Residential area						
Chi-square	13.26	9.80	13.36	6.35	7.71	9.95
Df	12	12	12	12	12	12
Probability	0.35	0.63	0.34	0.90	0.81	0.62
Gender						
Chi-square	9.19	5.25	9.22	7.97	6.27	7.66
Df	4	4	4	4	4	4
Probability	0.06	0.26	0.06	0.09	0.18	0.10
Age						
Chi-square	22.57	21.25	14.61	7.16	14.97	10.89
Df	20	20	20	20	20	20
Probability	0.31	0.38	0.80	1.00	0.78	0.95
Education						
Chi-square	17.31	25.38	19.04	24.14	13.80	14.47
Df	16	16	16	16	16	16
Probability	0.37	0.06	0.27	0.09	0.61	0.56
Income						
Chi-square	11.74	19.96	18.25	19.40	22.60	10.18
Df	16	16	16	16	16	16
Probability	0.76	0.22	0.31	0.25	0.12	0.86
Church						
Chi-square	39.20	31.96	33.64	27.71	48.92	33.92
Df	36	36	36	36	36	36
Probability	0.33	0.66	0.58	0.84	0.07	0.57
Ethnic group						
Chi-square	21.12	31.94	34.33	72.26	39.00	53.88
Df	32	32	32	32	32	32
Probability	0.94	0.47	0.36	0.00	0.18	0.01

Table 9.7 Influence of the type of pre-MC (religious and traditional)

Variable	Chi-square	*df*	Probability
Sexual relationship	4.4858	8	0.8109
Communication	10.6911	8	0.2198
Roles and responsibility	11.5841	8	0.1707
Leadership and authority	8.1331	8	0.4206
FP method	14.0822	8	0.0796
Family size	7.7564	8	0.4576

Contraceptive Prevalence Rate (CPR) in Zambia. This is because they offer a culturally appropriate and context fit platform (Airhihenbuwa et al. 2014) which can be deduced from some of the responses.

> Because my MC helped me to build a strong marriage so I can ask her when I am stuck **R2**, Age 40–49

> Marriage counselling has helped me and shaped me so much in my marriage **R14**, Age 30–39

9.6 Religion

Religion is one of the sociocultural factors used to understand health behaviour and practices because religion unites people of the same faith, beliefs, values and practices (Chatters 2000; Rumun 2014). However, it is difficult to define religion (Nath 2015; Niekerk 2018). Bruce (2011, p 133) explains that religion consists of beliefs, actions and institutions which assume the existence of supernatural entities with powers of action, or impersonal powers or processes possessed of moral purpose. Thus, religion influences people's behaviour and society at large.

For example, the influence of religion on reproductive health, specifically on fertility ideals, intentions and behaviour is evident (Mcquillan and Mcquillan 2019; Philipov and Berghammer 2007). This is because of religious ideologies about human procreation and family formation patterns. Although many studies show that this influence is common among developing and high fertility countries in Sub-Saharan countries (Agadjanian and Yabiku 2014), religion continues to be an important determinant of fertility in developed countries like France, Britain and Netherlands (Baudin 2007, 2015; Cohorts and Peri-rotem 2016; Philipov and Berghammer 2007). A study conducted in Zambia by Nyundo et al. (2020) in Lusaka district involving randomly selected respondents established that all respondents ($n = 149$) were Christian (see Table 9.8). The study among other things established that children are perceived to be a gift from God. They are the reason why people get married, as instructed by God in Genesis 9 v.7, to procreate

> "You must have many children, so that your descendants will live all over the earth." (Good News Bible).

The need to procreate is embedded as being common sense among Christians that when asked the question; why do you have/want children? Many respondents were surprised by this question because the Bible is clear on this issue and is reflected in the sample responses:

> "It is part of the gospel- the bible says go and fill the earth. So we wanted to have many children but my husband died…" (**R2**, Age: 50–59 years)

> "This is like an automatic thing, when you get married you are expected to have children" (**R5**, Age: 30–39 years)

> "Because it is mandatory for people who are married" (**R37**, Age: 30–39 years)

Table 9.8 Respondents' Church denomination profile

Church	Frequency	Percent
Anglican	1	0.67%
Baptist	3	2.01%
Catholic	28	18.79%
Jehovah's witness	9	6.04%
New apostolic	8	5.37%
Other	13	8.72%
Pentecostal	51	34.23%
Prefer not to say	1	0.67%
Seventh day adventist	20	13.42%
United Church of Zambia	15	10.07%
Total	149	100.00%

This religious conviction about procreation is common in Zambia where 95% of the population is Christian and characterised by different churches/denominations. Protestants account for 75%, followed by Catholics (20%). In the recent past, a new generation of protestant churches have emerged in the country (Cheyeka et al. 2014). For instance, in Bauleni compound of Lusaka district, there are over 82 different churches/denominations (Cheyeka et al. 2014). Although there are insignificant doctrinal differences among these new generation protestant churches, their charismatic approach and convincing messages are influencing the masses.

Christianity as a leading religion in Zambia has been reinforced by the declaration of Zambia as a Christian nation (Cheyeka 2016). The Ministry of National Guidance and Religious Affairs was created in 2016 to actualise Zambia's Christian identity. This has practical implications on social marketing programmes because Christianity has had both negative and positive impacts on African societies (Manala 2013). Some of the progressive impacts have been on education and health while the negative impact has been on social and cultural structures such as undermining the role of women in leadership, due to some of the teachings and church structures (Manala 2013). An example of Christian teachings that may negatively influence fertility behaviour is the importance of submission and leadership in marriage. The teaching is outlined in Ephesians 5:22–23:

> "Wives, submit yourselves to your husbands as to the Lord. For a husband has authority over his wife just as Christ has authority over the church; and Christ is himself the Saviour of the church, his body" (Good News Bible).

This has implications for couples that have different fertility desires because the teaching about submission and leadership place the woman in a weaker position during the negotiation and approval stage in the fertility decision-making process. Although communication remains key, negotiation in such cases becomes important for the couple to reach a consensus otherwise marital conflict may arise (Bankole and Audam 2011). Such marital conflict may force some women to opt for concealable contraceptives such as injections (Ashraf et al. 2014) or to not use contraception. This has implications on FP programmes in countries where men tend to have a slightly higher fertility desire compared to women (Westoff 2010) and practice Christianity.

Churches and faith-based organisations have been instrumental in promoting health and addressing epidemics such as HIV and AIDS (Campbell et al. 2007; Schmid et al. 2008). However, the proliferation of Pentecostal/charismatic churches in Africa, particularly in Zambia can be retrogressive in the acceptance of health programmes. This is because the new churches place emphasis on God's miraculous and mystical power that can change any life situation including healing of diseases and prosperity or wealth (Sakupapa 2018). This message appeals to the peri-urban and rural population where these churches are mushrooming. Therefore, FP programmes in Zambia must be sensitive to emerging Christian influence, which may reinforce a laissez-faire approach to family size decisions in spite economic and financial constraints as can be inferred from some of the responses to the question–what factors did you take into consideration when deciding the number of children to have?

> "They are a blessing from God so you can't refuse" (**R14**, **Age**: 20–29 years, **Income Level**: Low)

> "It is God's choice" (**R27**, **Age**: 40–49 years, **Income level**: Low)

Furthermore, it may be argued that such responses reflect Christians' support for birth spacing but not limiting of births. Barden-O'Fallon (2017) asserts that Faith-Based Organisations (FBOs) are less likely to offer FP services despite them being key providers of health care services in developing countries. In the case of the Catholic church which has many health facilities especially in peri-urban and rural areas, FP services may not even be offered due to the Ethical and Religious Directives for Catholic Health Care Services (Guiahi et al. 2017). In Zambia, the Catholic Church's health service provision at national and district level accounts for more than 30% (Schmid et al. 2008). This has practical implications because the Church through the Churches Health Associations of Zambia (CHAZ) is instrumental in health promotion, especially in peri-urban and rural areas. Therefore, the informality and flexibility of FBOs in delivering health services in Africa is a social marketing opportunity. Campbell et al. (2007) confirm that health programmes that are sensitive to spiritual and cultural contexts have been effective.

9.7 Unexpected Behaviours

9.7.1 Fertility Behaviour

Some populations exhibit behaviours that are contrary to the recognised best practice for healthy living in other countries. For example, in Bangladesh there are a number of folk-beliefs that are currently significant barriers to improving maternal and infant nutrition. These include beliefs in 'eating down' (i.e. women should eat less during pregnancy) and of avoiding meat, fish and eggs during pregnancy (World Bank 2005; Karim et al. 2016; White 2009). In Papua New Guinea, the introduction of solid foods considerably earlier than recommended is associated with an increased risk of malnutrition, anaemia, childhood infection diseases, diarrhoeal diseases and an increased likelihood of contracting pneumonia and developing food sensitivity disorders (Anga et al. 2008; Cripss et al. 2008; Manning et al. 2012).

Extant research on fertility behaviour in Sub-Sahara Africa shows that context, circumstances and method-related factors influence behaviour. For example, in Ethiopia it is reported that women whose spouse/partner disapproves of the use of contraceptives, prefers contraceptives which cannot be detected by the spouse/partner (Alvergne et al. 2017). Similarly, in Zambia women who have lower fertility desire than their spouses/partners prefer concealable contraceptives (Ashraf et al. 2014). Furthermore, reported side effects of hormonal contraceptives such as weight gain, headaches and irregular periods have become a widespread reason for discontinuation and an emerging strong barrier to the uptake of modern contraceptives (Chebet et al. 2015). Some of the claimed side effects have compounded myths, fears and misperceptions about modern contraceptives such as reduced sexual pleasure and infertility (Blackstone et al. 2017; Gueye et al. 2015). In some cases, Zambian women resort to the use of herbs such as cannabis seed as a form of natural contraceptive perceived to have negligible side effects (Lusaka Times 2009).

The use of herbs (which may include cannabis) in primary healthcare and fertility is not a new practice. Humans have used herbal contraceptives for centuries but the challenge has been their limited, if any, effectiveness, and efficacy. The reported side effects of hormonal contraceptives have spurred research in more ecofriendly or herbal contraceptives (Anand et al. 2015; Pradhan et al. 2012), driven by the need to make progress in herbal contraceptives—an alternative to hormonal contraceptive. Several studies predominantly in India have been carried out to identify plants that have the potential to regulate fertility (Bala et al. 2014; Kaur et al. 2011; Rai and Nath 2005).

While claims about herbal contraceptives have been reported in some population segments in Africa, the documented evidence is scant making it difficult to establish the extent of how widespread these practices and the use of herbs are. Partly this could be due to the fact that knowledge about medicinal plants or herbs is often in the form of indigenous knowledge peculiar to an ethnic group, resulting in a myriad of potential herbal contraceptives whose efficacy and effectiveness remain uncertain. While the studies in India (Bala et al. 2014; Rai and Nath 2005) have succeeded in identifying plants with anti-fertility properties and the specific parts of the plant (e.g. seeds, flowers, leaves, etc.); none of them identifies cannabis or marijuana (family name *Cannabaceae*) as one of the plants with anti-fertility properties. Recently, there has been a surge in evidence supporting medicinal properties of cannabis for different health conditions (Sznitman and Bretteville-jensen 2015). However, medicinal cannabis is highly debated, especially its legalisation (Larkin 2018).

Informal and casual chats on WhatsApp (in confidence) with women aged 20–35 about FP methods as part of the preliminary study by Nyundo et al. (2020) in

Zambia revealed that some women use cannabis seeds as a natural contraceptive. Zambian Law prohibits the cultivation, possession and use of cannabis. Consultations with the Zambia Drug Enforcement Commission (DEC) confirmed that the Law does not specifically address cannabis seeds but if the seed(s) test positive for Tetrahydrocannabinol (THC) or if they germinate, a criminal offence is charged on the person who was found in possession of the seed(s). Thus, the fear of women to openly discuss their experience of using cannabis seeds as a contraceptive is justified because of legal implications. According to Chileshe (2016), DEC is aware of three forms of cannabis namely herb, resin and oil and the wide use of these forms by persons for different reasons. Women are reported by the Zambia Drug Enforcement Commission to use it to grow their hair and to treat nausea and vomiting during the early stages of pregnancy.

Although the Zambia Dangerous Drugs Act of 1967 permits the licensing of the cultivation of medicinal cannabis, the Ministry of Health (2017) argues that there is no current need or gap that warrants the Ministry to issue such a license to any of the many unsolicited applicants because marijuana is the most abused drug in Zambia. The Court ruling on an appeal to legalise this has stifled the debate on medicinal cannabis in Zambia (Lusaka Times 2019), in spite, the several calls in the media by different sections of the society to reconsider cultivation of medicinal cannabis (Phiri 2017; Zambia Daily Mail 2018; Zimba 2018).

Claimed uses of cannabis for reproductive health like birth spacing remain scarcely documented. Therefore, the claimed contraceptive use of cannabis seeds as a natural method is peculiar and has practical implications for FP programmes in the face of the rising attention on medicinal cannabis in Zambian popular media and communities. The need for evidence to support the design of appropriate, and target specific communication about natural or traditional contraceptives is urgent. This is in view of the widespread fear of side effects of modern contraceptives amid myths and misinformation in many African countries which have the potential to inhibit the progress of fertility transition on the continent (Ajayi et al. 2018; Alvergne et al. 2017). Documented evidence can be used to train FP service providers on how best to counsel women and couples performing peculiar contraceptive practices in certain societies. It may also stimulate pharmaceutical research in herbs that exhibit anti-fertility properties.

9.8 Discussion

Identification of facilitators and barriers to health behaviour and practices is essential in developing behaviour change programmes. For example, establishing and overcoming barriers to the uptake of modern contraceptives is important in reducing fertility rates. Furthermore, grounding of behaviour change programmes in theory is important. However, the majority of behaviour change theories do not reflect the context or social factors that influence and shape health behaviour (Bruijn 2006a, b; Davis et al. 2014; Huinink et al. 2015). This has led to scholars like Morgan and

Bachrach (2011) to question the appropriateness of the Theory of Planned Behaviour (TPB)—a dominant theory, in understanding human fertility because of the soft factors such as social norms, religion, tradition, culture and social network influence, which are at play.

Practitioners, researchers and scholars agree that like other health behaviours, a mix of soft factors influences fertility behaviour, which may be unique to a population segment. Some social norms, which are peculiar to a society, may result in unusual behaviour. For example, desire for concealable contraceptives. Such behaviour—though unusual—is logically based on lived experiences or circumstances like side effects of hormonal contraceptives or marital conflict due to differences in fertility desires.

Therefore, people's lived experiences, culture and social norms necessitate the design and implementation of context-appropriate fertility and health behaviour change programmes. Furthermore, there is a need for evidence based, theory and frameworks that can ease the evaluation of such programmes. Although the Sub-Saharan region is characterised by cultural heterogeneity, varying fertility,and transition phases, the findings presented in this chapter have implications on design of effective FP programmes for countries in the region.

9.9 Conclusion

Social marketing relies on a mix of fields such as anthropology, psychology, sociology and economics to bring about change which cannot be achieved through mere awareness or information campaigns. This multi-faceted approach is made clear in the consoliadted definition of social marketing (iSMA, ESMA and AASM 2013).Thus, social marketing is appropriate for behaviour change because it recognises the different layers (i.e. individual, community and state) and context which are important in attaining holistic systemic change (Kennedy 2017).

Futhermore, social marketing recognises the demand and supply factors which influence health behaviour and decisions. For example, policies, quality of services, accessibility of services and availability of commodities affect the supply side of the 'health market'. While the demand side of this market is shaped by personal characteristics and the profile of the society, which shape people's health behaviours. Therefore, Airhihenbuwa et al. (2014, p. 78) assert that to achieve meaningful and sustainable health behaviour change, programmes that recognise the demand and supply factors, are context fit, and culturally logical, are required.

The dominant theories of change (Theory of Planned Behaviour, Ajzen 1991; Social Cognitive Theory, Bandura 1986; Transtheoretical Model or Stages of Change Model, Whitelaw, Baldwin, Bunton, and Flynn 2000; Health Belief Model, Sundstrom et al. 2015 and Diffusion of Innovation Theory, Rogers 2010) that are frequently used in social marketing may not be the most effective when applied in designing health-related behaviour change programmes for African countries or other Global South countries. This is because these theories do not adequately reflect the personal characteristics (e.g. stage of life, beliefs, values, religion, past experiences, social network influence, etc.) and the profile of the society (e.g. culture, traditions, practices, social norms) of population segments, yet these factors influence health behaviour.

Since the Global South remains a region that is comprised of countries that are culturally diverse and fall within the low and middle-income level, the need for evidence to support social marketing programmes for the region remains urgent. Thus, social marketing researchers need to uncover the factors that influence the demand side in population segments of the Global South in order to develop customised health programmes that are theory based, alive to the complexities and multi-faceted nature of behaviour change (Rundle-thiele et al. 2019).

References

Abel GJ, Barakat B, Kc S, Lutz W (2016) Meeting the sustainable development goals leads to lower world population growth. 113(50). https://doi.org/10.1073/pnas.1611386113

Agadjanian V, Yabiku CT (2014) Religious affiliation and fertility in a sub-sahran context: dynamics and lifetime perspectives. Popul Res Policy Rev 33(5):673–691. https://doi.org/10.1007/s1113-013-9317-2

Ahmed S, Choi Y, Rimon J, G, Alzouma S, Gichangi P, Guiella G, . . . Tsui A (2019) Trends in contraceptive prevalence rates in sub-saharan Africa since the 2012 London summit on family planning: results from reapeated cross-sectional surveys. The Lancet Global Health, 7:904–911. https://doi.org/10.1016/S2214-109X(19)30200-1

Airhihenbuwa CO, Ford CL, Iwelunmor JI (2014) Why culture matters in health programmes: lessons From HIV/AIDS stigma and NCDs. Health Educ Behav 4(1):78–84. https://doi.org/10.1177/1090198113487199

Ajayi AI, Adeniyi OV, Akpan W (2018) Use of traditional and modern contraceptives among childbearing women: findings from a mixed methods study in two southwestern Nigerian states. https://doi.org/10.1186/s12889-018-5522-6

Ajzen I (1991) The theory of planned behaviour. Organ Behav Hum Decision Process 50:179–211

Ajzen IDA, Hornik R (2007) Prediction and change of health behavior: applying the reasoned action approach. Lawrence Erlbaum Associates, New Jersey

Ajzen I, Fishbein M (1980) Understanding attitudes and predicting social behavior. Prentice Hall, New Jesrsey

Ajzen I, Klobas J (2003) Fertility intentions: an approach based on the theory of planned behavior. Demogr Res 29(8):203–232. https://doi.org/10.4054/DemRes.2013.29.8

Ali MM, Cleland JG, Shah IH, World Health Organization (2012) Causes and consequences of contraceptive discontinuation: evidence from 60 demographic and health surveys W. H. Organisation Ed.

Alvergne A, Stevens R, Gurmu E (2017) Side effects and the need for secrecy: characterising discontinuation of morden contraception and its causes in Ethiopia using mixed methods. Contracept Reprod Med 2(24):1–16. https://doi.org/10.1186/s40834-017-0052-7

Anand AK, Prasad V, Alam M (2015) Herbal or modern methods of contraception ! Choice is yours 4(4):947–953

Anga G, Vince JD, Kaupa M (2008) Early introduction of solids and pneumonia in young infacts in Papua New Guinea: a case control study. J Trop Pediatr 54(3):192–195

Armitage CJ, Conner M (2001) Efficacy of the theory of planned behaviour: a meta-analytic review. Br J Soc Psychol 40(4):471–499

Ashraf N, Field E, Lee J (2014) Household bargaining and excess fertility: an experimental study in zambia. Am Econ Rev 104(7):2210–2237. https://doi.org/10.1257/aer.104.7.2210
Bala K, Arya M, Katare DP (2014) Herbal contraceptive: an overview. World J Pharm Pharm Sci 3(8):1305–1326
Bank W (2005) Maintaining momentum to 2015? An impact evaluation programmes to improve maternal and child health and nutrition in Bangladesh
Bankole A, Audam S (2011) Fertility preferences and contraceptive use among couples in sub-Saharan Africa. Afr Popul Stud 25:173–188
Barden-O'Fallon J (2017) Availability of family planning services and quality of counseling by faith-based organizations: a three country comparative analysis. Reprod Health 14(1):1–11. https://doi.org/10.1186/s12978-017-0317-2
Baudin T (2007) A role for cultural transmission in fertility transitions to cite this version : HAL Id : halshs-00159595. Centre d' Economie de la Sorbonne Documents de Travail du
Baudin T (2015) .Religion and fertility: The French Connection. Demographic Research 32(February):397–420. https://doi.org/10.4054/DemRes.2015.32.13
Blackstone SR, Nwaozuru U, Iwelunmor J (2017) Factors influencing contraceptive use in sub-saharan Africa: a systematic review. Int Quart Commun Health Educ 37(2):79–91. https://doi.org/10.1177/0272684X16685254
Bongaarts J, Casterline J (2019) Fertility transition: is sub-saharan Africa different? Popul Dev Rev 38:153–168
Bray J (2008) Consumer behaviour theory: approaches and models. Bournemouth
Bruce S (2011) Defining religion: a practicle response. Int Rev Sociol 21(1):107–120. https://doi.org/10.1080/03906701.2011.544190
Bruijn BJDE (2006a) Fertility: theories, frameworks, models, concepts. In: Demographic analysis and synthesis, pp 549–569
Bruijn BJDE (2006b) Models, concepts and development, pp 549–569
Casterline JB, Agyei-Mensah S (2017) Fertility desires and the course of fertility decline in sub-saharan Africa. Population Dev. Rev. 84–111. https://doi.org/10.1111/padr.12030
Camargo K, Grant R (2015) Public health, science, and policy debate: being right is not enough. Am J Public Health 23(3):517–532
Campbell MK, Hudson MA, Resnicow K, Blakeney N, Paxton A, Baskin M (2007) Church-based health promotion programmes: evidence and lessons. Annu Rev Public Health 28:213–234. https://doi.org/10.1146/annurev.publhealth.28.021406.144016
Central Statistical Office C, Ministry of Health M, ICF International (2014) Zambia demographic and health survey (ZDHS). Lusaka
Chatters LM (2000) Public health research and practice
Chebet JJ, Mcmahon SA, Greenspan JA, Mosha IH, Callaghan-koru JA, Killewo J, Winch PJ (2015) Every method seems to have its problems: Perspectives on side effects of hormonal contraceptives in Morogoro Region, Tanzania. BMC Women's Health 1–12. https://doi.org/10.1186/s12905-015-0255-5
Cheyeka A, Hinfelaar M, Udelhoven B (2014) The changing face of zambia' christianity and its implications for the public sphere: a case study of Bauleni Touwnship, Lusaka. J South Afr Stud 40(5):1031–1045. https://doi.org/10.1080/0357070.2014.946228
Cheyeka AM (2016) Zambia, a 'Christian nation' in post Movement for Multiparty Democracy (MMD) Era, 2011–2016. 6(7), 2011–2016
Chileshe MF (2016) Cannabis in Zambia: who uses it and probably why? Retrieved 7 Sept 2019 from https://facebook.com/notes/drug-enforcement-commission/cannabis-in-zambia-who-uses-it-and-probably-why/765144903620973/
Cialdini R (2007) Descriptive social norms as underappreciated sources of social control. Pychometrika 72(2):263–268. https://doi.org/10.1007/s11336-006-1560-6
Cohorts A, Peri-rotem N (2016) Religion and fertility in Western Europe : trends. Eur J Popul 32. Springer Netherlands. https://doi.org/10.1007/s10680-015-9371-z

Cripss AW, Orczmy DC, Barker J, Lehmann D, Alpers MP (2008) The relationship between undernutrition and humoral immune status in children with pneumonia in Papua New Guinea. PNG Med J 51(3–4):120–130

Dados N, Connell R (2012) The global south. Contexts 11(1):12–13. https://doi.org/10.1177/1536504212436479

Davis R, Campbell R, Hildon Z, Hobbs L (2014) Theories of behaviour and behaviour change across the social and behavioural sciences : a scoping review. https://doi.org/10.1080/17437199.2014.941722

Davis R, Campbell R, Hildon Z, Hobbs L, Michie S, Davis R, Michie S (2015) Theories of behaviour and behaviour change across the social and behavioural sciences : a scoping review. Health Psychol Rev 9(3):323–344. https://doi.org/10.1080/17437199.2014.941

Donovan R (2011) Social marketing's mythunderstandings. J Soc Market 1(1):8–16. https://doi.org/10.1108/20426761111104392

Fang C (2018) Population dividend and economic growth in China, 1978–2018. China Econ J 11(3):243–258. https://doi.org/10.1080/17538963.2018.1509529

Firestone R, Rowe CJ, Modi SN, Sievers D (2017) The effectiveness of social marketing in global health: a systematic review. Health Policy Planning 32(1):110–124. https://doi.org/10.1093/heapol/czw088

Fishbein M (2008) A reasoned action approach to health promotion. Med Decis Making 28(6):834–844

Fishbein M, Ajzen I (2010) Predicting and changing behavior: the reasoned action approach. Taylor & Francis, New York

Fishbein M, Capppella JN (2006) The role of theory in developing effective health communications. J Commun 56. https://doi.org/10.1111/j.1460-2466.2006.00280.x

Fishbein M, Middlestadt SE (1987) Using the theory of reasoned action to develop educational programmes: applications to illicit drug use. Health Educ Res 2(4):361–371

Fishbein M, Yzer MC (2003) Using theory to design effective health behaviour programmes. Commun Theory 13(2):164–183

Gueye A, Speizer IS, Corroon M, Okigbo CC (2015) Belief in family planning myths at the individual and community levels and modern contraceptive use in Urban Africa. Int Perspect Sexual Reprod Health 41(4):191–199. https://doi.org/10.1363/4119115

Guiahi M, Hoover J, Swartz M, Teal S (2017) Impact of catholic hospital affiliation during obstetrics and gynecology residency on the provision of family planning. J Grad Med Educ 440–446. https://doi.org/10.4300/JGME-D-16-00496.1

Hollerbach PE (1980) Power in families, communication and fertility decision-making. Popul Environ 3(2):146–173

Huinink J, Kohli M, Ehrhardt J (2015) Explaining fertility: the potential for integrative approaches. Demogr Res 33(4):93–112. https://doi.org/10.4054/DemRes.2015.33.4

International Federation of Gynecology and Obstetrics (FIGO). (2018, October 14-19) Contraception: the key to achieving sustainable development goals by 2030. Paper presented at the XXII FIGO World Congress Brazil

iSMA, ESMA, AASM (2013) Consensus definition of social marketing, https://marketingsociale.net/isma-global-consensus-on-social-marketing-principles-concepts-and-techniques/

Kapambwe S, Parham G, Mwanahamuntu M, Chirwa S, Mwanza J, Amuyunzu-Nyamongo M (2013) Innovative approaches to promoting cervical health and raising cervical cancer awareness by use of existing cultural structures in resource-limited countries: experiences with traditional marriage counseling in Zambia. Global Health Promotion 20(57):57–64. https://doi.org/10.1177/1757975913502689

Karim MR, Mondal MNI, Rana MM, Karmaker H, Bharati P, Hossain MG (2016) Maternal factors are important predictors of low birth weight: evidence from Bangladesh demographic & health survey—2011. Malays J Nutr 22(2):257–265

Kaur R, Sharma A, Kumar R, Kharb R (2011) Rising trends towards herbal contraceptives. J Nat Prod Plant Resour 1(4):5–12

Kennedy A (2017) Macro-social marketing research: philosophy, methodology and methods. J Macromarket 37(4):347–355. https://doi.org/10.1177/0276146717735467
Kiene SM, Hopwood S, Lule H, Wanyenze RK (2014) An empirical test of the theory of planned behaviour applied to contraceptive use in Rural Uganda
Kraft PW, Lodge M, Taber CS (2015) Why people "Don't Trust the Evidence" motivated reasoning and scientific beliefs. Ann Am Acad Polit Soc Sci 658(1):121–133
Larkin PJ (2018) Introduction to a debate: "Marijuana: legalize, decriminalize, or leave the status quo in place ?" California: Regents of University of California. https://doi.org/10.15779/Z385H7BT6H
Lee NM, VanDyke SM (2015) Set It and forget it: the one-way use of social media by government agencies communicating science. Sci Commun 37(4):533–541 https://doi.org/10.1177/1075547015588600
Lefebvre C (2000) Theories and models in social marketing. In: Bloom PN, Gundlach GT (eds) Handbook of marketing and society. Sage, Newbury Park
Lopez LM, Grey TW, Chen M, Tolley E, Stockton L (2016) Theory-based programmes for contraception (Review). Cochrane Database Syst Rev 11. https://doi.org/10.1002/14651858.CD007249
Lusaka Times (2009) Mansa residents prefer traditional methods of child spacing. Retrieved from http://www.lusakatimes.com/2009/09/02/mansa-residents-prefer-traditional-methods-of-child-spacing/
Lusaka Times (2019) Court rejects appeal for medical marijuana use in Zambia. Lusaka Times. Retrieved from https://lusakatimes.com/2019/06/28/court-rejects-appeal-for-medical-marijuana-use-in-zambia/
Manala MJ (2013) The impact of christianity on sub-saharan Africa 1
Manning L, Laman M, Rosanas-Urgell A, Michon P, Aipit S, Bona C, Davis TM (2012) Severe anemia in papua new guinean children from a malaria-endemic area: a case-control etiologic study. PLoS Neglect Trop Dis 6(12):1972
May JF (2016) The politics of family planning policies and programs in sub-saharan Africa. Popul Dev Rev 43(1):308–329
May JF, Turbat V (2017) The Demographic dividend in sub-saharan Africa: two issues that need more attention. Demogr Econ 83:77–84
McEachan RRC, Conner M, Taylor NJ, Lawton RJ (2011) Prospective prediction of health-related behaviours with the theory of planned behaviour: a meta-analysis. Health Psychol Rev 5(2):97–144
Mcquillan K, Mcquillan K (2019) When does religion influence fertility? Popul Dev Rev 30(1):25–56
Ministry of Health (2017) Ministerial statement on the cultivation of cannabis for medicinal use in Zambia by the Minister of Health, Dr. Chilufya. National Assembly of Zambia, Lusaka. Retrieved from https://parliament.gov.zm/node/6442
Moore GS (2014) Crossing the chasm: marketing and selling disruptive products to mainstream customers (3 ed) HarperCollins, New York, NY
Morgan SP, Bachrach CA (2011) Is the theory of planned behaviour an appropriate model for human fertility? In: Philip Morgan S, Bachrach CA (eds) Stable model for human f. Vienna Yearbook of Population Research, vol 9. Austrian Academy of Sciences Press, pp 11–18. URL: https://www.jstor.org/stable/41342798
Nath S (2015) Religion and its role in society. J Human Soc Sci 20(11):82–85
Niekerk VB (2018) Region and spirituality: what are the fundamental differences? HTS Theog Stud 74(3). https://doi.org/10.4102/hts.v.74i3.4933
Nyundo L, Eagle L, Whittiker M, Low D (2020) Constructing the contraceptive decision-making process: a social marketing framework for family planning programmes in sub-saharan Africa. A zambian case study. James Cook University, Townsville, Australia
Oladeji D (2008) Communication and decision-making factors influencing couples interest in family planning and reproductive health behaviours in Nigeria. Stud Tribes Tribals 62:99–103

Philipov D, Berghammer C (2007) Religion and fertility ideals, intentions and behaviour: a comparative study of European countries. 271–305. https://doi.org/10.1553/populationyearbook2007s271

Phiri P (2017) In Zambia, old law ignites new debate on Marijuana Cultivation. GPJ Accurate. Retrieved from https://globalpressjournal.com/africa/zambia/zambia-old-law-ignites-new-debate-marijuana-cultivation/

Pradhan DK, Mishra MR, Mishra A, Panda AK, Behera RK, Jha S, Choudhury S (2012) A comprehensive review of plants used as contraceptives. Int J Pharm Sci Res 4(1):148–155

Rai R, Nath V (2005) Some lesser known oral herbal contraceptives in folk claims as anti-fertility and fertility induced plants in Bastar region of Chhattisgarh. J Nat Remedies 5(2):153–159

Rogers EM (1962) Diffussion of innovations. Simon and Schuster, London

Rumun AJ (2014) Influence of religious beliefs on healthcare practice. Int J Edu Res 2(4):3748

Rundle-thiele S, David P, Willmott T, Pang B, Eagle L, Hay R, Eagle L (2019) Social marketing theory development goals : an agenda to drive change. J Market Manage 35:1–22. https://doi.org/10.1080/0267257X.2018.1559871

Sakupapa TC (2018) Christianity in Zambia. In: Werner D, Phiri I (eds) Anthology of African Christianity. Regnum Books, pp 758–765

Schmid B, Thomas EJO, Cochrane JR (2008) The contribution of religious entities to health in sub-Saharan Africa

Schoumaker B (2019) Stalls in fertility transitions in sub-saharan Africa: revisiting the evidence. Studies in Family Plann 50(3):257–278

Seiler TB, Engwall M, Hollert H (2013) Lost in translation? Ways for environmental sciences to communicate about risk and research. Environ Sci Eur 25(1):1–3

Simis ML, Madden H, Cacciatore MA, Yeo SK (2016) The lure of rationality: why does the deficit model persist in science communication? Public Understand Sci 25(4):4000–4414

Starbird E, Norton M, Marcus R (2016) Investing in family planning: key to achieving the sustainable development goals. Retrieved from https://www.ncbi.nlm.nih.gov/pmc/articles/PMC4982245/pdf/191.pdf

Sznitman SR, Bretteville-jensen AL (2015) Public opinion and medical cannabis policies: examining the role of underlying beliefs and national medical cannabis policies. Harm Reduct J 12(46):1–9. https://doi.org/10.1186/s12954-015-0082-x

Truong VD (2014) Social marketing : a systematic review of research 1998–2012. 20(1):15–34. https://doi.org/10.1177/1524500413517666

Westoff CF (2010) Desired number of Children: 2000–2008. Calverton, Maryland

White H (2009) Theory-based impact evaluation: principles and practice. J Dev Effect 1(3):271–280. https://doi.org/10.1080/17439340903114628

World Health Organisation Department of Reproductive Health and Research and John Hopkins Bloomberg School of Public Health/Center for Communication Programs, Knowledge for Health Project (2018) Family planning: a global handbook for providers (2018 update), 3rd edn. World Health Organisation, Geneva

Zambia Daily Mail (2018) Doctors nod marijuana. Zambia Daily Mail Limited. Retrieved from https://daily-mail.co.zm/doctors-nod-marijuana/

Zimba CZ (2018) Reasons why Govt is right on legalizing cannabis, marijuana or "ganja" for medicines. The Mast. Retrieved from https://themastonline.com/2018/05/25/reasons-why-gvt-is-right-on-legalizing-cannabis-marijuana-or-ganja-for-medicines/

Lucy Nyundo is a Principal Consultant at the National Institute of Public Administration, Zambia. She holds a B.Sc. (Marketing) from the Copperbelt University, Zambia, M.Sc. Marketing Management from Liverpool Hope University, England and a Professional Diploma in Marketing from the UK Chartered Institute of Marketing. She is currently completing a Ph.D. at James Cook University, Townsville, Australia. Her thesis focuses on investigating family planning decision processes in Zambia.

Lynne Eagle is a Professor in Marketing at James Cook University, Townsville, Australia and an Adjunct Professor at Canterbury University, Christchurch, New Zealand. She holds a Ph.D. from the University of Auckland in her native New Zealand. Her research interests centre on marketing communication effects and effectiveness, including the impact of persuasive communication on children; the impact of new, emerging and hybrid media forms and preferences for the use of formal and informal communications channels, along with trans-disciplinary approaches to sustained behaviour change in social marketing/health promotion/environmental protection campaigns. She has published in a wide range of academic journals, including the Journal of Advertising and European Journal of Marketing, led the development of both Marketing Communications and Social Marketing texts and contributed several book chapters for other texts as well as writing commissioned expert papers and presenting numerous research papers at international conferences.

Maxine Whittaker is the Dean of the College of Public Health, Medical and Veterinary Sciences at James Cook University. Maxine Whittaker has lived and worked in Bangladesh, Zambia, Zimbabwe and PNG and worked extensively in Asia and the Pacific. She is the Co-Chair of the WHO Human Reproduction Programme-Research Project Review Pane. She has published more than 100 peer-reviewed publications and awarded the Royal Australasian College of Physicians International Medal in recognition of outstanding service in developing countries.

Chapter 10
Declare or Dispose: Keeping Biosecurity Threats Out of New Zealand Using Behaviour Change

Phill Sherring

Abstract The purpose of this chapter is to highlight the case study of how the Ministry for Primary Industries' (MPI) Border Compliance Social Marketing Programme follows a benchmarking criterion to deliver behaviour change. The programme is tasked with changing behaviour of international travellers to New Zealand to protect the country's environment and horticultural and agricultural industries from harmful pests and diseases. This is by either leaving biosecurity risk items at home, declaring items to border staff on arrival, or disposing them of in special amnesty bins. Advocacy from local communities to family and friends overseas is also targeted. The programme follows the UK's National Social Marketing Centre (NSMC) benchmark criteria in ensuring theory is followed, segmentation and insights are essential (gathered by ethnographic and quantitative research), and methods mix creates interventions throughout the passenger journey to New Zealand. The programme over its ten years has seen a clear reduction in potential biosecurity risk items seized at the New Zealand border. Quantitative research also points to significant change in the behaviour of leaving items behind pre-travel and declaring items on arrival. The programme contributes to the wider biosecurity system, which aims to protect New Zealand's horticultural and agricultural industries. The horticulture industry alone is worth $5.6bn to the country's economy. Biosecurity also guards New Zealand's native flora and fauna from pests and disease, which is a key attraction in the $12.9bn tourism industry of New Zealand.

Keywords Behaviour change · Biosecurity · Customs · New Zealand · Tourism · Agriculture

10.1 Introduction

Social marketing context

Ten years ago, to protect the country's horticultural and agricultural industries, the environment, and its native flora and fauna, the Ministry for Primary Industries (MPI)

P. Sherring (✉)
Ministry for Primary Industries, Wellington, New Zealand
e-mail: phill.sherring@mpi.govt.nz

R. Hay et al. (eds.), *Broadening Cultural Horizons in Social Marketing*,
https://doi.org/10.1007/978-981-15-8517-3_10

created a marketing programme with the strategic goal of reducing the risk of the introduction of pest and diseases to New Zealand by overseas visitors failing to declare or dispose of biosecurity items. To do this, it had to prevent overseas visitors from entering the country with items that may carry harmful pests and diseases. Risk items can include (but are not limited to):

- Food—cooked, uncooked, fresh, or preserved
- Animals and animal products—including meat, fish, and honey
- Plant or plant products like fruits, flowers, seeds, and plant products that may be part of native medicinal products or religious offerings
- Used outdoor equipment—such as hiking boots, camping, and hunting equipment.

The programme began with concentrating marketing efforts strictly at the main point of arrival, that is at New Zealand's international airports with a focus on signage and collateral. Over time, this expanded to include targeting New Zealand citizens with advertising and communications to promote local compliance and encourage advocacy to friends and family overseas.

In 2013, the programme moved to targeting higher risk overseas audiences, and at the same time, it moved to a more social marketing approach. In doing this, the programme mapped overseas visitors' journeys in an attempt to understand visitor barriers to compliance and to understand how the programme could encourage behaviour change. It also adopted a benchmark criterion to ensure that measurable behavioural change was achieved and not just "education or attitude change" (Andreasen 2002). This criterion is based on the programme theory, solid segmentation, and insights into target audiences, and interventions to create change.

New Zealand welcomed 3.8 million visitors to its shores in 2018 (Stats NZ 2019) and this is where the main focus of the border compliance programme sits. A secondary focus on locals remains to ensure compliance and advocacy from New Zealanders.

Biosecurity context

Biosecurity (or quarantine) laws exist in New Zealand to prevent unwanted pests and diseases, such as the brown marmorated stink bug or foot and mouth disease, from entering the country, and on how to respond should they do so. This is to protect New Zealand's primary industries, which are key contributors to the New Zealand economy. For example, the horticulture industry is worth $5.6b annually, exporting to 128 international markets (Horticulture New Zealand 2017). If an unwanted pest arrives into New Zealand, there is also the cost of eradication to also consider. The 2015 Queensland Fruit Fly incursion in Auckland cost the New Zealand Government $15.7 m to eradicate (RNZ 2016).

It is not just New Zealand's primary industries that biosecurity protects. These regulations also guard New Zealand's environment and native flora and fauna for locals and the country's 3.8 m international visitors to continue to enjoy. Further to this, the social costs cannot be discounted. Myrtle rust, a disease that entered New Zealand in 2017 has done considerable damage to the country's iconic Pōhutukawa and Mānuka tree population. Both these species are sacred to local Māori.

10.2 Programme Structure—Benchmark Criteria

The programme's structure was developed to closely follow benchmark criteria set by the National Social Marketing Centre (NSMC) in the United Kingdom (Hopwood and Merrit 2011) and also French and Blair-Stevens (French and Russell-Bennet 2015).

Both the NSMC's and French and Blair-Stevens' models are advancements on the original benchmarks written by Andreasen (2002). Andreasen (2002) argued that to be a legitimate social marketing programme, a set of criteria had to be met. These benchmarks include (Andreasen 2002):

1. A change in actual behaviour is the benchmark to evaluate success.
2. Audience research is used to understand the audience, pre-test interventions before implementation, and monitor these interventions as they are implemented.
3. Segmentation of target audiences to ensure effectiveness of interventions.
4. Creating an attractive and motivational "exchange" with the target audience.
5. Using the "four Ps" of the traditional marketing mix. This means having a product (i.e. an attractive benefit), while minimising costs (price), making the exchange easy (place), and communicating this to the target audience (promotion).
6. Paying careful attention to competition faced by the desired behaviour.

The NSMC and French and Blair-Stevens' updated criteria then go on to add two more benchmark criteria and expand on some of the existing. These are detailed in Table 10.1.

The first of the new criterion is the use of theory to understand audiences' behaviour and influence the interventions used. This recommends to find the most appropriate theory based on the behaviour of your audience and should be applied once you have done your audience research (Hopwood and Merrit 2011). This also reflects the amount of work done in the social and natural sciences to understand what influences behaviour (French and Russell-Bennet 2015).

The second additional criterion is "insight". This is focussed around gaining a deeper understanding of what emotional and physical barriers your audience face, and in turn, what motivates and influences their behaviour (Hopwood and Merrit 2011). Other differences between Andreasen's original criteria and that of the NSMC and French and Stevens are the audience research being expanded to become "Customer orientation". This considers ethnographic research methods, including qualitative and quantitative data sources.

Finally, the four Ps incorporate broader intervention methods (inform, educate, design, and control) in order to result in a more effective mix to change behaviour (French and Russell-Bennett 2015). The structure of this case study follows the NSMC's benchmark criteria to highlight how this has helped structure the Border Compliance Social Marketing programme.

The NSMC's planning process (Hopwood and Merrit 2011) also influences the border programme. A yearly evaluation following the steps of scope, develop, implement, evaluate, and feedback results in a strategy and intervention review and refresh.

Table 10.1 NSMC benchmark criteria (Hopwood and Merrit 2011)

Criteria	Description
1. Behaviour change	Actual specific and measurable change in behaviours should be the target for success
2. Customer orientation	Ethnographic research, interview research, and data should combine to gain a deeper understanding of the customer or target audience. This in turn should influence the methods mix (criteria number 8)
3. Theory	Understand behaviours and influence interventions through the use of behavioural theories
4. Insight	Insights are the key factors identified through customer research that motivate audiences to change behaviours, but may also present emotional or physical barriers that prevent change. This assists in developing an attractive exchange and methods mix
5. Exchange	For the audience, what are the benefits and costs involved for adopting (and maintaining) the desired behaviour? Customer orientation and insights should guide what the audience value and what incentives and rewards can be offered
6. Competition	Careful attention should be paid to any competition faced by the desired behaviour
7. Segmentation	Customer orientation and insights help identify the audience segments to target Focus on behavioural and psychographic data over demographic and geographic targeting Heavily influence the interventions included in the methods mix
8. Methods mix	Utilising methods to generate behaviour change. Includes the 4Ps of the marketing mix (product, price, place, and promotion) and intervention methods (inform, educate, support, design, and control)

10.3 Behaviour Change

The border compliance programme seeks to achieve a key objective of increasing voluntary compliance with biosecurity laws. There are two main behavioural goals that contribute to achieve this:

1. Consider biosecurity risk items when packing bags pre-departure (and leave them at home).
2. If carrying potential risk items, declare them to biosecurity staff or dispose of in special amnesty bins upon arrival.

The programme also has a secondary goal of increasing advocacy of biosecurity laws from local communities to friends and family overseas before they travel to New Zealand.

10.4 Measurement

MPI measures compliance annually through using operational data from airports, including the number of seized undeclared risk items and the number of items voluntarily declared for inspection. The annual compliance rate for each target segment is compared against previous years and the overall compliance rate for all passengers. Compliance during the New Zealand spring/summer holiday period is also closely monitored. This period is the busiest for inbound tourism, and includes culturally significant holidays such as Christmas, Chinese New Year, and Diwali. Quantitative research of target segments then tracks awareness, consideration of biosecurity items when packing, and their subsequent behaviour before travel and on arrival. It also measures what programme interventions may have (or have not) influenced their behaviours during their journey to New Zealand. This research occurs every two years.

10.5 Customer Orientation

10.5.1 Identifying Motivators and Barriers

The MPI Marketing team in 2013 began a piece of work with the Ministry's Service Design team to better understand their target segments. This included having a clearer understanding of both the motivators and the barriers that the audiences faced in complying with biosecurity rules. This began with using ethnographic research to learn how recently arrived visitors navigated the airport experience and interacted with biosecurity signage, amnesty bins, and frontline staff. Ethnographic research (sometimes called ethnography) is a form of observational research and anthropology to understand how people live their lives (Anderson 2009).

Passengers from selected flights were followed and observed as they disembarked the plane and proceeded through the airport experience to arrive at biosecurity screening. MPI staff were particularly interested if signage, bins, and other stimulus had any influence on behaviours. Following that, contextual interviews were then held with visitors to gain deeper insights. This was mostly done in the airport departures area, due to departing passengers having longer waiting periods for outbound flights and being more open to answering questions. Passengers in the arrivals area were discounted, as they were generally tired after long haul flights and simply wanting to leave the airport as quickly as possible. The interviews helped validate what was recorded in the observational research, and gained important information from visitors about behaviours before and during travel.

Frontline MPI staff were next interviewed for their everyday observations and experiences when dealing with those in the target segments. This included seeking their opinions on what tools may influence better compliance from passengers.

Finally, trial interventions were run and observed in the airport. This included additional signage, new positioning of amnesty bins, special benches for checking bags for risk items pre-biosecurity checks, loud speaker announcements, and staff assisting visitors before they arrived at the screening area.

10.6 Passenger Journey Mapping

Following the ethnographic research stage, MPI was then able to create a passenger journey map for travellers coming to New Zealand. This mapped out all the various common steps that travellers took before arriving in the country. This was very similar to customer journey maps used predominantly in service marketing. These illustrate the different touchpoints or steps taken by customers through products, online experiences, or services (Richardson 2010). The reason for mapping a customer journey is to improve the customer experience at each touchpoint (Rosenbaum et al. 2016). In service marketing the customer journey is split commonly into three stages: pre-purchase (the experience from the outset of a need recognition and consideration), purchase (all the customer interactions with the service or brand), and post-purchase (any experiences after the purchase including service requests) (Lemon and Verhoef 2016). MPI followed this approach in splitting their passenger journey into three stages:

(1) Pre-flight—includes steps such as researching and booking, obtaining a Visa, packing bags, checking-in, and boarding.
(2) Flight—includes briefings from cabin crew, meals, on-board entertainment, and completing the New Zealand arrival card.
(3) Arrival—exiting the plane, clearing customs, collecting baggage, and biosecurity checks.

From here, smaller individual steps could be identified and mapped out in more details.

Research of customer mapping has further identified how different touchpoints can be classified. These are broken down into four categories: brand-owned, partner-owned, customer-owned, and social/external touchpoints (Lemon and Verhoef 2016). Table 10.2 demonstrates the types of touchpoints in the passenger journey.

MPI's current and potential touchpoints were then able to be classified to illustrate how many the Ministry owned and how many required partnerships with others. It also highlighted where influence from friends and family was required as social touchpoints. This helped in developing intervention logic for the programme, see Table 10.3.

Table 10.2 Types of touchpoints in the Customer Journey (Lemon and Verhoef 2016)

Type of touchpoints	Description
1. Brand-owned touchpoints	Customer interactions designed and managed by the company/service provider. Includes all owned media such as website, advertising, and owned elements of the marketing mix
2. Partner-owned touchpoints	Any interactions that are designed and/or managed jointly by the company along with any partners. This can include marketing agencies, distribution partners, and shared communication channels
3. Customer-owned touchpoints	Any interactions included in the customer journey that the company and its partners have no influence or control over. Such as any specific needs or requirements from the customer
4. Social/external touchpoints	The role played by others in the customer experience. This can include any external touchpoints that could add influence including other customers, opinions of peers, and independent information sources such as reviews

10.7 Theory (and Models)

A number of theories, models, and processes influence the border compliance programme. The leading theory is Azjen's Theory of Planned Behaviour (Azjen 1991) to influence the segments visiting New Zealand. However, it also relies on the Social Norms Approach (Berkowitz 2004) and McKenzie-Mohr's Community Based Social Marketing (McKenzie-Mohr 2011) model to generate compliance and advocacy in local communities.

10.8 The Theory of Planned Behaviour

The theory of planned behaviour (TPB) has been one of the most dominant behaviour change theories of the last thirty years and works off the key belief that a person's intention to perform a behaviour directly relates to that behaviour occurring (Sniehotta et al. 2014). The stronger the intention, the more likely the behaviour will occur (Azjen 1991).

There are three factors that then directly affect and influence an individual's intention to perform a behaviour—attitudes, subjective norms, and perceptions of behavioural control (Azjen 2011). MPI tries to directly influence these factors through the programme's interventions and exchange, assisted by key insights from ethnographic and quantitative research. The idea of attitudes refers to how favourable or unfavourable the subject is to the consequences of performing the behaviour (de Leeuw et al. 2015). For example, many people have positive attitudes towards zero food waste and composting.

Table 10.3 Passenger journey steps and types of touchpoints

Journey stage	Individual steps	Type of touchpoint
Pre-flight	Planning and booking	• Brand-owned—advertising, website content, and collateral • Customer-owned—thoughts on where they should travel, knowledge of New Zealand, and biosecurity • Social/external—advice from peers, family, and travel review sites
	Obtaining a Visa	• Partner-owned—Immigration New Zealand manage Visas. MPI has partnership.
	Packing bags	• Brand-owned—advertising and web content • Partner-owned—information from partner travel agents and airlines • Customer-owned—have their own needs and desires for items they pack and bring • Social/extend—advice from friends and family on what to pack and bring
	Check in/boarding	• Brand-owned—advertising/signage at departure area • Partner owned—information given by airline staff at check in
In-journey (in-flight)	Relaxing and entertainment	• Partner-owned—inflight biosecurity video. Supplied by MPI, played by airlines
	Eating	• Customer-owned—hard to prevent passengers from taking food off flight
	Passenger arrival card	• Partner-owned—translated guide to completing arrival declaration card correctly
Arrival	Navigating airport	• Partner-owned—signage advising of biosecurity rules, including translated signs • Partner-owned—public announcements in the airport arrivals area • Partner-owned—wayfinding signage advising of biosecurity checks approaching • Customer-owned—wants to get through airport quickest way, blocks out signage/information
	Luggage retrieval	• Partner-owned—benches provided for passengers to check bags

(continued)

Table 10.3 (continued)

Journey stage	Individual steps	Type of touchpoint
	Risk assessment	• Brand-owned—translated materials (including guides/flashcards) • Brand-owned—translators to assist with language difficulties • Brand-owned—signage reminding what needs to be declared • Customer-owned—self-assesses, decides for themselves what they need to declare and what is "ok"

MPI's biosecurity focus is on fostering a positive attitude towards the behaviour and "doing the right thing" in order to protect the country's environment and industries. This stems from an audience insight that they wish to know why the biosecurity laws exist and why they cannot bring banned items into New Zealand. Therefore, many of the marketing interventions such as advertisements and web pages focus on explaining the country's reliance on biosecurity.

Subjective norms are how people feel social pressure to either perform or not perform the behaviour (Azjen 1991). The opinions of friends and family can influence the individual's intention of engaging in the behaviour (Kautonen et al. 2015). Using the zero food waste example, people who have friends that compost and actively reduce their waste may also be inclined in their intention to do so as well.

MPI uses subjective norms in the manner of advocacy from friends and family that live in New Zealand. To do this, MPI uses a mix of other models (expanded upon later)—the Social Norms Approach (Berkowitz 2004) and Community Based Social Marketing (McKenzie-Mohr 2011). The final contributing factor to intention is the perceived behavioural control. This relates to the level of ease or difficulty an individual feels there is to perform the behaviour (Michie 2014). Using the zero waste example again, someone with the facilities to compost (such as an outdoor compost bin) would have a greater sense of behavioural control to someone who would need to transport their food scraps to a green waste collection point.

It is at this point where MPI targets a large majority of their interventions. By creating awareness about the desired behaviour pre-travel and supplying tools during the travel journey to make declaring easier (for example, translated guides to the declaration form for those whose second language is English), MPI seeks to increase a visitor's perceived control in complying. Greater behavioural control combined with favourable attitudes and subjective norms should strengthen the intention to perform the actual desired behaviour. This is why MPI focusses on all three factors to generate a strong intention to comply with border laws, creating positive attitudes by showing the importance of biosecurity, making advocates of friends and family to generate strong subjective norms, and finally aiming tools to result in enhanced behavioural control.

10.9 VADE Compliance Model

The programme subscribes to a New Zealand Government compliance model called the VADE (Voluntary, Assisted, Directed, Enforced) compliance model. This was first developed by a former government agency (The Ministry of Fisheries) and was strongly influenced by theory from Dr. Edward Maibach (NZ Department of Internal Affairs 2011; Maibach and Parrot 1995). People purposely not complying with laws are punished with enforcement such as fines and deportation. Those wishing to do the right thing but needing assistance to do so are guided by social marketing interventions to result in positive behaviour.

10.10 Social Norms Theory and Community-Based Social Marketing

When targeting local communities within New Zealand to create advocacy to family and friends offshore, the programme uses approaches such as community-based social marketing (McKenzie-Mohr 2011) and the social norms approach (Berkowitz 2004).

Social norms theory proposes that a person's behaviour is influenced by inaccurate perceptions of thoughts and behaviours of others that are in that person's social group (Michie et al. 2014). In fact, peer influences have a greater impact on behaviour than biological, religious, cultural, and personality influences (Berkowitz 2004).

Berkowitz (2004) states that peer influences are more based on what individuals think others believe and do (called the "perceived norm") than on real beliefs and behaviours (the "actual norm"). The difference between perceived and actual norms forms what is called "the misperception". From this theory stems the social norms approach (SNA), which aims to challenge the misperceptions and norms held by audiences, and motivates behaviour on tendencies to conform (Schuster et al. 2016). SNA mostly does this through educating the target audience about behaviours of those that are similar to them (Schuster et al. 2016), usually via mass media campaigns.

There are three forms of misperception—pluralistic ignorance, false uniqueness, and the one most applicable to border compliance—false consensus (Berkowitz 2004). False consensus occurs when an individual in the minority believes their behaviour is that of the majority (Berkowitz 2004). In MPI's case, false consensus occurred when visitors held the incorrect belief that the majority of their peers thought it was acceptable to bring in items not available in New Zealand. Those in the minority therefore felt they could bring traditional food and medicines with them. However, MPI's quantitative research showed this was not the case, with 62% of surveyed visitors stating that they believed it was important to leave items behind or declare to officials upon arrival. Hence, MPI uses mass media and social media advertising to local communities to stress the importance of advising friends and family overseas of the misperception and they should comply with the majority's belief and leave items at home. In addition, MPI utilises principles of community-based social marketing (CBSM) to further assist with local community behaviour and advocacy. CBSM developed by Doug McKenzie-Mohr focusses more on engaging the community to foster change, not using mass media like SNA (Schuster et al. 2016).

CBSM follows five stages when developing a community-based social marketing approach: selecting behaviours, identifying barriers, developing the strategy, piloting, and broad implementation (McKenzie-Mohr 2011). Selecting behaviours (to change), finding barriers, and piloting (in particular with interventions) have been vital steps in both engaging with communities and in the broader programme. Other aspects such as prompts to act (McKenzie-Mohr 2011), in particular with signage at arrival airports, have also been borrowed from the CBSM model.

Using direct contact between people rather than advertising is a cornerstone of CBSM (Schuster et al. 2016) and MPI has adopted this by having frontline staff active in key local communities. Quarantine officers give regular talks about the importance of biosecurity in community halls, churches, mosques and temples, and schools. Officers chosen are often part of the communities they speak in as well, in keeping with CBSM's recommendation of using credible sources to deliver behaviour change messages in the community (McKenzie-Mohr 2011).

10.11 Insight

Quantitative research

The programme uses quantitative research of visitors to New Zealand, along with New Zealand residents and citizens within local communities. Visitor research is conducted via face-to-face interviews at Auckland International Airport. Local residents and citizens are questioned via online panels. The visitor research tracks awareness of and attitudes towards biosecurity as a whole, knowledge of prohibited items, and the laws around compliance. It also engages respondents with the many interventions used through the passenger's journey and tracks visitors recall of these and rates their effectiveness in prompting behaviour intention and action. This can include asking whether interventions prompted visitors to not pack certain items before they travelled or if anything made them declare goods when arriving. Local respondents were questioned to establish their own biosecurity compliance, but more importantly their levels of advocacy to friends and family overseas.

The most recent study was completed in 2017 (the next study is due for completion late 2019). In total 1,531 visitors to New Zealand were interviewed. This consisted of what MPI considers high risk ($n = 731$) and high volume ($n = 800$). Interviews were conducted in the departure lounge at Auckland International Airport, and data was collected electronically on tablets instead of paper questionnaires to assist with ensuring a high level of accuracy of answers. All respondents from overseas had entered or returned to New Zealand in the past 12 months. An independent research provider commissioned by MPI carried out the fieldwork and also managed the analysis and reporting. Post fieldwork, all open-ended questions were coded. Along with giving a result for each segment, a single dataset from all target groups was formed and data weighted, so each group contributed equally to a "total average" score.

10.12 Identifying Motivators for Behaviour

Through the customer orientation criterion, a key motivator to compliance with biosecurity laws was established. The main motivator for compliance is seen as avoiding enforcement (and subsequent punishment) by those visiting the country. A desire to comply with the law, "be on the safe side", and avoiding a fine were the main reasons given by visitors who had declared items upon arrival.

Understanding barriers

Alongside learning the motivators to compliance, an understanding of the barriers to the behaviour was also required. These insights to the barriers that passengers faced were derived from the ethnographic research, in particular the contextual interviews; and the quantitative research. It was useful for MPI to line up the barriers to the individual steps within their passenger mapping to illustrate what was preventing compliance when arriving in New Zealand. Table 10.4 shows the passenger journey, the barriers mapped against it, and the types of touchpoints used to alleviate the barriers.

10.13 Key Insights

Because of both the ethnographic and quantitative research, a list of key insights of passengers was developed. This was used to help inform the mix of interventions, messaging, and creative approaches.

These insights were:

- Passengers have an all-time high perception of biosecurity risk, although some passengers still do not perceive a risk in bringing small items for personal use. Perception has been trending upwards since 2013.
- Increasingly more passengers are considering high-risk items when packing bags pre-departure.
- Passengers want more information in their native languages. Low understanding of English is a major barrier to compliance.
- They want to know why biosecurity rules exist and why they cannot bring items from home.
- A main motivating factor is the desire to avoid penalties and fines, and a wish to do "the right thing".

Table 10.4 Passenger journey steps, identified barriers, and types of touchpoints

Journey stage	Individual steps	Barriers identified	Type of touchpoint
Pre-flight	Planning and booking	Cannot find information about biosecurity laws/process Travel agents/tour companies lack information about biosecurity rules	Brand-owned—advertising, website content, and collateral Customer-owned—thoughts on where they should travel, knowledge of New Zealand, and biosecurity Social/external—advice from peers, family, and travel review sites
	Obtaining a Visa	Visitors see their holiday and its entirety, not a series of interactions Received no information with their Visa about biosecurity rules	Partner-owned—Immigration New Zealand manage Visas. MPI has partnership.
	Check in/boarding	Lack of information at departing airports and at check in	Brand-owned—advertising/signage at departure area Partner-owned—information given by airline staff at check in
In-journey (in-flight)	Relaxing and entertainment	Biosecurity video isn't in my native language and no subtitles are provided No biosecurity video played on the flight	Partner-owned—inflight biosecurity video. Supplied by MPI, played by airlines
	Eating	Given food items (e.g. fruit) during flight Not told by cabin crew that the food cannot be taken off the plane	Customer owned—hard to prevent passengers from taking food off flight

(continued)

Table 10.4 (continued)

Journey stage	Individual steps	Barriers identified	Type of touchpoint
	Passenger arrival card	Cannot read the card as it's only in English Crew cannot assist through lack of time/understanding/language Visitors unsure how to complete form correctly and struggle with the nuances of the questions	Partner-owned—translated guide to completing arrival declaration card correctly
Arrival	Navigating airport	Wayfinding through the airport is not clear Little signage in languages other than English Difficulties in communicating with airport and/or biosecurity staff No knowledge of what biosecurity checks entail	Partner-owned—signage advising of biosecurity rules, including translated signs Partner-owned—public announcements in the airport arrivals area Partner-owned—wayfinding signage advising of biosecurity checks approaching Customer-owned—wants to get through airport quickest way, blocks out signage/information
	Luggage retrieval	Nowhere to check bags for any biosecurity items that may be in bags	Partner-owned—benches provided for passengers to check bags
	Risk assessment	Visitors unsure why bags are being searched by border officers (especially when visitor thinks they've done everything right) Language difficulties between visitors and biosecurity officers Lack of explanation why some items are allowed and others aren't – doesn't increase understanding or acceptance	Brand-owned—translated materials (including guides/flashcards) Brand-owned—translators to assist with language difficulties Brand-owned—signage reminding what needs to be declared Customer-owned—self-assesses, decides for themselves what they need to declare and what is "ok"

- Target audiences respond to messaging that details consequences for non-compliance.
- Visitors have a fear that food from their home country is not available in New Zealand hence carrying it with them.
- Passengers have a resistance to declaring items they think may not be available in New Zealand, for example traditional medicines, and religious items.
- Many passengers do not consider food (a major risk) to be classed as "food" until it is cooked and prepared as a meal.
- Passengers are heavily influenced by any relatives of friends that already live in New Zealand.

10.14 Exchange

Exchange is one of the most important concepts of both social and commercial marketing (Vinnikainen 2017). Exchange is the cost/benefit undertaken by a person when deciding if the cost of performing the desired behaviour is worth the reward, they receive in return (Hopwood and Merrit 2011).

With the border compliance programme, the exchange revolves around avoiding the punishment if caught with biosecurity items. The cost/benefit here is whether avoiding a $400 NZD fine is worth going without their favourite foods and other items such as herbal medicines from their home country. To many people this may sound like an easy choice, but it can be difficult for many who believe these items will not be available in New Zealand. In addition, some items may have cultural or religious significance, for example offerings from local temples or gifts given for good luck when travelling by family at home.

10.15 Competition

MPI also has to give consideration to competition for the attention of the passenger and the ability to get its messages and interventions across. This can come from many sources including travel agents, in-flight entertainment on planes, and even distractions (or "noise") at the airport like duty-free. Instead of competing with these, MPI has actually tried to place interventions within these environments. For example, digital advertising on travel sites, providing documents in-flight, and biosecurity focussed in-flight videos (similar to safety videos airlines play at take-off). This way the intervention is complementary to the "noise" and passengers see it as important to pay attention to the information.

Then there is the competition within the desire of the audience to bring their own food and items from home when they visit New Zealand. This desire to have their familiar goods may outweigh the want to comply with biosecurity law or perceived risk these items can cause. MPI's advertising, collateral, and online information detail the damage that risk items pose, plus inform future visitors that many items can be purchased in New Zealand through Asian supermarkets and specialty stores. Through their local marketing activities, MPI reminds local audiences that if visiting friends and family are found with prohibited items, they will be confiscated and fines will also be issued if they do not declare.

10.16 Segmentation

Because of the large diversity and ever-growing number of visitors to New Zealand, MPI has to segment the programme towards a selection of priority audiences. The programme therefore focusses on those that present the highest identified risk. Risk is established by visitor numbers, how they are arriving in the country (air or sea), the types of items they carry, and their awareness of and compliance with biosecurity laws. Operational data contributes the most to establishing these levels of risk. All items that are not declared by visitors to New Zealand are confiscated and the information recorded along with passport and flight details. This meaningful data combined with frontline staff experiences and observations can form a clear picture of what visitor segments need targeting to improve their compliance.

The two main audiences that pose the highest biosecurity risk are Chinese and Indian passport holders. These segments both have a limited knowledge of biosecurity, a propensity to bring food and other high-risk items, and finally a higher rate of being found with undeclared risk items. To use New Zealand passport holders as a baseline, those holding Chinese passports are four times higher to be found with undeclared items, and those with an Indian passport, six times higher. The number of Chinese visitors has increased from 228,928 in 2013 to 417,872 in 2017 (Stats NZ 2018). This makes them the third highest visitor group to New Zealand behind New Zealand and Australian passport holders. Chinese are not generally familiar with the idea of biosecurity and have limited knowledge of the risks (though this is improving). They can be both frequent and infrequent travellers.

Visitors from India is New Zealand's new growth market, with an 18% growth in visitors in 2017 (Stats NZ 2018). Tourism New Zealand is targeting India as one of three key emerging markets, and most Indian holidaymakers are coming to New Zealand for the first time (Tourism NZ 2016). The ethnographic research with Indian visitors also identified that visiting family already settled in New Zealand was the main reason for over half of visitors coming to New Zealand. In addition to marketing to passport holders offshore, a secondary audience of local communities is targeted. This links to the insight that word of mouth from friends and family is a big influence on the compliance of visitors. Reminding local communities, helps achieve this, as well as acts as a reminder to residents who may head offshore to visit family when returning to New Zealand.

10.17 Methods Mix

The methods mix combines the traditional 4Ps of marketing with intervention methods to not only raise awareness, but to deliver behaviour change from the target audiences

10.18 Marketing Mix

The Ps of "product", "price", "place", and "promotion" came originally from commercial marketing. They were later adapted for use in social marketing (Kotler and Zaltman 1971), and feature in both Andreasen's original benchmark framework, and the expanded "methods mix" criterion of the NSMC.

Product. What a "product" is specifically within social marketing can be hard to quantify, and is commonly considered to be the achievement of the behavioural goals set out in a programme (Vinnikainen 2017). This is true for the border compliance programme, with the product considered to be visitors to New Zealand complying with biosecurity laws.

Price. Within social marketing, it is argued that price is perceived as the cost for the audience in carrying out the desired behaviour change (Vinnikainen 2017). In MPI's programme, the cost of non-compliance is the main price consideration. The minimum fine for being caught with undeclared risk items is $400 NZD. If caught with deliberately hidden items, this can increase to $10,000 NZD or in extreme cases deportation.

An additional price consideration for overseas visitors is how much items from their home country can cost to purchase in New Zealand. Many items (in particular food) can be found in specialty stores in New Zealand, having gone through proper importing procedures. However, these items are generally substantially cheaper to buy in their country of origin. MPI reminds visitors in their collateral that certain items are allowed to be carried into the country (for example, commercially processed food) with the intention to create a psychological trade-off for the banned items. Price aligns very closely to the benchmark criterion of "Exchange", which was detailed earlier in this case study.

Place. In commercial marketing, the idea of place would be centred on outlets for purchase and distribution channels (Kotler and Zaltman 1971). In delivering a change programme, "place" should instead refer to the importance of ensuring resources and interventions to encourage changes are accessible to the target audience (Peattie and Peattie 2003).

All interventions and marketing for the border compliance programme are aligned to the passenger journey mapping. Interventions are placed where they are most relevant within the journey. For example, receiving biosecurity reminders with flight itineraries, and providing translated guides to the New Zealand arrival card at the time when it is required to be completed. Also carefully considered are prompts such as signs and announcements leading up to and at the assessment area in airports where any declarations of potential risk items need to be made. This is also a compulsory stopping point before any visitor can exit the international terminal.

Promotion. Promotion is the most aligned and transferable P between commercial and social marketing disciplines. Many social marketing interventions can be dependent on promotion to ensure they are visible, recalled, and effective (Vinnikainen 2017). As with the place component, promotion aligns closely with the passenger journey, but also the quantitative research in guiding the messages delivered. For example, MPI focusses their offshore advertising indicating why it is important to leave items behind and why the rules exist—as that is a key motivator pre-travel. Messages then change through the journey to focus on compliance upon arrival along with the enforcement and penalties. MPI utilises a wide range of channels for promotion including digital and traditional advertising, brochures, translated guides, staff tools, signage, and word of mouth.

10.19 Social Marketing Intervention Mix

MPI's interventions sit alongside the marketing mix and are in three stages: pre-travel, in-journey, and upon arrival. These consist of a mix of inform, educate, support, and control.

Pre-journey interventions. MPI starts creating awareness of biosecurity at the very beginning of a passenger's journey—when they start to research trips to New Zealand. This begins with digital advertising in both India and China on popular travel booking sites such as Skyscanner and Ctrip. These ads are purposely targeted only to searches on New Zealand. They help get biosecurity front of mind and click through to translated pages (such as www.nzquarantine.com) and resources on the MPI website to inform and educate potential visitors.

Search engine marketing is also prominent, and MPI targets the most common search terms regarding travel and entry requirements to New Zealand on Baidu (China) and Google (India). Messaging at this stage becomes important as well. A key insight from MPI's quantitative research is that visitors want to know the reasoning why they cannot bring items with them to New Zealand. Therefore, MPI not only focusses on the desired behaviour, but more importantly the reasons behind the behaviour. Explaining the damage that pests and diseases in items such as fruit and wood becomes a key message in both the advertising, and the websites and collateral that are presented to the audience.

Partnerships were a key type of touchpoint through the passenger journey, and one of MPI's most important partnerships is in the pre-journey phase. MPI has a close relationship with the country's immigration service. They assist in disseminating information at the key time of issuing Visas to residents of China and India. Until 2016, all Visas were issued in physical processing centres. MPI provided posters and brochures in these centres. More importantly when a Visa was issued, Immigration New Zealand staff would also provide a small booklet to the new Visa holder, which explained biosecurity and why it was important, and detailed what wasn't allowed to be bought into New Zealand.

Post 2016, digital e-Visas have become the norm, meaning the booklets have become redundant. To counter this, Immigration New Zealand now provides biosecurity advice on their website and e-Visa application portal. More importantly, information is included in Immigration New Zealand's emails when an e-Visa is confirmed, and reminder emails two weeks before the e-Visa officially commences. These are highly targeted, down to specifying common biosecurity items that come from a visitor's region of their country (based on seizure data at the New Zealand border). MPI also works with travel booking agents and tour companies, alongside Tourism New Zealand to ensure staff at these companies have the required knowledge and training to pass onto clients.

Advocacy and "word of mouth" from friends and family in New Zealand are also extremely important during this phase. MPI reaches out to local Chinese and Indian residents to encourage offshore friends and relatives to know and comply with the rules before they travel. Local advertising on Chinese and India media (as per the social norms approach) and using members within the communities (in keeping with community-based social marketing methods) are deployed as interventions at this point. "Tell your friends and family so they avoid a fine and don't have items confiscated" is the message relayed to local communities.

In journey interventions. Now that the opportunity for passengers to not pack items has passed for those on their journey to New Zealand, MPI's interventions change from an educate focus to an "inform" (of the compliance behaviours) and "support" function. Messaging changes from the "why" must a passenger comply, to explaining what is required upon arrival. Interventions explain how a passenger has to either "declare" items to MPI staff at the biosecurity checkpoint, or "dispose" of any potential risk items in specially marked "amnesty bins" found in the airport. MPI also begin to include messaging around consequences for non-compliance, mainly the $400 NZD fine.

The in-journey interventions occur mainly on the flight to New Zealand. The most important tool during this stage, and of the whole journey, is the New Zealand passenger arrival card that must be completed by all arrivals into the country. This has been identified as a major barrier to complying and correctly declaring items to MPI officers. As a legal document, it has to be completed and signed in English. This meant that those for whom English is a second language struggled with understanding what was required of them. This was compounded by poor wording of questions on the form that did not consider cultural differences. For example, the form asked the audience to declare any "food". However, for many cultures, including Indian, "food" is only referred to as such when its ingredients are cooked and prepared as a meal. The questions have now been changed to ask about "ingredients" instead.

More importantly, MPI has worked with key airlines to issue translated guides that wrap around the arrival cards on all direct flights from China. Note this intervention is not currently possible for Indian travellers as there are no direct flights from India and passengers have to go through hub airports, making this difficult. The guides have a simplified Chinese copy of the arrival card and explanations of what is required when completing it and why it is important to do so correctly. It also details any nuances that may confuse passengers. The airlines see this an important support function to not only assist passengers in understanding their compliance obligations, but to also free up cabin crew from explaining the requirements to confused customers. Finally, an in-flight biosecurity video is screened on all inbound flights, explaining the rules and what passengers must do upon arrival to comply. For Chinese airlines, a Chinese spokesperson delivers the video in Mandarin, and it is subtitled into 10 different languages (including Hindi and Punjabi) for all other airlines flying into New Zealand.

On-arrival interventions. Upon arrival, the infrastructure of the airport and the promotion within it become a key focus for the interventions. Airport signs and bins are placed in carefully considered key locations such as the baggage claim throughout the airport journey to maximise the desired behaviour. This includes piloting (as per the community-based social marketing approach of McKenzie-Mohr) various messages and locations within the airports. Translated signs and public address announcements for both Chinese and Indian passengers are commonplace to remind arriving passengers of upcoming biosecurity checks and the appropriate behaviour. Tools for staff have also been developed, especially to break down language barriers including flash cards, translated guides, and having translators available during peak travel times.

At this point, the messaging focusses on the key motivators for the audiences as per the programme's insights. The desire to obey the law, avoid punishment, and "do the right thing" becomes the main focus throughout the airport journey. All signs, bins, and collateral all refer to the punishment for non-compliance with the phrase "declare, dispose or pay the fine" prevalent.

10.20 Evaluation and Results

10.20.1 Compliance Measures

The New Zealand Government's target for border compliance is stopping 98.5% of risk goods from entering the country. The Border Compliance Social Marketing programme is a key tool in ensuring the Ministry achieves this target, which is consistently met year-on-year.

Based on operational data, the number of items seized from Chinese passengers over the lifetime of the programme (it began in 2008) has reduced by 67%. More importantly, the biggest jump in behaviour came in 2014, when compliance by Chinese improved by over 20%. This was the year after the passenger journey mapping first took place and new interventions were introduced. Similarly, the seizure rate for Indian passengers has been cut by half in the years the programme has run. This result is more impressive given New Zealand has become a new major destination for Indian passport holders with visitor numbers doubling in the last five years.

The quantitative research also reinforces the compliance results, with significant changes achieved in attitudes, intention, and behaviours in the three stages of the passenger journey (as detailed below).

10.21 Pre-journey Consideration and Behaviour

At the first stage of the passenger journey, MPI has seen statistically significant increases in consideration of biosecurity risk items when packing bags. Ninety percent of Chinese visitors considered biosecurity items when packing according to the last quantitative survey (from 2017) before travelling. This was an increase of 15% from the 2015 survey. This was comparable to Indian audiences as well, with 92% considering what went into their bags. The previous survey had consideration of items at 63% of visitors.

After consideration, comes actual behaviour. 78% of Chinese and 73% of Indian visitors followed through and left items that could pose a risk at home. In 2015, this figure was 61% and 55%, respectively.

As per the insights, friends and families were the most responsible for prompting pre-journey behaviour, with search engine marketing, Visa information (from Immigration New Zealand) and brochures the most successful introduced interventions. Recall and effectiveness of interventions can be seen in Fig. 10.1.

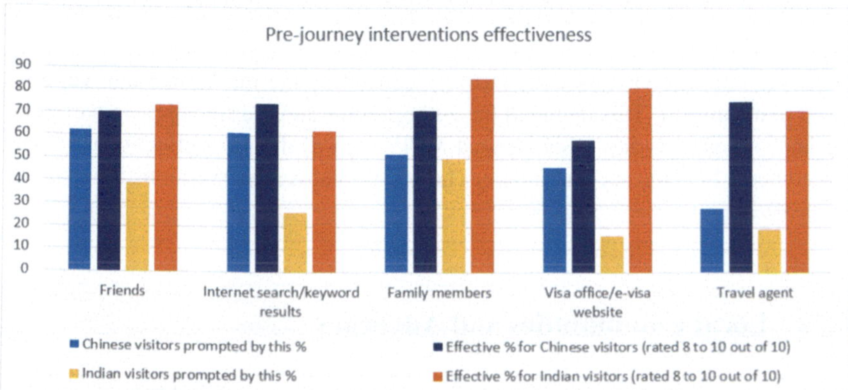

Colmar Brunton - Border Monitor. Base 2017 - visiting Chinese passengerswho considered risk items pre-journey, n=158
Colmar Brunton - Border Monitor. Base 2017 - visiting Indian passengers who considered risk items pre-journey, n=159

Fig. 10.1 Effectiveness of pre-journey interventions to Chinese/Indian visitors to consider biosecurity items

10.22 In-Journey Consideration and Behaviour

At this point passengers may or may not have considered items before departure and are now on route to New Zealand. However, MPI has seen that consideration of items while travelling to New Zealand significantly improved for both Chinese and Indian passengers. Of those surveyed in 2017, 73% of Indian passengers responded that they considered items in their luggage. This is an increase of nearly 40% from the 2015 survey. Chinese passengers did not record as high an increase, but 75% of this audience responded to in-flight stimulus and considered what was in their bags, up from 60% two years earlier. The compulsory passenger arrival card obtained a high recall from both audiences, with Indian passengers rating it the most effective tool to make them think about risk items. For the Chinese passengers, the translated guide of the arrival card (that wraps around the card) was considered the most effective intervention.

10.23 Post-journey (on Arrival) Consideration and Behaviour

Matching MPI's results of meeting their 98.5% compliance target, nearly all of those surveyed said they believed they were not carrying risk items into the country or they declared or disposed of items upon arrival. Ninety-seven percent of Chinese and 98% of Indian passengers stated they carried out one of MPI's desired behaviours of either not bringing items in the first place or "doing the right thing" at the airport. On arrival, Chinese visitors were most influenced by the intentional interventions placed in the airport. This included the amnesty bins and the signage—in particular translated signage. Indian visitors rated MPI's staff as the most effective influence with amnesty bins the second choice. The most effective interventions are detailed in Fig. 10.2.

10.24 Local Communities and Advocacy

In keeping with the audience's importance on word of mouth and MPI using social norms and community-based social marketing to foster this; local communities have responded and advocated to friends and family overseas. Local Chinese and Indian communities had a 52% and 65% advocacy rate, respectively. Advocacy from New Zealand-based relatives and friends is indicated as the most important overall influence on behaviour change as validated by overseas visitors. In addition, the interventions aimed at local communities ingrain the desired behaviours when they travel. Eighty percent of local Chinese who were survey considered or did not pack items when returning to New Zealand from overseas. Local Indian respondents were similar on 84%.

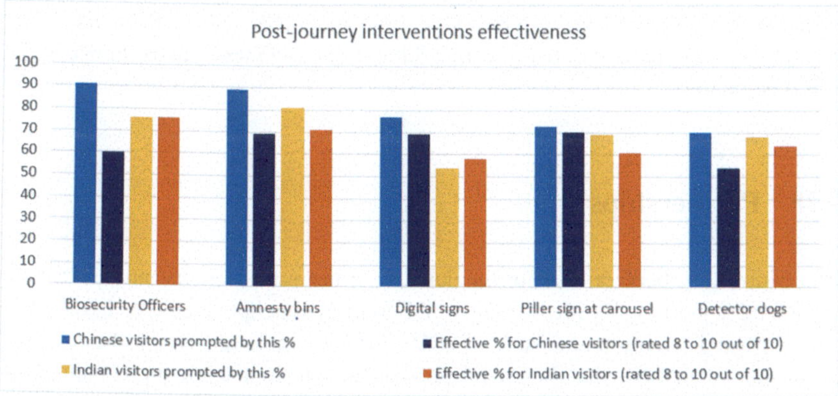

Colmar Brunton - Border Monitor. Base 2017 - visiting Chinese passengers who considered risk items post-journey, n=166
Colmar Brunton - Border Monitor. Base 2017 - visiting Indian passengers who considered risk items post-journey, n=165

Fig. 10.2 Effectiveness of post-journey interventions to visitors to declare or dispose

10.25 Discussion and Lessons Learned

The border compliance social marketing programme has been an integral part of the New Zealand biosecurity system for a decade. While steady improvements occurred year-on-year in the programme's early years, ever since the steps to really focus on fully understanding the customer, their journey and link the programme to theory; the results have dramatically improved to create real behaviour change.

10.26 Understanding the Passenger Journey and Barriers

Mapping the passenger journey has been a vital step in the development of the programme. By mapping all the steps in the journey, it helped MPI identify the barriers that passengers faced to comply. With these barriers now known, ways to relive those barriers in the form of interventions could be developed, piloted, and implemented. It also identified which potential touchpoints in the journey were owned by MPI that required partnerships and which were customer or externally owned (and harder to influence).

10.27 Insight

The insights delivered by the quantitative and ethnographic research were key to making the programme a success. The insights ensured a thorough understanding into the audiences' cultural nuances, barriers to compliance, and in turn motivators for change. By understanding the reasons for the audiences' current behaviour, they were able to work more effectively with interventions and messages to change to the desired behaviour.

10.28 Partnerships

Partnerships have been critical for the delivery of the various interventions through the passenger journey. By identifying which journey touchpoints were classified as "partner owned" (Lemon and Verhoef 2016) MPI was able to develop the required relationships with those that they required assistance from in using their channels or expertise. Those interventions that are delivered via partnerships have generally been the most successful, in particular those with Immigration New Zealand and airlines. It has also ensured that the programme's channel mix is relevant, targeted, varied, and cost-effective.

10.29 Social Norms

Outside of the passenger journey, that MPI's main intervention mix focusses on the social norms created by friends and family within New Zealand cannot be understated. These have a major influence on the audiences' awareness, intentions, and resulting behaviours. Both Chinese and Indian segments rated word of mouth from friends and family as the biggest reason for considering biosecurity when packing bags pre-departure. While placing great importance on the passenger journey, MPI has learnt not to discard their local audience, and implement measures based on social norms and community-based social marketing approaches. This is why one of the key next steps for MPI is to begin working collaboratively with local communities. It is aiming to implement a programme of co-designing tools, messaging and possible interventions to help New Zealand residents deliver the message of biosecurity even more effectively to friends and family back offshore.

10.30 Exchange

MPI understanding the cost of compliance behaviour versus the benefit of doing so has assisted greatly in framing relevant messages to the target audiences. However, MPI has mainly focussed on one key piece of exchange, the desire to do the right thing and avoid punishment. It needs to investigate if other benefits may offer an even stronger exchange for the audiences. This could include faster passage through airports on arrival, or the feel-good factor of protecting New Zealand's environment.

10.31 Use of Theory in the Programme

The programme places a high value on the use of theory. In doing so, the programme has had a clear reference point against which to measure interventions. By basing the majority of the programme on the Theory of Planned Behaviour, MPI knew that interventions either had to meet the three key factors of either changing attitudes, influencing a social norm, or enhancing behavioural control. These would then contribute to intention and lead to behaviour change. If an intervention was not meeting one of these factors, then a lesser focus would be placed on it. The NSMC's benchmarking criterion is also crucial as a sense check for the programme's overall strategy. By striving to meet the criterion set by the NSMC (and Andreasen 2002 as well), it ensured the programme had clear behavioural goals, carefully considered segmentation and insights, and appropriately delivered interventions. This gave MPI confidence that the programme would not just raise awareness and knowledge but deliver meaningful and measurable behaviour change.

10.32 Notes

1. More information about biosecurity and what is allowed into New Zealand can be found at www.biosecurity.govt.nz/declare.
2. Information in Hindi for Indian passengers can be found at https://www.biosecurity.govt.nz/travel-and-recreation/arriving-in-new-zealand/items-to-declare/items-from-india/hindi/.
3. And Punjabi at https://www.biosecurity.govt.nz/travel-and-recreation/arriving-in-new-zealand/items-to-declare/items-from-india/punjabi/.
4. Chinese biosecurity information can be found at www.nzquarantine.com.
5. Colmar Brunton is an independent research company contracted by Ministry for Primary Industries to carry out research and analysis on its behalf.
6. For Figs. 1 and 2, the "effectiveness" of interventions is based on percentages of those who answered 8–10 (from a scale of 1–10).

References

Ajzen I (1991) The theory of planned behavior. Organ Behav Hum Decis Process 50:179–211

Azjen I (2011) The theory of planned behaviour: reactions and reflections. Psychology & Health 26(9):1113–1127

Anderson K (2009) Ethnographic research: a key to strategy. Harv Bus Rev. https://hbr.org/2009/03/ethnographic-research-a-key-to-strategy. Accessed 25 September 2018

Andreasen A (2002) Marketing social marketing in the social change marketplace. J Public Policy Mark 21:3–13

Berkowitz A (2004) The social norms approach: theory, research and annotated bibliography (August 2004). http://alanberkowitz.com/articles/social_norms.pdf
deLeeuw A, Valois P, Azjen I, Schmidt P (2015) Using the theory of planned behaviour to identify key beliefs underlying pro-environmental behavior in high-school students: Implications for educational interventions. J Environ Psychol 42:128–138
French J, Russell-Bennett R (2015) A hierarchical model of social marketing. J Soc Mark 5(2):139–159
Hopwood T, Merritt R (2011) Big pocket guide to using social marketing for behaviour change. The NSMC, London. http://www.thensmc.com/sites/default/files/Big_pocket_guide_2011.pdf
Horticulture New Zealand (2017). HortNZ annual report. http://www.hortnz.co.nz/assets/HortNZ AnnualReport17.pdf. Accessed 25 September 2018
Kautonen T, van Gelderen M, Fink M (2015) Robustness of the theory of planned behavior in predicting entrepreneurial intentions and actions. Entrep Theory Pract 39(3):655–674
Kotler P, Zaltman G (1971) Social marketing: an approach to planned social change. J Market 35(3):3–12
Lemon K, Verhoef P (2016) Understanding the customer experience throughout the customer journey. J Market 69–96. AMA/MSI Special Issue
Maibach E, Louiselle Parrot R (1995) Designing health messages. SAGE Publications Ltd, California
McKenzie-Mohr D (2011). Fostering sustainable behavior—an introduction to community-based social marketing, 3rd edn. http://www.cbsm.com/pages/guide/preface/. Accessed 28 November 2018
Michie S, West R, Campbell R, Brown J, Gainforth H (2014) ABC of behaviour change theories. Silverback Publishing, Great Britain
NZ Department of Internal Affairs (2011) Achieving compliance—a guide for compliance agencies in New Zealand. https://www.dia.govt.nz/diawebsite.nsf/Files/Achieving%20Compliance%20-%20A%20Guide%20for%20Compliance%20Agencies%20in%20New%20Zealand/$file/AchievingComplianceGuide_17July2011.doc. Accessed 10 September 2018
Peattie S, Peattie K (2003) Ready to fly solo? reducing social marketing's dependence on commercial marketing theory. Market Theory 3(3):365–385
Richardson A (2010) Using customer journey maps to improve customer experience. Harv Bus Rev. https://hbr.org/2010/11/using-customer-journey-maps-to. Accessed 11 April 2019
Rosnbaum M, Losada Otalora M, Contreras Ramirez G (2016) How to create a realistic customer journey map. Bus Horiz 60:143–150
RNZ (2016) Fourteen flies cost $15.7 m to eradicate. https://www.radionz.co.nz/news/national/300805/fourteen-flies-cost-$15-point-7m-to-eradicate. Accessed 25 September 2018
Schuster L, Kubacki K, Rundle-Thiele S (2016) Community-based social marketing: effects on social norms. J Soc Market 6(2):193–210
Sniehotta F, Presseau J, Araũjo-Soares (2014) Time to retire the theory of planned behaviour. Health Psychol Rev 8(1):1–17
Stats NZ (2018) International Visitor Arrivals to New Zealand: December 2017. https://www.stats.govt.nz/information-releases/international-visitor-arrivals-to-new-zealand-December-2017. Accessed 25 September 2018
Stats NZ (2019) International Visitor Arrivals to New Zealand: December 2018. https://www.stats.govt.nz/information-releases/international-travel-december-2018. Accessed April 2019
Tourism New Zealand (2016) About the industry. https://www.tourismnewzealand.com/about/about-the-industry/. Accessed 15 February 2018
Vinnikainen K (2017) The role of Andreasen's benchmark framework in social marketing interventions. Bachelor of Science in Economics and Business Administration, Espoo, Finland

Phill Sherring is a marketing practitioner, with a career spanning 20 years across private and government sectors. The Ministry for Primary Industries is the government agency tasked with protecting New Zealand from biosecurity threats to the country's horticulture and agriculture industries, and environment. Phill manages programmes influencing the behaviour of international visitors to not bring items that could carry biosecurity risk, and mobilizing New Zealanders to look out for pests and diseases in their communities. Phill has delivered award-winning campaigns in both banking and government sectors, and has a Bachelor of Business Information, specialising in Marketing and Communication Management.

Chapter 11
Technology Acceptance, Social Marketing and the Design of a Mobile Health App to Support Active Ageing Amongst Senior Citizens in the Asia-Pacific Region

Shailey Chawla, Jagdeep Kaur Sabharwal, Breda McCarthy, and Rene Erhardt

Abstract This chapter presents the findings from a usability study on health apps on older members of the population in the Asia-Pacific region. Life expectancy continues to increase around the world, and this has long-term implications for society in terms of managing the health care needs of an ageing population. Mobile health apps are increasingly seen as a mechanism to help improve the health, independence and quality of life of older individuals. However, health professionals highlight that the natural process of ageing has, in general, an adverse impact on people's physical and cognitive functions; hence, closer interactions between health care and technology researchers are necessary to ensure that health apps correctly address the needs of this target population. This study shows that several design aspects of health apps in the marketplace require modification to produce robust, reliable, meaningful and usable technologies. The research draws on the concept of usability and the technology acceptance model to explain the factors that predict the successful adoption of health apps amongst senior members of the population in Singapore and Australia.

Keywords Health apps · Ageing population · Quality of life · Usability · Technology acceptance model · Technology adoption

---

S. Chawla (✉) · J. K. Sabharwal
James Cook University, Singapore, Singapore
e-mail: Shailey.chawla@jcu.edu.au

B. McCarthy
James Cook University, Townsville, Australia

R. Erhardt
Deakin University, Melbourne, Australia

R. Hay et al. (eds.), *Broadening Cultural Horizons in Social Marketing*,
https://doi.org/10.1007/978-981-15-8517-3_11

11.1 Introduction

Information and communication technologies (ICT) have evolved and developed rapidly in the last few years, allowing ease of access to users. Mobile technology, such as smartphones, now has a broad penetration and acceptation within society. It gives user groups, 50+ and older, more opportunities to monitor their health and live an autonomous life (Årsand et al. 2008; Blake 2008; Lorenz and Oppermann 2009). Mobile technology, such as health applications ("apps"), has become an important tool for information transfer. It may form part of a behavioural change strategy developed to nudge people to be proactive and make better choices. It may support patient care and be used as a mobile "self-help" tool for people with health conditions, which is useful particularly where treatments depend on individual change such as modifying one's diet and lifestyle, taking steps to stop smoking and reduce the risk of cancer, obesity and diabetes (McCurdie et al. 2012; Gilliland et al. 2015). Currently, there are more than 165,000 mHealth apps, including free and paid, publicly available to download and use. These apps hold enormous potential to enhance access to quality health care at reduced and affordable cost. However, most apps have not yet lived up to their potential. This is due to multiple barriers, including lack of regulation, privacy and security concerns, and limited support in evidence-based literature for the efficacy and clinical utility of health apps (Kao and Liebovitz 2017).

It is argued that groups of people aged 50+ show greater diversity in their cognitive, sensory and motor skills than younger people (Lorenz and Oppermann 2009). Therefore, it is imperative that such health apps are created taking into account the user needs using adaptive technology and incorporating design features that meet the needs of older consumers (García-Peñalvo et al. 2014). Despite the growing availability of health apps on the market, academic research on the development and evaluation of such apps is in relatively early stages (Dennison et al. 2013). There appears to be little research on older age groups, their experiences with and views on health apps. Given this context, the objectives of this study are to examine the usability of health apps from an older person's perspective. This study seeks to describe people's experiences with health apps and views on features that might assist them and support a healthy lifestyle. It also examines how usage of health apps might vary when cognitive and demographic characteristics are taken into account.

The research reported in this section is broken down into three sections: firstly, it outlines the ageing of the population and implications for society in relation to health care management. Secondly, it examines the concept of usability and outlines the technology acceptance model and thirdly, it examines the influences of new technologies, such as mobile health apps, on decision-making around health, including health monitoring, food, eating habits and exercise.

11.2 Literature Review

11.2.1 The Ageing of the Population and Health Implications for Society

The rapid ageing of the world's population is arguably one of the important demographic shifts facing the developed world (United Nations Population Fund 2012). Between 2015 and 2050, it is predicted that the proportion of the world's population over 60 years will nearly double from 12 to 22% (WHO 2018). Certain forces are responsible for this global ageing: sharp increase in life expectancy due to medical advances, better nutrition, sanitation, success in dealing with fatal childhood disease, declining fertility rates and fall in maternity mortality rates (WHO 2017). A longer life brings with it the opportunities, but the extent of these opportunities and contributions to society depends on health. Ageing can be defined as the decline and deterioration of functional properties at the cellular, tissue and organ level. This loss of functional properties yields a loss of homeostasis and decreased adaptability to internal and external stress, yielding an increased vulnerability to disease and mortality (Holliday 1997).

Ageing is also associated with change in social roles and life transitions, such as retirement, relocation to safer or more appropriate housing and the death of friends and partners. All countries are facing major challenges to ensure that their health and social systems are ready to make the most of this demographic shift (WHO 2018). The definition of an "older person" varies depending on culture and there is great diversity in terms of ageing; for instance, some 70-year-old may enjoy very good physical and mental capacity, others may be frail and require significant support to meet their basic needs (WHO 2018). Terms used to describe older people are mature-age consumers, retirees, elderly people or senior citizens. There is no general agreement on the age, at which a person becomes "old", yet pension schemes mostly use age 60 or 65 years for eligibility.

The UN Sustainable Development Goals (2014) make it clear that the right to a healthy life is applicable to all ages (Goal 3: Good Health and Well-Being—Ensure healthy lives and promote well-being for all at all ages) (UN 2014). According to the WHO, healthy ageing is a key goal for the World Health Assembly, and it is defined by the *World Report on Aging and Health* as "the process of developing and maintaining the functional ability that enables well-being in older age" (WHO 2017). Fostering healthy ageing depends on many factors, including lifestyle changes and the fostering of innovation (WHO 2017).

In the context of mobile technology, the most important user group is arguably the 50+ cohort (Lorenz and Oppermann 2009), since technology may help foster healthy habits. It is argued that nutrition and lifestyle choices in younger years influence health in older age. Lifestyle changes can ameliorate health declines such as eating a balanced diet, engaging in regular physical activity and refraining from tobacco use (WHO 2018). Furthermore, health conditions can be effectively managed if detected and treated early (WHO 2017). Globally, 32 million people died in 2016 due to cardiovascular disease, cancer, diabetes or chronic respiratory disease. The probability of dying from these causes was about 18% in 2016 for people between 30 and 70 years of age (United Nations 2018). In older age groups, the biggest underlying risk factor for chronic disease is high blood pressure (UNFPA 2012); other key determinants are smoking and high blood glucose levels. Heavy drinking, insufficient nutrition, obesity, unhealthy waist-hip ratio, hypertension and low physical activity are other risk factors among males and females aged 50 or over (UNFPA 2012). It is argued that type 2 diabetes is one of the greatest challenges to human health, and the risk factors include genes, early life events (such as famine), lifestyle and behaviour, with almost two-thirds of the Australian population in 2000 being overweight or obese (Zimmet 2017). Hence, the promotion of knowledge on the prevention and control of disease is critical to healthy ageing. The next section examines mobile technology and its implications for public health promotion and healthy ageing.

11.2.2 Impact of Mobile Technology and Health Apps on Healthy Living and Healthy Ageing, Limitations and Risks

Mobile technology can open new possibilities for older people to acquire health-related information, monitor their health, improve their independence and enhance quality of life (Plaza et al. 2011; Abascal and Civit 2001; Gao and Koronious 2010). Visual and hearing impairment, dementia and osteoarthritis are the main causes of disability in older persons. Mental health and wellbeing are core health issues at all ages, but depression, social isolation and loneliness are likely problems in older age. Injuries, particularly those caused by falls, are frequent events for older persons, and falls could be prevented through a number of interventions (UNFPA 2012). Technology may help ensure more safety in the home and perhaps alleviate the sense of isolation experienced by some older persons.

Health apps are playing a role in educating people about the role of diet and exercise in maintaining a healthy lifestyle. In 2016, the global market for health-related self-monitoring technologies was valued at approximately USD16.7 billion and is predicted to reach around USD72 billion by 2022 (Reportlinker 2017). The use of health and fitness apps increased by 330% between 2015 and 2018 (Medical Director 2018). In addition to the increasing array of wearable technology, mobile applications and electronic food diaries allow consumers to more closely monitor what they eat. Advances in technology may facilitate greater consumer involvement and digital wellness through self-monitoring of dietary intake and calorie tracking (Ho et al. 2017). Self-monitoring involves the use of "electronic devices and software technologies to collect, process and display a wide range of personal data", which allows consumers to monitor their health, including food consumption and physical activity (Reportlinker 2017). The main benefit of health apps is convenience (Liu et al. 2019). Consumers can search for information at any time, at any place, even when they are on the go and they do not necessarily have to travel to an outlet or sit in front of a computer (Verhoef et al. 2015). However, consumers can only use apps to search for information on websites in situations with network coverage (mobile apps will only work if the internet is available).

Regarding mobile apps repositories, Google Android and Apple Apps Store have thousands of health apps. There are numerous apps that promote healthier food decisions (e.g., "FoodSwitch") (Flaherty et al. 2018; Sucharitha and Muruganantham 2017). It is reported that apps are useful and novel; they help users capture real-time food intake and reduce participant burden and time taken to report eating habits over long periods, and hence people's satisfaction tends to be higher for use of smartphone platforms than for conventional platforms (Pendergast et al. 2017). Health apps do exist in the main mobile app repositories that are related to the needs of older people, such as quit smoking and diabetic recipes, but they are not specifically designed for older people (García-Peñalvo et al. 2014). There has been an increase in apps for specific medical purposes such as type II diabetes, high blood pressure and other medical conditions (Cho et al. 2014). There is a growing number of studies on specialised apps that are designed for patients, and assessments have been made on their effectiveness as a tool in health care interventions, such as whether they promote positive behavioural changes towards healthy eating (Gilliland et al. 2015).

While health apps have emerged as a popular category in apps overall, they have their limitations. There has been a rapid proliferation of health information on the internet and some concerns have been raised about the quality of the information, the risk of patients being misinformed, the tendency towards self-diagnosis or self-treatment and its impact on people's interactions with their physicians (Tan and Goonawardene 2017). However, new technologies in the health sector, such as mobile health apps, wearables and sensors, are promising. Further research is needed to validate their use and long-term impact and to identify possible risks, from physiological harm, to breaching of privacy and confidentiality with insecure devices (Ho et al. 2017), as well as the quality of the health information delivered to people.

11.2.3 Usability and the Technology Acceptance Model

With the increase in the ageing population worldwide, m-health apps are increasingly seen as essential for elderly users of health services (Guo et al. 2015; Iwaya et al. 2013). However, very few m-health apps have involved usability testing during their development in older adults (Hong et al. 2014; Grindrod et al. 2014; Pak et al. 2009). Scholars have used the technology acceptance model (TAM) as a theoretical lens, focusing on the factors driving the adoption of health apps, since many researchers consider m-health as a technology (Guo et al. 2013). Two key components of the model are "perceived ease of use" and "perceived usefulness", which are treated as two enablers, or factors that encourage adoption, of new technologies (Davis et al. 1989; Venkatesh et al. 2003). Venkatesh and Davis (2000) define perceived ease of use as the degree to which a person believes that he/she will be able to use a given technology in a convenient manner without much effort. Perceived usefulness is defined as the extent to which a person believes that the use of technology will help improve his/her performance (Venkatesh and Davis 2000).

Along with TAM, usability models have been used to explain technology acceptance (Abran et al. 2003). However, very few m-health apps involve usability testing in older adults, apart from a few notable exceptions (Hong et al. 2014; Grindrod et al. 2014; Pak et al. 2009). The concept of usability is well-established in the literature on software development. Usability is seen as a basic determinant of product acceptance. Two definitions of usability, offered by the International Standards Organisation (ISO), are as follows: "the extent to which a product can be used by specified users to achieve specified goals with effectiveness, efficiency and satisfaction in a specified context of use" (ISO 9241-11 2001); and "the capability of the software product to be understood, learned, used and attractive to the user, when used under specified conditions" (ISO/IEC 9126-1 2000). The most common dimensions that determine usability are effectiveness, efficiency, satisfaction and learnability; some scholars have added security to the framework, which refers to the ability of software to prevent unauthorised access, which is particularly important in an industrial context (Abran et al. 2003). Rubin and Chisnell (2008) describe a usability test plan and propose that it should be comprehensive, detailed and should document exactly what the researcher is going to do, how the test will be conducted, what metrics will be captured, the number of participants to be tested and the kinds of scenarios that will be used.

Since mobile health apps are a new entrant into IT, usability frameworks are particularly relevant to this study. Chan et al. (2015) provide the following guidelines and framework to evaluate m-health apps, which cover three key dimensions, usefulness, usability and integration and infrastructure dimensions. The guidelines, based on a literature review, are comprehensive.

(1) Usefulness dimension:

   a. *Validity and accuracy*: Does the app work as advertised?
   b. *Reliability*: Will the app consistently function from session to session?
   c. *Effectiveness*: Is the app clinically effective—with demonstrated improved outcomes—for the target population, disease, or disability?
   d. *Time and number of sessions*: What time is required for the user to derive some benefit from the app?

(2) Usability dimension:

   a. *Satisfaction and reward*: Is the app pleasurable and enjoyable to use, or does it discourage repeat use?
   b. *Usability*: Can the user easily—or with minimal training—use and understand the app?
   c. *Disability accessibility*: Is the app usable by those with disabilities (e.g., incorporates screen readers for blind users, closed captions for the hard-of-hearing and deaf communities)?

d. *Cultural accessibility*: Does the app work effectively with the user's culture (as defined by factors such as ethnicity and language)?
e. *Socioeconomic and generational accessibility*: Does the app take into account socioeconomic status and the user's age, with potential implications for the user's digital health literacy?

(3) Integration and infrastructure dimension:

a. *Security*: Are the app's data encrypted on the device and/or in transmission? Are they anonymised, or do they contain personal health information? If so, what do they do?
b. *Workflow integration*: Does the app work within its user's workflow?
c. *Data integration*: Does the app share data with other apps, networks and medical record systems?
d. *Safety*: Does the app take into account patient safety, such as suicidality or homicidality?
e. *Privacy*: Does the app contain a robust privacy policy addressing the type of information collected, rationale for collecting information, sharing of information and user controls?

It is beyond the scope of this study to incorporate all of the above dimensions in this usability study on m-health usage by older age groups, but the first two dimensions are particularly relevant to this study.

11.3 Research Aim and Research Questions

The aim of this study is to examine the usability of m-health apps for older adult users. This also includes an analysis of key factors, such as cognitive efficiency and demographic variables that are likely to affect usability, along with incorporating user perspectives into the study. Hence, this study identifies important variables that may act as enablers or inhibitors to the acceptance of mobile health apps by older age groups. This study is, to our knowledge, the first to incorporate cognitive testing into a usability study on older adults and the first to conduct such testing with a cross-cultural sample. This study is important since there are research gaps in the existing literature. According to Guo et al. (2015: 1), "m-health is a new emerging phenomenon; empirical studies on this issue are still inadequate".

In order to study the factors influencing seniors' acceptance of m-health apps, we developed an integrated model (Fig. 11.1) and drew on a popular usability framework (Fig. 11.2).

Figure 11.2 was drawn from the work of Abran et al. (2003) on usability testing. We included variables such as cognitive ability and demographic variables such as gender, age and country of residence. The framework also draws on two key components (perceived ease of use and perceived usefulness) of TAM. The following literature review explains and supports the framework presented in Fig. 11.2.

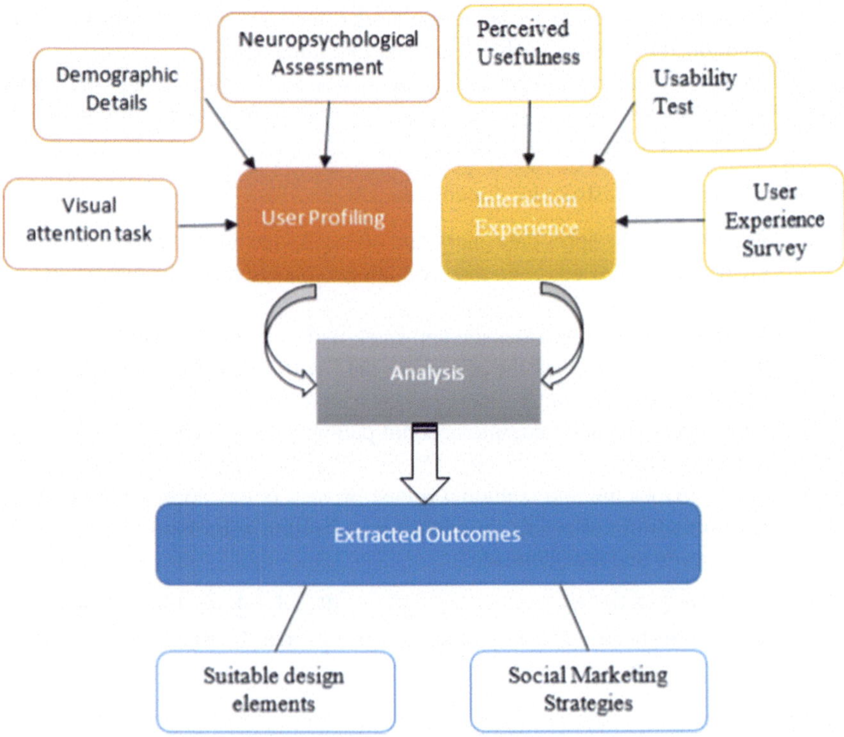

Fig. 11.1 Framework for usability study of m-health apps for seniors

11.4 Cognitive Efficiency and Usability of m-Health Apps

Ageing is associated with biological changes such as a gradual decline in physical and mental capacity and a growing risk of chronic disease (WHO 2018). As people age, their decision-making around physical activity also changes (Reuter et al. 2010). When a person ages, his or her mental and/or physical health conditions deteriorate so this may affect usability of mobile health apps and inhibit adoption. Studies drawing on TAM (Bhattacherjee and Hikmet 2007; Guo et al. 2013) show that any effort to learn new technology can increase resistance to adoption.

Earlier studies on mobile phone technology suggest that older people are likely to face multiple challenges in using technology due to ageing. Hence, designers need to incorporate "age-friendly" features into devices to aid memory, vision, auditory and haptic functions (García-Peñalvo et al. 2014; Urdaibay-Villaseca 2010). Usability in mobile phones for older people tends to be linked to their requirements, such as memory aids (i.e., appointments with doctors, reminders); visual aids (i.e., ability to increase the size of text); haptic aids (i.e., a rubber grip making the phone easier to hold); features to minimise error (i.e., keypad auto-lock) and safety features such as a dedicated panic button (García-Peñalvo et al. 2014). An earlier study by Kurniawan

Usability of mHealth apps for seniors

Effectiveness	Efficiency	System Usability	Satisfaction	Learnability
Percentage of tasks accomplished Usability tasks a) Key in any two food items that you had in your last meal (Breakfast/Lunch). b) Set up your goals: Starting weight, current weight, weekly goal and activity level.	Repetitions / number of failed attempts. Time spent on a single task Percentage of errors Number of cues needed	System Usability Scale (Brooke, 1996) Perceived ease of use. Perceived usefulness	Number of favourable comments Net Promoter Score: number of promoters / people willing to recommend the app to friends. Valuable feature recalled by the users Recommendations for improvement	Time to learn: taking the psychological context of the user into consideration Test of cognitive and physical functions (Mini-Cog and Trail Making Test A and B)

Fig. 11.2 Usability test dimensions. *Source* Adapted from Abran et al. (2003)

(2008) on mobile phone usage recommended adaptation to suit older user's special requirements. These included menus and navigation, personal safety features and age-friendly features (i.e., larger buttons or device size, larger text and stronger backlights for persons with vision problems). Although an old study, current studies on m-health app design reveal similar design issues, such as poor navigation, small button size and inadequate data visualisations (Morey et al. 2019). Other studies have found that the app's interface design needs to be adjusted to suit the needs of older adults (Grindrod et al. 2014; Lee et al. 2014).

The first research question is as follows: does cognitive decline or inefficiency (which can be objectively assessed by cognitive and psychomotor tests) lead to poor efficiency on app navigation and hence decreased ease-of-use for older age groups.

11.4.1 Demographics and Usability of m-Health Apps

Some studies regard gender and age as important indicators of usability and these factors are discussed below. Furthermore, this study attempts to address the call by Cho et al. (2014: 867) for a greater focus on demographics to explain disparities in health app use.

11.4.2 Gender

Although there is growing concern about gender stereotypes, there are some studies that regard gender as an important indicator of technology acceptance (Chung et al. 2010; Guo et al. 2015). In the field of evolutionary psychology, it is posited that as women and men are different biologically, they think and behave differently (Saad and Gill 2000). One study found that males feel they are better at using technology than females (Bain and Rice 2006) and another study posited that males are more task-oriented and are more inclined to exert effort to overcome difficulties to pursue their goals than women (Morris and Venkatesh 2000). One study found that males enjoy a higher level of m-health adoption intention compared to females, thus gender differences were found in m-health adoption (Zhang et al. 2014). Hence, incorporating gender into a study on mobile health apps make sense.

11.4.3 Age

Incorporating age into a study on mobile health apps make sense given that scholars tend to view ageing as a negative factor in technology acceptance studies. A study by Guo et al. (2013) identified two typical characteristics of elderly users, which tend to act as barriers in the adoption of mobile health services: technology anxiety and dispositional resistance to change. Guo et al. (2013: 49) explain why these factors are two critical barriers for older people:

> First, it is always difficult and costly for the elderly to learn how to operate new devices. They are worried about the negative consequences induced by their wrong operations and thus avoid using new devices. Secondly, they would like to keep to their routines and do not like any activity that can change their lifestyle. Although they know they could expect benefits from using mobile health care services, changes in their life scare them.

Other scholars posit that the elderly are often slower than younger people, are in technological progress and many of them have a sceptical attitude towards new technologies (Harrison and Rainer 1992). On the other hand, when considering health threats, age differences may be manifested in different attitudes. Since older people have a higher possibility of getting ill, an aged person generally pays more attention to health-related issues, may have positive attitudes towards health apps and be more inclined to accept apps (Guo et al. 2015). Therefore, older adults may be motivated to complete tasks efficiently on apps. A US report from PEW research also addressed notable differences in the use of m-health apps among different demographic groups based on differences in age, sex, educational level and annual household income (Fox and Duggan 2012).

Hence, the second research question is as follows: to what extent do demographic factors (i.e., gender and age) predict both cognitive efficiency as well as efficiency on app navigation?

11.4.4 Attitudes of Users Towards m-Health Apps

In addition, this study seeks to capture people's attitudes towards health apps, their perspectives on perceived usefulness and ease of use, and whether the individual would, or would not, recommend it to others. Mobile apps for health exist in large numbers today, but there is limited research on user perspectives of health apps (Peng et al. 2016), particularly the views of elderly consumers. Consequently, this study sought to incorporate the voice of the older persons into the study. According to the Theory of Planned Behaviour (Ajzen 1991), attitudes refer to personal feelings towards adopting a behaviour and attitudes influence behavioural intentions. Hence, favourable attitudes towards health apps are likely to facilitate the adoption of health apps and negative attitudes are likely to be an inhibitor—although the intention-behaviour gap is well recognised in the health literature (Reuter et al. 2010).

Thus, the third research question is as follows: what are older people's attitudes towards a mainstream m-health app, and would they recommend it to others?

11.4.5 Cultural Background and Context

We focus on two countries, Australia and Singapore, for a variety of reasons. A cross-cultural sample may provide some insights into international differences in usability, how users perform on cognitive tests and their perception of m-health apps. The promotion of digital technology in general, and the promotion of e-health services to the older age groups, is particularly important in Singapore. The Singaporean government places a strong emphasis on active ageing and are actively training seniors to empower them in using digital devices and thus becoming more self-reliant (IMDA 2019). Not surprisingly, governments and the health service in both Singapore and Australia recognise the need to embrace digitalisation. Singapore is well-regarded as a triple-A rated economy and Singapore's "Smart Nation" Strategy is designed to harness digital technologies to drive economic growth (Smart Nation 2018). Likewise, the Australian government recognises that digital technologies will transform the economy (Austrade 2016). Understanding how contextual forces affect e-health usage will provide guidance on improving the user experience with m-Health apps across two different markets. Recommendations will be made to assist m-health service providers, the government and aid the creation of social marketing campaigns.

11.5 Materials and Methods

This study analysed the experiences of older age groups with m-health apps through a series of tests and a survey administered in a face-to-face format. The Human Research Ethics Committee of James Cook University approved this study. The study was advertised via posters at community centres that attract the over 60s, food and hawker centres, exhibitions and community events aimed at seniors, and other organisations that attract the over 60 in demographics. We believe that this sampling procedure is appropriate given the characteristics of older age groups who are not easily accessible. Information about the study was shared with participants before they gave their consent to participate in the study. Only those participants who could read and speak English, and who were familiar with smartphones, were selected to participate in the study. Once the participants agreed to participate, they were given a consent form to sign. An incentive, a gift voucher, was utilised to encourage participation. The data was collected over the period August 2019 to September 2019, in Singapore and Australia.

The participants were requested to give their demographic details (age, gender, education profession) since the literature review indicates that demographic factors contribute to technology acceptance. This study is limited to a mainstream and popular m-health app, MyFitnessPal app, for a couple of reasons. Firstly, it is available in both Singapore and Australia, which facilitates cross-comparison of data. Secondly, it is a mainstream app, designed to help users maintain a record of their diet and weight, and it also offers articles and blogs on healthy eating and exercise. Hence, the participants were presented with an opportunity to elaborate on their day-to-day experiences, without restricting them to disease-specific apps or old-age specific apps.

11.5.1 Cognitive Tests and Usability Testing

The participants performed paper-and-pen tasks for assessment of their neuropsychological efficiency. Neuropsychological tests are useful for assessing behaviour and cognitive changes resulting from normal ageing (Douglas et al. 2012). The first cognitive test used in the study was Mini-Cog (Borson et al. 2000), which is a validated 3-min instrument that can increase detection of cognitive impairment in older adults. It can be used effectively in community settings. It consists of two components, a 3-item recall test for memory, and a simply scored clock-drawing test. The participants also completed two trail-making tests (A and B) (Reitan and Wolfson 1985). The trail-making test (TMT) is similar to "connect-the-dots" in which individuals connect a set of 25 dots as quickly as possible. TMT-A requires an individual to draw lines sequentially connecting 25 encircled numbers distributed on a sheet of paper. Task requirements are similar for TMT-B except the person must alternate between numbers and letters (e.g., 1, A, 2, B, 3, C, etc.). The score on each part represents the amount of time required to complete the task. TMT A and B make good use of executive functioning such as memory, going back and forth while doing the tasks, planning and execution (Tombaugh 2004).

After completion of the Mini-Cog and TMT tasks A and B, the user was directed to the chosen m-health application on a phone or tablet where he/she was requested to complete two tasks requiring app navigation. The user session clicks and navigation path was recorded to evaluate the usability of the app. The two tasks were as follows:

(a) Keying in any two food items consumed in the last meal (Breakfast/Lunch).
(b) Setting up of goals: Starting weight, current weight, weekly goal and activity level.

For each task, the following data were collected:

(a) Task completion—success/failure/abandon.
(b) Number of cues given.
(c) Time to task (in seconds), meaning the time taken for completion of the task.

11.5.2 *Attitudes and Perspectives of Users*

We examined seniors' perceptions of health apps in three different ways:

- A system usability scale (SUS), based on a Likert scale, where 1 = strongly agree and 5 = strongly disagree; respondents are asked ten questions and the answers indicate if they think the app is easy to use. This is a standard usability scale for software applications (Brooke 1996).
- The net promotor score (NPS), where respondents are simply asked to indicate, on scale from 1 to 10, how likely they are to recommend the app to other people. The NPS sorts customers into three groups, promoters, passive or detractors, based on their response to a single question. The NPS is simply the percentage of promoters minus the percentage of detractors and is used as tool to learn about the customer experience and improve that experience (Reichheld 2003).
- A short user experience survey. The survey included the following open-ended questions: What is the most valuable feature in the app? What are the three things you would like to improve in the app? What impact, if any, will the app have in improving the wellbeing or quality of life of the users? The latter was designed to measure perceived usefulness and was drawn from Bhattacherjee and Hikmet (2007).

11.6 Sample

A sample of 15 adults from Singapore and Australia aged 60 and above was recruited for the study. The average age of participants was 65.33 years (s.d. = 9.7). A total of eight males and seven females participated in the study. Out of the total sample, seven participants were from Australia and eight participants were from Singapore.

11.7 Data Analysis

The data was collected from 15 participants; however, one participant's data was removed due to the participant's inability to complete any of the usability tasks and TMT-B. The 14 participants were able to complete MiniCog, TMT-A, TMT-B and task 1 but only 13 participants could complete task 2. The data was analysed to examine differences due to gender, country of residence and age. All of qualitative or textual data was entered into an excel sheet and analysed manually. We identified the most frequently used phrases or keywords, and delineated a range of responses for each open-ended question on people's overall experience with the health app.

The quantitative analysis was done using the software package IBM SPSS (version 25). Prior to the analysis, the data was checked for assumption violations. The assumption of homogeneity of variance was satisfied; however, the assumption of normality was violated for two variables. Thus, the data was analysed using non-parametric statistical method. Mann-Whitney U test was used to examine the differences of group characteristics and outcomes, and Spearman's Rho was used to examine the relationships.

11.8 Results

11.8.1 Analysis by Gender

The Mann-Whitney U test was used to evaluate the gender differences for the different tasks and activities. Results showed that males took longer (Mdn = 75 s) to complete TMT A than females (Mdn = 47 s). A similar trend was noticed for TMT-B, with males taking a longer time (Mdn = 114 s) than females (Mdn = 93 s).

For task 1, females took a shorter time (Mdn = 140 s) than males (Mdn = 150 s). For task 2 as well, females were faster to navigate the app (Mdn = 128 s) as compared to males (Mdn = 180 s).

Despite the differences in median scores for males and females, there were no significant gender differences noticed for the trail-making tasks and usability tasks.

11.8.2 Analysis by Country of Residence

The Mann-Whitney U test revealed significant differences in performance on TMT-A for seniors from Australia and Singapore. Data analysis revealed that the time taken to complete TMT-A was shorter for Australian seniors (Mdn = 39 s) than for Singaporean seniors (Mdn = 94 s), $U = 5.00$, $p = 0.013$, $r = 0.66$. Though Australians were faster on TMT-B (Mdn = 93 s) compared to Singaporean seniors (Mdn = 130 s), the difference was not statistically significant.

Singapore Seniors were faster to navigate task 1 (Mdn = 120 s) and task 2 (Mdn = 127 s) as compared to Australian seniors' performance on task 1 (Mdn = 169 s) and task 2 (Mdn = 230 s). However, the differences were not statistically significant.

Despite performing efficiently on tasks 1 and 2, Singaporean seniors' Net Promoter Score was lower (Mdn = 5) than that of Australian seniors (Mdn = 7). The same result was observed for the System Usability Scale score, where Singaporeans rated their usability score lower (Mdn = 40) as compared to Australians (Mdn = 65).

There was no difference by nationality for the Mini-Cog scores of Singapore and Australian seniors.

11.8.3 Age and Time Taken for the Usability Task on the m-Health App

A Spearman's rank-order correlation was run to determine the relationship between the participants' age (n = 14) and time taken to complete the navigation task, as well as TMT-A and B. The results showed that there was a moderate, positive correlation between age and time to task 1, which was statistically significant (r_s (14) = 0.56, p = 0.037).

Statistically significant, moderate and positive correlation was also observed between age and time to task 2 (r_s (13) = 0.68, p = 0.011). As age increased, time taken to complete usability tasks 1 and 2 also increased.

11.8.4 Attitudes and Users' Perspectives

Perceived ease of use was addressed by the system usability scale (SUS). The average score on the system usability scale was 50.8 (Mdn = 50). Tullis and Stetson as cited in Zaphiris and Ioannou (2016), state that SUS score above 68 would be considered above average and anything below 68 is below average. The mean score on the Net Promoter Score (NPS) was four (Mdn = 4), which puts responses in the detractor category. Anything below six is considered to be in the detractor category and above nine is considered to be promoters (Reichheld 2003).

One open-ended question from the User Experience Survey addressed perceived usefulness, such as its contribution to quality of life. A total of four respondents responded with a definite "yes" and outlined the benefits of the app, such as helping people maintain a healthy lifestyle; helping with nutrition, exercise, weight control and keeping a diary (n = 4). Other respondents (n = 6) outlined benefits, such as good for raising awareness of daily activities and what's happening to the body; effective for offering tips and reminders; a useful tool for visualisation of health information,

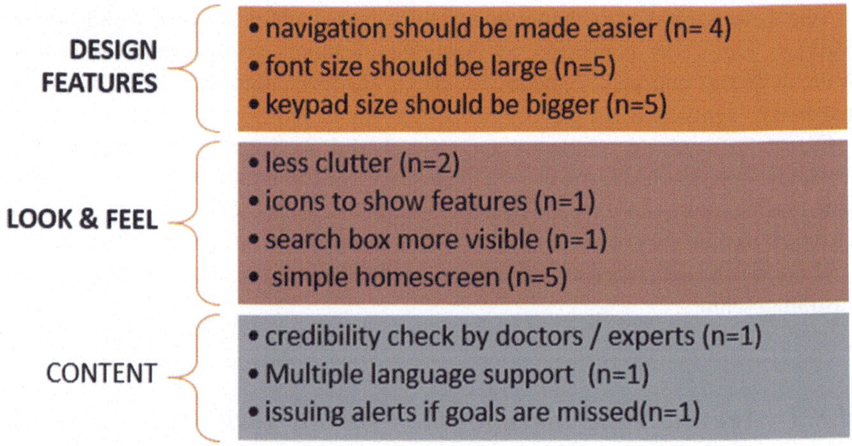

Fig. 11.3 Recommendations for improvement by participants

health monitoring and creating interest in health management, and the convenience of a mobile phone was mentioned.

Five people did not agree or were unsure about the usefulness of the app. One respondent stated that it depended on level of technical awareness. One respondent cited no motivation to use the app and another stated a preference for self-management. One person stated language was a major concern, along with poor navigation and restricted use of phones for calls only. Another respondent was highly critical of health apps and the quality of health information from government sources in general, stating that it was up to the individuals and their will power to manage their personal health.

When asked about the most valuable feature of the app, respondents commented primarily on the information and articles ($n = 6$), as well as ability to measure weight and monitor calories and diet ($n = 3$), and the visual presentation of current status and progress ($n = 1$).

In relation to improvements that could be made to the app, participants' feedback (in order of frequency) is shown in Fig. 11.3.

11.9 Discussion

This is the first study, to our knowledge, that tests whether cognitive efficiency, along with demographic variables, is related to m-health app navigation (i.e., usability) in a cross-cultural sample of seniors. Furthermore, qualitative data is gathered to evaluate seniors' perceptions of a health app after undergoing a trial of the app's basic features.

Results suggest that overall, the participants performed the cognitive tests efficiently (apart from one respondent who was dropped from the study due to suspected cognitive impairment) and they had few problems with using the apps. The average time taken to complete TMT task A and TMT task B was well within the acceptable limit of 90 and 180 s, respectively.

The proposition (research question 1) that efficiency on cognitive tests was associated with higher proficiency in app navigation was not confirmed. Despite Australian participants outperforming Singaporean participants in TMT A and B by taking a shorter time, Singaporean seniors were more economical in their time taken to navigate the apps. This leaves us with a conclusion that perhaps efficiency on the cognitive and psychomotor task does not predict efficiency on app navigation. To interpret this finding, we need to consider government policy and contextual factors in Singapore. The Singaporean government places a strong emphasis on active ageing, the promotion of e-health services and demonstrates a commitment to training seniors on how to use technology devices (IMDA 2019); hence, the participants from Singapore may have a greater capacity to use technology and may benefit from greater exposure to technology, which is perhaps the "game changer" in this case.

Furthermore, despite performing efficiently on task 1 and task 2, the Net Promoter Score from Singaporean seniors was lower (Mdn = 5) than that of Australian seniors (Mdn = 7). Likewise, the System Usability Scare score was lower for Singaporeans (Mdn = 45) as compared with Australians (Mdn = 65). To interpret this finding, we need to consider that Singaporean seniors come from a population known to be technologically savvy, are familiar with technology and this perhaps raised their expectations of an app, and hence that is why they gave it a lower rating than Australian users. Concisely, familiarity breeds contempt. Australians, who are perhaps less technologically savvy or novel users, place more value on the health apps, since they have no yardstick with which to judge m-health apps. The implication of this finding is that designers of m-health apps need to continually improve their apps when dealing with sophisticated users or with people who are accustomed to using digital technologies and familiar with a breadth and depth of digital media.

The practical implication of this research for governments, health service providers and social marketers is that Australia can learn from Singapore. In order to improve Australian seniors' efficiency on app navigation, education and training are needed to encourage seniors to use digital technology in general and m-health apps in particular. Recommendations are to provide free internet (WIFI) in public areas where senior members of the population are likely to gather (i.e., libraries, community centres, food courts and shopping centres), run technology-training programmes or foster the development of a virtual community, where elderly members provide tips and guidance to their peers.

Despite some studies which suggest a link between gender and technology acceptance (Chung et al. 2010; Guo et al. 2015) and between gender and intentions to use m-health apps (Chang et al. 2013), we found no evidence that gender is an indicator of usability or cognitive efficiency. In relation to the cognitive tests, results showed that males took longer to complete TMT A and B than females and the results were not significant. With regard to the usability tests, for tasks 1 and 2, females took a shorter time than males, and there were no significant gender differences noticed for the usability tasks. The practical implication of this research is that governments, designers of social marketing campaigns and creators of m-health apps do not need to segment the market according to gender.

Age was related to efficiency on app navigation. The results showed that there was a moderate, positive correlation between age and time to tasks 1 and 2. As age increased, time taken to complete usability tasks 1 and 2 also increased. This was not surprising given that studies suggest that ageing is associated with biological changes such as a gradual decline in physical and mental capacity (WHO 2018; Reuter et al. 2010). As noted in the literature review, elderly users have particular characteristics, which tend to act as barriers in the adoption of mobile health services, notably technology anxiety and dispositional resistance to change (Guo et al. 2013). Several studies on usability highlight that age-appropriate features need to be incorporated into technology, including mobile phones (García-Peñalvo et al. 2014; Kurniawan 2008; Urdaibay-Villaseca 2010) and m-health apps (Morey et al. 2019; Grindrod et al. 2014; Lee et al. 2014). In addition, the qualitative data revealed areas for improvement. A few of the participants highlighted design elements, such as font size and keyboard size, that needed to be improved in order to make the app easier to use, suggesting that long sightedness or manual dexterity might have been an issue.

The practical implication of this research is that the creators of health apps need to ensure that health apps are easier to use if they wish to target people who are fall into the older age categories. Health app designers need to turn their attention to design and incorporate "age-friendly" features into m-health apps. It is recognised that, when developing mobile programs for older adults, it is important to involve end users from the early stage of design and conduct on-going usability testing (Hong et al. 2014; Grindrod et al. 2014; Lorenz and Oppermann 2009; Pak et al. 2009).

The qualitative data revealed both positive and negative attitudes towards m-health apps. In relation to its "perceived usefulness", the survey revealed that some respondents (n = 5) were either not satisfied with the m-health app or unsure about its value. One respondent stated that it depended on level of technical awareness. One respondent cited no motivation to use the app; another stated a preference for self-management. One person stated language was a major concern, along with poor navigation and restricted use of phones for calls only; another respondent was highly critical of health apps and the quality of health information from government sources in general, stating that it was up to the individuals and their will power to manage their personal health. Only four respondents (26.6% of the sample) responded very positively to the m-health app, and they outlined the benefits of the app, such as helping people maintain a healthy lifestyle; helping with nutrition, exercise, weight control and keeping a diary (n = 4). Other respondents (n = 6) also outlined benefits, such as good for raising awareness of daily activities and what's happening to the body; effective for offering tips and reminders; a useful tool for visualisation of health information, health monitoring and creating interest in health management, and the convenience of a mobile phone was mentioned. These results are not surprising. The main functional strength of health apps is their distribution of quality medical information, which originates from highly credible sources and is often customised for the users (Cho et al. 2014). Taken from the perspectives of seniors, we can infer that most of them will not use health apps, without considerable encouragement. Furthermore, the mean score on the Net Promoter Score (NPS) was 4 (Mdn = 4), which puts responses in the detractor category. The low score given to usability overall, combined with some lack of support for the app in the open-ended questions, suggests weak intentions to adopt the app.

The practical implication of this study is that governments must work hard to encourage seniors to use m-health apps, if they wish to promote m-health apps as a tool for healthy and active ageing. The government and m-health providers could use social marketing and persuasive communications to dispel the attitude that m-health apps offer no real value to people in their daily lives. For example, marketers could first impress upon seniors that m-health apps are easy to use and that they help with the monitoring of diet and exercise and add value to people's lives. After this assurance, marketers could demonstrate how the m-health apps actually work. Primary care doctors, nutritionists and alternative health practitioners may be another important communications source and they could recommend m-health apps to seniors, who may be clients. This addresses the needs of users who might have some concerns about the credibility of the information. As a society ages, preventive medicine becomes increasingly important. Much research in the area of preventive medicine focuses on diet and exercise, and addressing healthy behaviours, not just in elderly populations but also in young-adulthood and middle age in order to prevent the onset of disease (WHO and Consultation 2003). In the long run, an improvement in the general health of the population will benefit individuals and society, if it results in a more efficient allocation of scarce health care resources.

11.10 Limitations

We close by pointing out several limitations. Firstly, the sample size was small; the study focused on a special population (e.g., seniors and elderly users) and one particular m-health app (e.g., MyFitnessPal). This limits the generalizability of our findings to other populations or to other types of m-health apps. Secondly, we used a free app, which is a popular app that appeals to young populations, but it may not be attractive to seniors, potentially reducing the "perceived usefulness" score of the app. Finally, simple cognitive tests were used. Hence, future research should include a larger sample size and more advanced cognitive tests, which may help to advance theory linking cognitive efficiency and demographics to usability.

11.11 Conclusions

Although m-health apps have become popular in the contemporary world, empirical studies on their usability, especially as to elderly users, are rarely conducted. This study contributes to contemporary research on m-health apps and helps address a gap in the literature by examining whether cognitive efficiency on psychological tests predicts usability. Furthermore, we also analyse the role of demographic factors (gender and age) in influencing usability and explore senior users' perceptions of health apps in a cross-cultural sample. Testing usability and acceptability is a very important step in developing age-appropriate and user-friendly m-health apps, especially for older adults, and our study represents an effort to incorporate the voice of older adults into the rapidly growing field of mobile health.

Acknowledgements This study was supported by a grant from James Cook University.

References

Abascal J, Civit A (2001) Mobile communication for older people: new opportunities for autonomous life. In: Proceedings of EC/NSF workshop on universal accessibility of ubiquitous computing: providing for the elderly, vol 487

Abran A, Khelifi A, Suryn W, Seffah A (2003) Usability meanings and interpretations in ISO standards. Softw Qual J 11(4):325–338

Ajzen I (1991) The theory of planned behavior. Organ Behav Hum Decis Process 50(2):179–211

Årsand E, Tufano JT, Ralston JD, Hjortdahl P (2008) Designing mobile dietary management support technologies for people with diabetes. J Telemed Telecare 14(7):329–332

Austrade—Australian Trade and Investment Commission (2016) Investment opportunities in digital technologies in Australia. file:///C:/Users/jc203156/Downloads/Investment-opportunities-in-Australian-Digital-technologies.pdf. Accessed 29 Sept 2019

Bain CD, Rice ML (2006) The influence of gender on attitudes, perceptions, and uses of technology. J Res Technol Educ 39(2):119–132

Bhattacherjee A, Hikmet N (2007) Physicians' resistance toward healthcare information technology: a theoretical model and empirical test. Eur J Inf Syst 16(6):725–737

Blake H (2008) Innovation in practice: mobile phone technology in patient care. Br J Community Nurs 13(4):160–165

Borson S, Scanlan J, Brush M, Vitaliano P, Dokmak A (2000) The mini-cog: a cognitive 'vital signs' measure for dementia screening in multi-lingual elderly. Int J Geriatr Psychiatry 15(11):1021–1027

Brooke J (1996) SUS: A "quick and dirty" usability scale. In: Jordan PW, Thomas B, Weerdmeester BA, McClelland AL (eds) Usability evaluation in industry. Taylor and Francis, London

Chan S, Torous J, Hinton L, Yellowlees P (2015) Towards a framework for evaluating mobile mental health apps. Telemed e-Health 21(12):1038–1041

Cho J, Quinlan MM, Park D, Noh GY (2014) Determinants of adoption of smartphone health apps among college students. Am J Health Behav 38(6):860–870

Chung JE, Park N, Wang H, Fulk J, McLaughlin M (2010) Age differences in perceptions of online community participation among non-users: an extension of the Technology Acceptance Model. Comput Hum Behav 26(6):1674–1684

Davis FD, Bagozzi RP, Warshaw PR (1989) User acceptance of computer technology: a comparison of two theoretical models. Manag Sci 35(8):982–1003

Dennison L, Morrison L, Conway G, Yardley L (2013) Opportunities and challenges for smartphone applications in supporting health behavior change: a qualitative study. J Med Internet Res 15(4):e86

Douglas A, Letts L, Eva K, Richardson J (2012) Use of the cognitive performance test for identifying deficits in hospitalized older adults. Rehabil Res Pract 2012:1–9

Flaherty SJ, McCarthy M, Collins A, McAuliffe F (2018) Can existing mobile apps support healthier food purchasing behaviour? Content analysis of nutrition content, behaviour change theory and user quality integration. Public Health Nutr 21(2):288–298

Fox S, Duggan M (2012) Mobile health 2012. PEW Research Center, DC, Washington, DC

Gao J, Koronios A (2010, July) Mobile application development for senior citizens. In PACIS, p 65

García-Peñalvo FJ, Conde MÁ, Matellán-Olivera V (2014) Mobile apps for older users—the development of a mobile apps repository for older people. In: International conference on learning and collaboration technologies. Springer, Cham, pp 117–126

Gilliland J, Sadler R, Clark A, O'Connor C, Milczarek M, Doherty S (2015) Using a smartphone application to promote healthy dietary behaviours and local food consumption. BioMed Res Int 2015

Grindrod KA, Li M, Gates A (2014) Evaluating user perceptions of mobile medication management applications with older adults: a usability study. JMIR mHealth uHealth 2(1):e11

Guo X, Sun Y, Wang N, Peng Z, Yan Z (2013) The dark side of elderly acceptance of preventive mobile health services in China. Electron Mark 23(1):49–61

Guo X, Han X, Zhang X, Dang Y, Chen C (2015) Investigating m-health acceptance from a protection motivation theory perspective: gender and age differences. Telemed e-Health 21(8):661–669

Harrison AW, Rainer RK Jr (1992) The influence of individual differences on skill in end-user computing. J Manag Inf Syst 9(1):93–111

Ho K, Yao C, Lauscher HN (2017) Health apps, wearables, and sensors: the advancing frontier of digital health. B C Med J 59(10)

Holliday R (1997) Understanding ageing. Philos Trans R Soc Lond Se B: Biol Sci 352(1363):1793–1797

Hong Y, Goldberg D, Dahlke DV, Ory MG, Cargill JS, Coughlin R, Hernandez E, Kellstedt DK, Peres SC (2014) Testing usability and acceptability of a web application to promote physical activity (iCanFit) among older adults. JMIR Hum Factors 1(1):e2

Infocomm Media Development Authority—IMDA (2019) Digital clinics. https://www2.imda.gov.sg/programme-listing/digital-clinics. Accessed 9 Oct 2019

ISO 9241 (2001) Ergonomics requirements for office with visual display terminals (VDTs). International Organization for Standardization, Geneva

ISO, IEC 9126 (2000) Software product evaluation—quality characteristics and guidelines for the user. International Organization for Standardization, Geneva

Iwaya LH, Gomes MA, Simplício MA, Carvalho TCMB, Dominicini CK, Sakuragui RR, Håkansson P (2013) Mobile health in emerging countries: a survey of research initiatives in Brazil. Int J Med Inform 82(5):283–298

Kao CK, Liebovitz DM (2017) Consumer mobile health apps: current state, barriers, and future directions. PM&R 9(5):S106–S115

Kurniawan S (2008) Older people and mobile phones: a multi-method investigation. Int J Hum Comput Stud 66(12):889–901

Lee JA, Nguyen AL, Berg J, Amin A, Bachman M, Guo Y, Evangelista L (2014) Attitudes and preferences on the use of mobile health technology and health games for self-management: interviews with older adults on anticoagulation therapy. JMIR mHealth and uHealth 2(3):e32

Liu H, Lobschat L, Verhoef PC, Zhao H (2019) App adoption: the effect on purchasing of customers who have used a mobile website previously. J Interact Mark 47:16–34

Lorenz A, Oppermann R (2009) Mobile health monitoring for the elderly: designing for diversity. Pervasive Mob Comput 5(5):478–495

McCurdie T, Taneva S, Casselman M, Yeung M, McDaniel C, Ho W, Cafazzo J (2012) mHealth consumer apps: the case for user-centered design. Biomed Instrum Technol 46(s2):49–56

Medical Director (2018) How wearables will revolutionise health care. https://www.medicaldirector.com/news/future-of-health/2018/06/how-wearables-will-revolutionise-healthcare. Accessed 25 Aug 2019

Morey SA, Stuck RE, Chong AW, Barg-Walkow LH, Mitzner TL, Rogers WA (2019) Mobile health apps: improving usability for older adult users. Ergon Des. 1064804619840731

Morris MG, Venkatesh V (2000) Why don't men ever stop to ask for directions? Gender, social influence, and their role in technology acceptance and usage behavior. MIS Q 24(1):115–139
Pak R, Price MM, Thatcher J (2009) Age-sensitive design of online health information: comparative usability study. J Med Internet Res 11(4):e45
Pendergast F, Ridgers N, Worsley A, McNaughton S (2017) Evaluation of a smartphone food diary application using objectively measured energy expenditure. Int J Behav Nutr Phys Activity 14(30). https://doi.org/10.1186/s12966-017-0488-9
Peng W, Kanthawala S, Yuan S, Hussain SA (2016) A qualitative study of user perceptions of mobile health apps. BMC Public Health 16(1):1158
Plaza I, MartíN L, Martin S, Medrano C (2011) Mobile applications in an aging society: status and trends. J Syst Softw 84(11):1977–1988
Reichheld FF (2003) The one number you need to grow. Harvard Bus Rev 81(12):46–55
Reitan RM, Wolfson D (1985) The Halstead-Reitan neuropsycholgical test battery: therapy and clinical interpretation. Neuropsychological Press, Tucson, AZ
Reportlinker (2017) Health self-monitoring: technologies and global markets to reach $71.9 billion by 2022. https://www.prnewswire.com/news-releases/health-self-monitoring-technologies-and-global-markets–to-reach-719-billion-by-2022-300479733.html
Reuter T, Ziegelmann JP, Wiedemann AU, Lippke S, Schüz B, Aiken LS (2010) Planning bridges the intention–behaviour gap: age makes a difference and strategy use explains why. Psychol Health 25(7):873–887
Rubin J, Chisnell D (2008) Handbook of usability testing: how to plan, design, and conduct effective tests, 2nd edn. Wiley Publications, Indianapolis, IN
Saad G, Gill T (2000) Applications of evolutionary psychology in marketing. Psychol Mark 17(12):1005–1034
Smart Nation (2018) Transforming Singapore. https://www.smartnation.sg/why-Smart-Nation/transforming-singapore. Accessed 29 Sept 2019
Sucharitha ST, Muruganantham J (2017) FoodSwitch: can a smart phone app helpconsumer switch to healthier food choices? A pilot study. Natl J Res Community Med 6(2):155–160
Tan SL, Goonawardene N (2017) Internet health information seeking and the patient-physician relationship: a systematic review. J Med Internet Res 19(1):e9
Tombaugh TN (2004) Trail making Test A and B: normative data stratified by age and education. Archiv Clin Neuropsychol 19(2):203–214
Tullis TS, Stetson JN (2004) A comparison of questionnaires for assessing website usability. In: Usability professional association conference, vol 1
United Nations (2014) Sustainable development goals. https://www.un.org/sustainabledevelopment/sustainable-development-goals/
United Nations (2018) The sustainable development goals report 2018. https://unstats.un.org/sdgs/report/2018/overview/
United Nations Population Fund (UNFPA) (2012) Ageing the twenty-first century: a celebration and a challenge. https://www.unfpa.org/sites/default/files/pub-pdf/Ageing%20report.pdf
Urdaibay-Villaseca PT (2010) Usability of mobile devices for elderly people. Department of Information Technology, vol Master of Science in Software and Information Systems. National University of Ireland, Galway, Ireland
Usability (n.d) System usability scale (SUS). https://www.usability.gov/how-to-and-tools/methods/system-usability-scale.html. Accessed 8 Oct 2019
Venkatesh V, Davis FD (2000) A theoretical extension of the technology acceptance model: four longitudinal field studies. Manag Sci 46:186–204
Venkatesh V, Morris MG, Davis GB, Davis FD (2003) User acceptance of information technology: toward a unified view. MIS Q 425–478
Verhoef PC, Kannan PK, Inman JJ (2015) From multichannel retailing to omni-channel retailing: introduction to the special issue on multi-channel retailing. J Retail 91(2):174–181
WHO (2017) Global strategy and action plan on ageing and health. https://www.who.int/ageing/global-strategy/en/ and https://www.who.int/ageing/WHO-GSAP-2017.pdf

WHO (2018) Aging and health. https://www.who.int/news-room/fact-sheets/detail/ageing-and-health

Who J, Consultation FE (2003) Diet, nutrition and the prevention of chronic diseases. World Health Organ Tech Rep Ser 916(i–viii)

Zaphiris P, Ioannou A (eds) (2016) Learning and collaboration technologies: third international conference, LCT 2016, held as part of HCI International 2016, Toronto, ON, Canada, July 17–22, 2016, proceedings, vol 9753. Springer

Zhang X, Guo X, Lai KH, Guo F, Li C (2014) Understanding gender differences in m-health adoption: a modified theory of reasoned action model. Telemed e-Health 20(1):39–46

Zimmet PZ (2017) Diabetes and its drivers: the largest epidemic in human history? Clin Diabetes Endocrinol 3(1):1

Shailey Chawla is an academic in Information Technology at James Cook University Singapore. Her research interests include human–computer interaction, software engineering and business process management. Research and practice focus is on the development of technological solutions for social good.

Jagdeep Kaur holds a Ph.D. in Social Psychology. She is presently working as a Lecturer in Psychology at James Cook University, Singapore. As well as teaching undergraduate students and supervising postgraduate research projects, Jagdeep's research focuses on the social, relational and cultural influences on well-being; stress management and mindfulness; psychosocial gerontology and intergroup processes.

Dr. McCarthy is a Senior Lecturer of Marketing at James Cook University (JCU). She has over 80 publications, including articles spanning A, B and C rated journals that are ranked by the Australian Business Dean's Council (ABDC). Her research interests include consumer behaviour, sustainable food consumption, food waste and networks. She has published papers on consumer decision making around food purchasing, shopping and food waste behaviours in journals such as the *Australasian Marketing Journal, Rural Society and the British Food Journal.*

Chapter 12
Effectiveness of Public Advertisements to Influence Perceptions Towards Public Transport Among Young Adults in Singapore

Abhishek Bhati

Abstract Singapore, with a strong motive to go green, is focused on ways to improve social perceptions towards the use of public transport (PT) in order to become a "car light" community. Thus, the Land Transport Authority (LTA) is encouraging the general public to use PT. This research paper aims to explore the effectiveness of public information advertisements by the LTA to influence perceptions of young adults towards the use of public transport. The LTA is encouraging the general public to use PT by introducing advertisements near bus stops and train stations that highlight initiatives taken to create more park connectors, cycling paths, and covered walkways to make use of PT more convenient to access. This helps to promote the use of pedestrian pathways and the convenience of PT by showing the ease of MRT and bus connectivity. It is noteworthy that the young adults between 18 and 35 years are the most frequent user of PT and trendsetters in Singapore. Thus, this target segment was selected for the study. The primary aim of this study was to use the foundations of social marketing theories to explore the reach of LTA's advertisements. Three theories on community readiness model, social cognitive theory, and protection motivation theory provide the framework to analyse the current LTA adverts. A subsequent survey assessed the readiness stage of the target population and to identify ways to make the message more relevant. The paper concludes highlighting the ineffectiveness of the advert to reach targeted audience and proposes a set of recommendations and actions to enhance communication reach and to improve outcomes of LTA's social marketing efforts.

Keywords Public advertising · Singapore · Public transport · Perception

---

A. Bhati (✉)
James Cook University, Townsville, Australia
e-mail: Abhishek.bhati@jcu.edu.au

R. Hay et al. (eds.), *Broadening Cultural Horizons in Social Marketing*,
https://doi.org/10.1007/978-981-15-8517-3_12

12.1 Introduction

As a small island nation, Singapore is pressed for space and the need to curb traffic congestion and other environmental impacts from the use of vehicle and its emissions. Climate change fear and carbon emissions are critical issues for governments worldwide (Peattie and Peattie 2008). With a strong motivation to go green, the state is focusing on enhancing the use of public transport (PT) whilst discouraging private transport ownership. The statutory body for the land transportation developments in Singapore is the Land Transport Authority (LTA). The LTA is responsible for designing, planning, developing, and sustaining the nation's transport systems and infrastructure. As indicated on their website, LTA aims to provide "*a greener and more inclusive public transport system, complemented by convenient options to walk and cycle from their homes or to their destinations*" (LTA 2018). Hence, it encourages the use of PT with the intention of becoming a "car light" community through its marketing efforts.

Dentsu (nd) is a marketing service agency that handles all types of marketing process. This includes consulting, planning, implementation, and execution. Dentsu aims to become a trusted marketing partner to its clients, creating the best solutions and the most unforgettable experience for the people (Dentsu nd). Dentsu was chosen as the marketing partner for the Land Transport Authority (LTA) and the National Environment Authority (NEA), after collaborating for two years on the Graciousness campaign and the Clean and Green Singapore campaign.

The "Graciousness on PT" campaign by LTA encourages thoughtfulness and consideration for others while using the public transport through colourful cartoon "thoughtfulness characters" (LTA 2018). However, the "Clean and Green Singapore" campaign by the NEA inspires to maintain clean green environment and common spaces by caring, protecting, and leading a sustainable lifestyle (NEA 2018). Hence, Dentsu provides strategic communication planning, using different media channels to provide public awareness and promote living an environmentally responsible and sustainable life (Manjur 2016). It is currently responsible for creating the family of advertisements that encourage PT use by means of displaying public information on the connectivity of Mass Rapid Transport (MRT) and covered pedestrian walkways. Thus, Dentsu undertakes certain social marketing aspects for LTA and NEA.

Social marketing has the prospect of creating better sustainable communities and societies using commercial marketing, thus bringing about essential behaviour change in society (Peattie and Peattie 2008). As per Andreasen (2002), behaviour change is one of the key characteristics of social marketing and the effectiveness or success of social marketing efforts depends on achieving the proposed behaviour change. Social marketing just like commercial marketing is a strategic managerial and social function (Kotler et al. 2005), which is often misunderstood and reduced to some its key activities which are components of marketing such as selling and advertising. Thus, advertising and promotion are communication activities within the marketing function, which are used to inform and influence targeted audience for the exchange (Brassington and Pettitt 2003).

In social marketing, the root cause of conducting these marketing communication campaigns is the same; that is, to change people's behaviour. Advertising is only one of the components of social marketing despite being one of the most important (Thackery et al. 2007). The right advertisement helps to reach the targeted audience faster because advertisements are attractive compared to any other medium of sharing. Advertisements can be customized according to the target audience. It can have a greater impact on the thinking of the audience and a long-lasting impact on the behaviour change of the audience. Thus, to solve growing health issues, less availability of land, and convenient active mobility, the Singapore government started the Walk Cycle Ride campaign with the vision of developing walking and cycling as possible methods of green mobility to lessen the country's dependence on cars. This includes enhancing infrastructure so that people can focus on opting for other means of transport regardless of ability or age.

To make Singapore healthier and greener, Singapore's LTA is taking the initiative for building and publicizing more skywalks for pedestrians, park connectors, and walking and cycling paths. The publicity part involves multiple advertisements in print and on social media. The effectiveness of the advertisements needs to be analysed to understand how the message reaches the public. Hence, the aim of the research was to identify whether the specific family of advertisements by LTA is effective in reaching out to the young adults between 18 and 35 years of age in Singapore. It involved applying the concepts and theories of social marketing to judge the thoughts of the people and the way they interpret the advertisements. It also involved determining the reach of the advertisements.

The factors influencing an individual to accept the idea or thought suggested in LTA's public information advertisement are evaluated. This helps to answer questions relating to whether people see the advertisements, whether the idea behind the adverts is effective, and to check whether the advertisements are understood by Singapore society and followed.

12.2 Literature Review

Earlier approaches to influencing public behaviour were mostly from an educational context (Lefebvre and Flora 1988). On the contrary, social marketing utilizes concepts and tools of commercial marketing to achieve social goals (Peattie and Peattie 2003). Furthermore, there exists a lot of confusion around the term social marketing often generalized with advertising (Almestahiri et al. 2017) and the lack of full utilization of the features of social marketing. As per Andreasen (2002), social marketing has six common unique features such as consumer research, behaviour change, target and segmenting, marketing mix, exchange, and competition faced. Almestahiri et al. (2017) go on to propose another unique feature to the six features, namely "theory".

Behavioural theories provide the framework for developing and assessing social marketing (Almestahiri et al. 2017). Therefore, the seven features formulate the major components of social marketing. Social marketing is apt when promoting a social cause or change in behaviour, and most social marketing activities are undertaken for the purpose of bringing about behavioural change (Almestahiri et al. 2017). It is applicable to different kinds of social issues. Some of the benefits of social marketing include emphasis on customer orientation, maintaining the desired change, flexibility in being applicable to different stakeholders, creating new partnership opportunities, and open opportunities to create awareness and discourage unsustainable behaviours (Peattie and Peattie 2008). As in the case of commercial marketing, social marketing also needs to consider target market, competition, and marketing communication to achieve the benefits.

12.3 Understanding the Social Marketing Mix

In social marketing, the marketer is aware of the core idea to be marketed and must create tangible products and services to make the target interested in the product/services. This indirectly attracts the target interested in the cause. The difficulty lies in creating appropriate products and motivating individuals to buy the product (Kotler and Zaltman 1971). According to Wiebe (1951), the challenge in social marketing is in promotion, especially in terms that are broader than specific marketing terms. Furthermore, according to Wiebe's (1951) theory, the more resemblances of a social marketing campaign to a product campaign, the more successful it will be. Place is equal to adequacy, compatibility, and direction in the social factors mentioned above. After a social message is conveyed, for the most part, people do not know how to respond or act, as there is no clear "buy". There is a need for appropriate action channels. Therefore, place needs to provide adequate outlets to translate motivation into action (Kotler and Zaltman 1971). Pricing of social marketing is based on the consumer performing a cost/benefit analysis before investing money, time, and resources. Therefore, the focus should be on how the reward of the product can be increased in comparison to the cost. It is important to find a mix of product, promotion, place, and price that will increase rewards and control costs.

The convenience and gratification of the solution should be clearly presented to the buyers (Kotler and Zaltman 1971). Thus, social marketing mix also involves manipulating the marketing mix by applying the traditional four "Ps" by tailoring it as needed against the present behaviour and advocated to be changed (Peattie and Peattie 2008). Although the naive manner of applying the commercial marketing mix to social causes can create problems and debates, it highlights the need for a better and new social marketing approach to achieve behaviour change (Peattie and Peattie 2008). Moreover, in most cases, the cost of achieving the behaviour change may be social not economical and the fact that the traditional might not be apt in the social marketing context (Peattie and Peattie 2008). As discussed earlier, behaviour change is one of the unique features of social marketing (Andreasen 2002); hence, the effectiveness of any social marketing campaign depends on the achievement of the needed behaviour change (Almestahiri et al. 2017).

12.4 The Social Learning Models

Behavioural theories provide the structures from which social marketing could be developed and assessed (Almestahiri et al. 2017). However, there is a debate around the role of theory in social marketing with few authors advocating for the inclusion of theory as a key characteristic of social marketing (French and Blair-Stevens 2006; Almestahiri et al. 2017). Bandura (2002) suggests that apart from direct and experiential learning, learning also happens by observing models who convey cognitive skills, values, and knowledge and behaviour styles. Some of the theories that provide structure or frameworks to develop social marketing are discussed below.

Community readiness model: Transforming the community to bring about behavioural changes is a complex process involving different stakeholders, and community readiness can contribute to planning and implementing (Behar and Hydaker 2009). Community readiness marketing can be used as a framework to identify the different stages of marketing strategy at different stages of a client and change agent relationship. As a community-based process, it classifies people on their ability to handle the issue. It is successful by appropriate use of segmentation techniques that can target the marketer to the most suitable community-based strategies (Kelly et al. 2003).

According to Edwards et al. (2000), communities can be at various stages of readiness to implement programmes, and the challenge is that they are groups and not individuals. Assessing community readiness levels helps identifying the readiness level, the problem awareness, readiness for change among people, understanding the link between strategies and psychology of the people, and it has been used as a background framework to choose theories (Kelly et al. 2003). The stages of community readiness can be classified as lower, intermediate, and advanced (Edwards et al. 2000). At lower stage, the customer or target has no awareness, a denial, resistance, or a vague awareness of the situation or social message (Edwards et al. 2000; Kelly et al. 2003).

The awareness stage introduces the product; the interesting part focuses on making the buyer research the product/service. In social marketing, the desire stage involves a shift from liking to want; it is loosely linked to motivation and taking action (Edwards et al. 2000; Kelly et al. 2003). At the intermediate stage, people who are aware of the message are at the preplanning and preparation stage. A person evaluates the cost benefits and considers action as it is important to focus on the appropriate channels (Edward et al. 2000; Kelly et al. 2003). At the advance stage, the person/buyer has confirmed the cause and stabilized on the idea (Edwards et al. 2000; Kelly et al. 2003). There is a high level of community ownership at this stage. It is better for social marketers that people are at this stage for better acceptance of social messages.

Singapore is expected to be at the intermediate stage as the advertisements displayed are seen to focus on the cost/benefits model, facilities, and advantages of using PT instead of focusing on the awareness, reason, and need for public transport. At the pre-planning/preparation stage, people understand that there needs to be action regarding the social message/product. There is recognition of the problem and agreement that something needs to be done (Donnermeyer et al. 1997). Assessing community readiness is done by interviewing the community who has seen the advertisements.

Social cognitive theory: Behaviour and attitudes are mostly ruled by cognitions (Shaw 1979). This theory suggests that behavioural, personal, and environmental factors are reciprocal. The environment includes society, family, and the structural environment. Based on this, observational learning is used at this stage. That is, observing the models can influence an individual behaviour. Behavioural change is encouraged at this stage by boosting self-confidence, feedback from others, or self-efficacy. The combination of behavioural changes and positive outcomes is effective in social marketing interventions (Eagle et al. 2013). Yet, it is essential to note that not all observed behaviours are followed, and social cognitive theory suggests attention, retention, production, and motivation are four cognitive processes overseeing observational learning (Moyer-Guse 2008).

Motivation is actually influenced by self-efficacy and outcome expectancies or the observer's perception of behavioural consequences (Moyer-Guse 2008). Theory of perceived self-efficacy, which is a part of the theory of planned behaviour, plays a vital role in the decision-making of the person or customer towards the social message/product. If a person feels they need to improve self-efficacy, they work towards a goal. The need for self-efficacy comes from observing others. It serves as a dynamic motivational belief system that varies from task to task (Stajkovic and Luthans 2002). Therefore, if the advertisement creates a sense or need for self-efficacy, people tend to follow the advertisement. This can be done by showing examples of how people are successful by implementing the particular social message. Looking at examples that work well encourages the majority to follow.

Psychosocial model: Social marketing is successful only if it catches the eye of the consumer; hence, it is important to understand the consumer's psychological behaviour. Theories of behavioural change play a crucial role in developing, implementing, and evaluating social marketing strategies (Eagle et al. 2013). The messages in public information advertisements can be derived, using a psychosocial theory such as protection motivation theory, one of many theories of behavioural change, to make them more effective. Protection motivation theory provides a framework to design and assess persuasive communications (Moyer-Guse 2008). Hence, the current study focused on identifying whether the message that represents the components of the protection motivation theory could be effective in influencing people to use PT (Cathcart and Glendon 2016).

Protection motivation theory (PMT): Protection motivation theory (PMT) maps the impact of concerns about how people will react to a specific threat and select responses in order to cope with the consequences pertaining to those threats (Tanner et al. 1989). PMT, as shown in Fig. 12.1, explains that the variables proposed within this model (i.e., vulnerability, severity, efficacy, and costs) impact the convincing measures, though the ways in which they influence the consumer are unclear as they depend on the adaptive and maladaptive behaviours to the response (Cismaru 2006). These variables act as persuasion measures and are considered by consumers

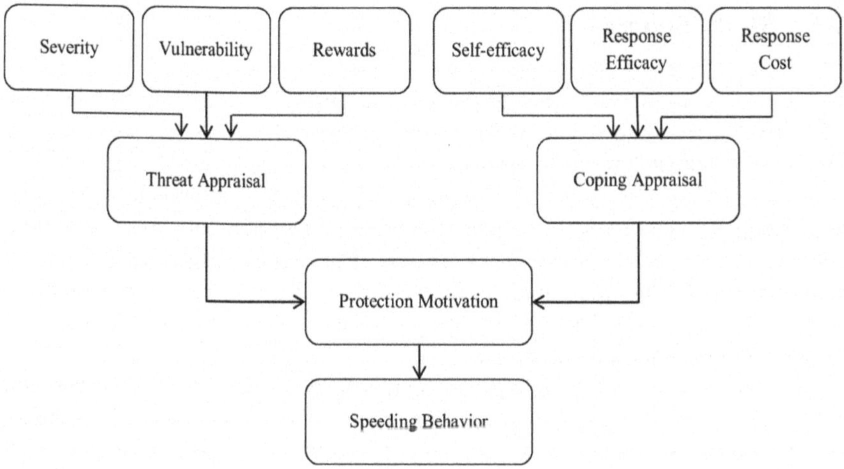

Fig. 12.1 Protection motivation theory. *Source* Cathcart and Glendon (2016)

in the decision-making process using different sources of information, namely, environmental and interpersonal. Environmental sources of information include verbal persuasion and observational learning, and interpersonal sources of information include personality variables and prior experience. These sources of information help to decide how decisions are based on appraisals of both a threat and the potential effects of coping strategies (Cismaru 2006).

In Fig. 12.1, the threat appraisal consists of the threat severity, perceived vulnerability to the threat, and the perceived rewards with respect to maladaptive response (Cathcart and Glendon 2016). The main process of threat appraisal is to evaluate the severity and vulnerability of a problem by assessing the personal impact of using a private vehicle rather than public transport. PMT suggests that when people are exposed to the threat of using a private vehicle, they experience the cognitive process of threat appraisal (Zhao 2016). The coping appraisal consists of perceived self-efficacy to adapt behaviour, response efficacy in controlling or minimizing the threat, and the cost of the adaptive response, which would reduce the possibility of adopting that response (Cathcart and Glendon 2016). In a coping appraisal, self-efficacy is the major component that defines the degree to which people are effectively able to perform pro-environmental measures. Response efficacy is the next factor that influences the perceived effectiveness of the pro-environmental behaviour (Zhao 2016). Response cost represents the effect of coping with the consequences due to engagement in the recommended behaviour (Cismaru 2006). Convenience and cost effectiveness are the response costs, which would motivate people to use PT over a private vehicle, focusing more on factors such as severity of the problem with respect to the threat appraisal due to the usage of private vehicles. Thus, how effective a social marketing campaign is depends on the extent to which it achieves its intended purpose (Wymer 2011).

12.5 Methodology

The study aims to identify whether the specific family of advertisements by LTA is effective in reaching out to young adults between 18 and 35 years of age in Singapore. A total of 200 youths between 21 and 35 years age group were approached for the purpose of this study. The response rate was 57% (complete and usable questionnaires). The survey was conducted in different areas of Singapore, including commercial, residential, and business districts. The sample covered an adequately equal number of male and female candidates. The 21 questions derived from the above discussed theories, subjected to the candidates were monitored for quality of response, and the data were analysed.

The questionnaire had three sections: demographic or personal, awareness, and questions based on the theories discussed previously. Variables included the inability to cover all places in Singapore, that the answers given by the candidates might be not truthful, the availability of an equal number of respondents within each age group, and misunderstanding of the questions by the candidates on reading the survey. The following steps were taken to ensure fair representation of the target population:

- Firstly, respondents were shown the relevant advertisement photograph or the actual advert at the field location to improve reliability of responses.
- Secondly, a range of geographical locations were selected in different parts of the city, including commercial, residential, and business districts, to cover different contexts and to ensure variety of responses.
- Thirdly, questions were kept to a minimum by the simple framing of responses to a five-point Likert scale. In this study, a questionnaire was used to collect the sample of data. The questionnaire had structured questions, with a series or set of options given and closed-ended answers.

This helped to identify patterns in responses and to make comparisons between groups. These types of question were primarily multiple choice, rating scales, and open-ended. The questions were easy to complete but did not have any room for the respondents' remarks. Therefore, two to three open-ended questions in the theory section were added to enable the remarks, which were kept to a minimum because it can be hard to classify and qualify varied responses (Cohen et al. 2000).

The data chosen for this report are from young adult respondents. This is because the predominant users of PT are young adults between 18 and 35 years, as confirmed by the survey findings with 89% respondents using public transport. They are also the key influencers within families and among friends. Therefore, a study of this age group will have a direct impact on society in Singapore.

Random sampling was used for the collection of data, and during the survey, every third person was asked to complete the survey to keep the results accurate. Images of advertisements were presented to the respondents, with advertisements near MRT and bus stops used as live examples. The data collected comprised a variety of individuals from different walks of life aged 18–35 who use different modes of transport. In this survey, 52% of respondents were students, 21% were homemakers, 9% were employed, and 6% were unemployed. An equal split of male and female candidates was considered, with 49% male and 51% female.

12.6 Findings and Discussion

The survey findings reveal that 68% of the young adult respondents felt the advertisements were ineffective. In addition, a strong percentage (86%) of all respondents felt that the government advertisements were not appealing. According to the respondents, the advertisements are not attractive, so they did not pay attention to them. They did not know about the campaign because the advertisements are not easy to understand. According to respondents, the advertisements do not explain their meaning and purpose. About 41% had not seen the advertisements. Most of them were unclear of what message the advertisements are conveying, and they are not attractive enough to grab their attention.

Print advertisements are all about content creation, and advertising should be engaging and persuading. However, from the survey results, 46% of respondents rarely pay attention. Around 60% were confused because the advertisement seemed irrelevant. One of the respondents to the survey stated, "The government needs to focus more on content and attractiveness which will persuade people". The concern is also about the placement of advertisements, whereby they were not very visible.

The second finding is that the majority of respondents understand the seriousness of social messages with 82% respondents agreeing to the comfort of public transport, and 76% agreeing that government initiatives are important in improving public transport. The confirmation of these points highlights the general understanding of the issue. According to the community readiness model, at the intermediate stage, the community understands there is a need for action to be taken regarding the social message.

Recognition of the problem and agreement that something needs to be done is evident from the sample, with the majority agreeing on government initiatives and the advantages of public transport. At this stage, society tends to evaluate the cost and benefits before choosing the path to follow (Donnermeyer et al. 1997).

The LTA initiative currently displays public information to encourage its use, but there is no actual use of advertisements highlighting the benefits to people. Adding benefits generates a push for the public to take steps, encouraging action. From the theories discussed under the social cognitive theory section, self-efficacy is understood to be a key motivator that encourages society to move towards a goal (Stajkovic and Luthans 2002).

The third finding is that a strong percentage of the respondents are willing to convince and convey the advantages of PT to their friends and relatives. Around 88% of the respondents are happy to convey their choice to use public transportation to relatives. In the community readiness model, at the advanced stage, society takes ownership of the message and initiative (Kelly et al. 2003).

Fig. 12.2 Word cloud to show perception towards public transport

It is evident from the survey that the majority are willing to translate the initiative to their friends and relatives. This proves that the community of young adults in Singapore is slowly moving towards the advanced stage. The "word cloud" of the respondents' replies is shown in Fig. 12.2. It highlights an understanding of the link between society's awareness of conservation and environment protection and the use of public transport. The respondents strongly correlated PT use with lower emissions and population, which proves that society is ready to accept the core message's advantages.

This shows that the sample of young adults are aware of the advantages of using PT and are ready to accept LTA's initiatives to encourage PT for the sake of environmental protection. Understanding the core message makes it easier to translate the public message via social marketing. Social marketing, as seen from the literature review, needs to create tangible products to reach the target's attention (Kotler and Zaltman 1971). In this case, the LTA may use the understanding of its core message (i.e., reduce emissions, pollution, savings) as a product to promote the message of PT use to the target audience.

The target audience, after internalizing the message, will then potentially promote it to friends and relatives. The fourth finding is that the majority (79%) feel that fitness is important to them. Further, 36% of respondents agree that they are not fit. This confirms that the respondents are aware of their health and wish to improve self-efficacy. According to social cognitive theory, the need to improve self-efficacy plays an important role in an individual's decision to make a choice (Stajkovic and Luthans 2002). The efficiency of PT and its cost effectiveness were not conveyed to the respondents by LTA's public information advertisements. This core message of self-efficacy can be used as a tangible product to attract the target audience. Social cognitive theory also suggests that individuals, while witnessing self-efficacy in others, tend to follow the crowd. This can be linked to the promotion of the message to friends and relatives at the advanced stage.

Furthermore, this result highlights the fifth finding from the data collected that about 79% of the respondents agreed that fitness is important to them, but only about 64% of the respondents agreed that they are fit. The difference in percentage between these two survey responses, regarding how fit and whether fitness is important, can be explained using the threat appraisal of PMT. According to this, when people are exposed to a threat, they become more involved in finding the severity and vulnerability of the problem. They tend to focus on finding the reasons for the threat (Zhao 2016). The severity of the problem, as found in the results, is due to various environmental hazards such as global warming, pollution, carbon emissions, and so on.

The sixth finding notes a preference by commuters for PT over private vehicles. According to the analysis, the majority of respondents prefer PT to private transport. Most respondents prefer public transportation such as bus, MRT, and so on because, according to the coping appraisal of PMT (Zhao 2016), when people understand the threat and the severity of using more private transport, such as higher carbon emissions, they respond effectively to the situation by performing various pro-environmental measures, which increases their self-efficacy. It was also learned that many of the respondents understand the benefits of using PT as they are at the intermediate stage, according to the community readiness model (Kelly et al. 2003).

The seventh finding based on analysis of the survey is that about 45% of the respondents understand that government advertisements promote the use of public transport. The majority of the survey respondents agreed that people are aware of the main motives of these advertisements. This explains the degree to which people would be able to perform pro-environmental measures effectively, as there is an increase in self-efficacy. Since there is an increase in the self-efficacy of a problem, the response efficacy to the threat appraisal is higher, which helps in taking safety measures with respect to emerging concerns (Zhao 2016) such as health and environmental issues.

12.7 Recommendations

Eight key benchmark criteria had been identified based on successful social marketing projects and building on Andersen's (2002) features or characteristics of social marketing interventions (National Social Marketing Centre (NSMC), nd). The eight criteria can be noted through the LTA campaign.

- Firstly, the campaign aims to change people's actual behaviour in using PT over private transportation which will also be a sustainable initiative towards less carbon emissions. However, the messages conveyed through public advertisements are confusing and unclear.
- Secondly, the campaign focuses on the betterment of the community by encouraging the use of PT considering the limited space available and how communities can be impacted if private transportation increases. The covered walks and cycling options also offer an easy connection. Yet, it is essential to create more awareness of the initiatives.

- Thirdly, while the campaign uses behavioural theories to understand how to motivate people to change behaviour, based on the findings, this is an area that could improve more.
- Fourthly, with regards to insights, the campaign does attempt to be customer-oriented in getting the combination of walk cycle and ride, having the intention of increasing the public transport and in highlighting how it would be more convenient for customers; however, the findings indicate that the message in the advertisement could improve and be more easily comprehendible.
- Fifthly, while the campaign highlights some benefits, the findings indicate the health benefits and environmental benefits could be emphasized more and the cost of using private transport or rather impacts could be highlighted.
- Sixthly, the need for more public transport, flexibility of owning a car, and the social image conscious "want" and not "need" of having own car can compete with the desired behaviour.
- Seventhly, segmentation is an aspect that could improve, not only geographically but also in terms of purpose of travel and increased connectivity.
- Eighthly, the advert by itself may not be able to bring about the transformational behaviour change and needs to be supported with activities, events, mascots to boost and promote the campaign.

Based on the findings and discussions, the following recommendations have been suggested:

Improving visibility of the advertisements: For a successful campaign to grab people's attention, the government must create advertisements with creative content. If the content of advertisements concerns or can be useful to the public, then the public is more likely to pay attention to it. The advertisement must have headlines and slogans, which helps the government to spread the message clearly and achieve the advertising objective. After advertising, the government should undertake proper monitoring of advertisements to find whether they are effective enough to change the behaviour of the target community. Proper execution is required to make government advertisement campaigns successful, so that any deficiencies found can be improved upon or replaced. Mascots can leave a great impression on the public's memory as they can help the public to remember an advertisement's message.

Mascots can be placed at key government events for people to approach them. For example, a social media campaign called Pride used species mascots as a marketing tool for marine conservation. This campaign is about protecting marine life from growing environmental threats (Hayden and Dills 2015). For any country or city, it is important to ensure streets are clean and safe for making it attractive and sustainable. For example, in Mississauga, a city situated in the Canadian province of southern Ontario, the Clean City campaign was implemented to manage waste, keep the city safe, protect the environment, and to encourage the move towards a sustainable approach. To market the campaign, they selected a mascot called Super Kitty. Clean City Kitty was painted on the side of trash bins. Therefore, this campaign was communicated through a mascot, which became popular throughout the city. The campaign was successful due to the mascot Super Kitty, who keeps watch, cleans the city, and keeps the city safe. Mississauga's residents became more aware of their responsibility towards their environment (Kennedy 2004).

Conserving a long-term relationship with local communities, with resources, changes a community's behaviour towards making small initiatives to establish their connection with natural resources and focus on sustainability. The Pride campaign targets local communities to help to protect the biodiversity around them. The campaign engages these communities by educating them about the future benefits, explaining the sense of ownership, letting citizens feel that it is a matter of pride to protect the local environment, inspiring them to bring change. By running a social marketing campaign, they aimed to increase the participation of communities who are at the intermediate stage (Hayden and Dills 2015).

Big conservation organizations such as the World Wildlife Fund (WWF) use mascots to connect with and communicate to society on an emotional level. The WWF has a giant panda as its mascot as the big, furry animal with black patched eyes is appealing to the public. In fact, it has now become a celebrity, and people recognize the organization more by the mascot than the name (WWF 2008). When a mascot gains enough fame and popularity, the mascot can be identified just by the design, even without mentioning the organization's name. The panda became one of the official mascots of the 1990 Asian Games and the 2008 Summer Olympic Games in Beijing. It has now become an icon of the global conservation effort. This has also helped in saving the giant panda population because due to human expansion, pandas are being driven to the edge of extinction. Having a mascot can help to spread the word more effectively (Liu et al. 2016).

Mini cases to show effective communication strategy: In January 2014, Hamburg became the first city in Europe to declare its ban on private cars by 2034. In October 2015, Oslo, the capital of Norway, followed suit, declaring a ban on all private vehicles from its centre by 2019 (Nieuwenhuijsen and Khreis 2016) and announcing plans for pedestrian and bicycle routes that will cover about 40% of the total city area. Its commitment to reduce air pollution to provide a healthy environment for its population followed asthma and respiratory complaints from its residents, particularly children, elderly, and pregnant women who are most affected by air pollution. A study conducted by the Norwegian government shows that road traffic is the primary source of air pollution, adversely affecting the population and natural environment. Moreover, due to acid rain, the quality level of water bodies such as lakes and rivers has started to deteriorate, resulting in contaminated water and depletion of aquatic plants and animals. National targets were made to reduce the volume of traffic and reduce emissions.

Starting with Oslo, officials decided to invest in bicycles and public transportation and ban cars permanently from the city. To promote awareness of the campaign, the Norwegian government first launched an inspiring video to make the goal a reality. Secondly, they created awareness through social media marketing, posting videos and articles related to the environment and sustainability and discussing on an online forum about what steps they are taking to make Oslo car-free. The Norwegian government is even giving away grants/subsidies to people to buy bicycles.

Similarly, Madrid in Spain is banning cars by 2020, which the Spanish government began to implement in 2015 by launching Europe's first fully electric bicycle scheme, actively encouraging people to use e-bikes and creating 42 new stations with 486 extra bikes (Sullivan 2017). In addition, increased bus services and pedestrian walkways are making the change convenient for the public. Common among these cities' governments is that they are providing suitable alternatives first, before implementing the change.

While conducting the survey for this research at bus stops, it was discovered there were no advertisements about the campaign around the bus stops. The equal presence of advertisements near MRT, bus stops, and the mall is important to reach LTA's target audience; only then can the effectiveness of the campaign be observed. The concern is for the public to at least know about the government campaign, seeing the advertisements consistently in different places, as repeatedly observing them will make positive changes in behaviour and thinking. This campaign aims not only to influence individuals, but also the community, which can actively take part in bringing a social change for the good of those living in Singapore and its environment. Therefore, for better results, rather than only displaying advertisements near MRT or in residential areas, the advertisements must cover the whole city.

Organizing large events and marathons: The LTA could organize large sponsored health events to attract the public by showcasing benefits. Events such as marathons can attract a large population where the message of benefits can be conveyed. The majority of the population at the intermediate stage can be motivated to accept the message. It would also serve as a platform on which those at the advanced stage of the community readiness model could meet those at other stages, serving as a medium to convey the advantages of public transport.

Using testimonies that showcase real-life experiences: Testimonies and experiences may have a direct impact on young adults. The benefits of cost effectiveness and faster travel can be showcased by real-life examples, as the public tends to look at the advantages and adapt the idea, according to social cognitive theory (Stajkovic and Luthans 2002). The publishing of advertisements and media banners that display the effectiveness of PT and the monetary savings to be gained by society will encourage the participation and following of both the intermediate and advanced stage communities of Singapore. Self-preservation and self-efficacy serve as motivational products to enable the social marketing strategy to reach the consumer.

Adverts showing the discomforts of health and environmental hazards: To make the LTA advertisements more appealing and effective, it is important for their message to be highly motivated to the real cause. The behaviour of the public can be analysed using theories of behavioural change. Hence, through psychosocial models and the components of PMT, it is understood that people are motivated through observational learning and by displaying the severity of problems in the adverts (Cismaru 2006).

Factors of convenience as response cost to influence the public: For social marketing advertisements to be effective, it is important to focus on two factors; that is, benefits and response cost. Social marketing is about focusing on changing people's behaviour to benefit the public (University of Kansas 2017). Choices to adopt PT are disrupted if the benefits and the convenience described as response costs are not properly addressed, as clearly understood from the 4Ps of social marketing (Kotler and Zaltman 1971). Therefore, convenience factors should be clearly listed to influence the public to adopt PT for commuting. Convenience factors include covered walkways, bicycles such as O-Bikes placed at every point, convenient skywalks, and so on, which benefit people in terms of health by helping them to get fit and encouraging them to use PT instead of personal vehicles.

12.8 Conclusion

This research report studied the effectiveness of LTA's family of advertisements in Singapore to reach young adults between 18 and 35 years of age. The results reveal that the effectiveness of the advertisements is low, with 68% of respondents finding the advertisement ineffective in influencing the target audience. The research team used current knowledge in the field in identifying the relevant stage of acceptance in Singapore society and to arrive at an understanding of the different stages of acceptance of public information. The models involving psychosocial and social cognitive theory are referred to show that principals of social marketing are not utilized to their full advantage. Several advertisements are not relevant to society's current stage of acceptance. The discussion on these points presents several recommendations and action points which may serve as a starting point for the LTA to introduce new integrated marketing strategies to implement their public information initiatives about the use of public transport. Future research projects could use a larger sample to arrive at deeper understanding of the influences of advertisements on the different sections of the society. The action points could prove useful in enhancing the effectiveness of the LTA's advertisements in reaching out to the desired target population, especially the young adult in Singapore.

Acknowledgements The author would like to acknowledge the support of research students Charvi Sharma, Vineeta Thomas and Asha Surendran, without whom this research could not have taken place.

References

Almestahiri R, Rundle-Thiele S, Parkinson J, Alri D (2017) The use of the major components of social marketing: a systematic review of tobacco cessation programs. Social Market Q 23(3):232–248

Andreasen AR (2002) Marketing social marketing in the social change marketplace. J Public Policy Market 21(1):3–13

Bandura A (2002) Social cognitive theory of mass communication. In: Bryant J, Zillmann D (eds) Media effects: advances in theory and research. Lawrence Erlbaum, Mahwah, NJ, pp 121–154

Behar LB, Hydaker WM (2009) Adm Policy Mental Health 36:81–392. https://doi.org/10.1007/s10488-009-0227-x

Brassington F, Pettitt S (2003) Principles of marketing, 3rd edn. FT/Prentice Hall
Cathcart RL, Glendon AI (2016) Judged effectiveness of threat and coping appraisal anti-speeding messages. Acci Anal Prevent 96:237–248
Cismaru L (2006) Marketing communications and protection motivation theory: examining consumer decision making. Int Rev Public Non-Profit Market 3:9–24
Cohen L, Manion L, Morrison K (2000) Research methods in education, 5th edn. Taylor & Francis Group, New York, NY
Dentsu (nd) About us, Dentsu Singapore. https://dentsu.com.sg/about-us/#.WkHq_N-WZPZ. Accessed 26 Dec 2017
Donnermeyer JF, Plested BA, Edwards RW, Oetting G, Litterthunder, L (1997) Community readiness and prevention programs. J Comm Develop Soc 28(1)
Eagle L, Dahl S, Hill S, Bird S, Spotswood F, Tapp A (2013) Social marketing. Pearson, Harlow
Edwards RW, Thurman PJ, Plested BA, Oeting ER, Swanson L (2000) Community readiness: research to practice. J Comm Psychol 28(3):291–307
French J, Blair-Stevens C (2006) From snake oil salesmen to trusted policy advisors: the development of a strategic approach to the application of social marketing in England. Social Market Q 12(3):29–40
Hayden D, Dills B (2015) Smokey the bear should come to the beach. Social Market Q 21(1):3–13. https://doi.org/10.1177/1524500414558126
Kelly K, Edwards RW, Plested BA, Thurman PJG, Gomello ML, Slater MD (2003) The community readiness model: a complementary approach to social marketing. Market Theory 3(4):411–426
Kennedy RD (2004) Pretty city kitty. Alternat J 30(1):52–53. https://search-proquest-com.elibrary.jcu.edu.au/docview/218756933?pq-origsite=summon. Accessed 26 Dec 2017
Kotler P, Zaltman G (1971) Social marketing: an approach to planned social change. J Market 3–12
Kotler P, Wong V, Saunders J, Armstrong G (2005) Principles of marketing, 4th edn. Pearson
Lefebvre C, Flora JA (1988) Social marketing and public health interventions. Health Educ Q 15(3):299–315
Liu J, Hull V, Yang W, Vina A, Chen X, Ouyang Z, Zhang H (2016) Pandas and people: coupling human and naturals systems for sustainability. https://books.google.com.sg/books?id=OPIyCwAAQBAJ&printsec=frontcover&source=gbs_ge_summary_r&cad=0#v=onepage&q&f=false. Accessed 22 Jan 2018
LTA (2018) Walk, ride & cycle. https://www.mot.gov.sg/Transport-Matters/cycling/detail/encouraging-walking-and-cycling
Manjur R (2016) Dentsu Singapore bags two government accounts: marketing interactive. https://www.marketing-interactive.com/dentsu-singapore-bags-two-government-accounts/. Accessed 26 Dec 2017
Moyer-Guse E (2008) Toward a theory of entertainment persuasion: explaining the persuasive effects of entertainment-education messages. Communication Theory 18(2008):407–425
National Social Marketing Centre (NSMC) (nd) https://www.thensmc.com/about-us
NEA (2018) https://www.nea.gov.sg
Nieuwenhuijsen MJ, Khreis H (2016) Car free cities: pathway to healthy urban living. Environ Int 94:251–262. https://doi.org/10.1016/j.envint.2016.05.032
Peattie K, Peattie S (2003) Ready to fly solo? Reducing social marketing's dependence on commercial marketing theory. Mark Theory 3(3):365–386
Peattie K, Peattie S (2008) Social marketing: a pathway to consumption reduction? J Business Res
Shaw EF (1979) Agenda-setting and mass communication theory. Int Comm Gazette 25–96
Stajkovic A, Luthans F (2002) Social cognitive theory and self-efficacy: implications for motivation theory and practice. In: Steers RM, Porter LW, Bigley GA (eds) Motivation and work Behaviour, 7th edn, pp 126–140
Sullivan FO (2017) Why does New York ban e-bikes? It could learn from Europe, CityLab. https://www.citylab.com/transportation/2017/11/why-does-new-york-city-ban-ebikes/544829/. Accessed 22 Jan 2018

Tanner JF, Day E, Crask MR (1989) Protection motivation theory: an extension of fear appeals theory in communication. J Busin Res 19:267–276
Thackeray R, Neiger BL, Hanson CL (2007) Developing a promotional strategy: important questions for social marketing. Health Promot Pract 8(4):332–336
University of Kansas (2017) Section 1. Understanding social marketing: encouraging adoption and use of valued products and practices. Community Tool Box. https://ctb.ku.edu/en/sustain/social-marketing/overview/main. Accessed 25 Jan 2018
Walk Cycle Ride (2017) Walk cycle ride: land transport authority. https://www.lta.gov.sg/content/ltaweb/en/walk-cycle-ride.html. Accessed 26 Dec 2017
Wiebe G (1951) Merchandising commodities and citizenship on television. Public Opin Q 15. https://doi.org/10.1086/266353
World Wildlife Fund (2008) The Nike + human race—panda style! https://www.wwf.sg/?145801/The-Nike-Human-Race-Panda-Style. Accessed 22 Jan 2018
Wymer W (2011) Developing more effective social marketing strategies. J Social Market 1:17–31
Zhao CZ (2016) A protection motivation explanation of base-of-pyramid consumers' environmental sustainability. J Environ Psychol 45:116–126

Professor Abhishek Bhati is the Campus Dean of James Cook University Singapore campus. As Campus Dean JCU Singapore, Abhishek contributes to JCU's initiatives in the Tropical Asia. Prof Abhishek Bhati's research investigates resilience planning, sustainable development of cities and scholarship of learning and teaching. In particular, he is interested in technology and the role it has as a catalyst for industry changes; "Smart City" as a mechanism for future sustainable development; and the need to deliver United Nations Sustainable Development Goals (SDGs).

Chapter 13
Marketing Social Change: Fixing Bush Internet in Rural, Regional, and Remote Australia

Rachel Hay and Lynne Eagle

Abstract Cattle producers in Australia have turned to social media to highlight deficits in internet access in rural, regional, and remote Australia. This chapter provides a case study about how a group of Australian cattle producer women used social marketing and advocacy to improve internet access and educational data allowances in rural, regional, and remote Australia. A content analysis of Facebook posts and comments from the Better Internet for Rural, Regional, and Remote Australia (BIRRR) evidenced a connection between advocacy and the principles of marketing. The results of the analysis highlighted deficits in access to internet connectivity and lead to a change in policy giving rural, regional, and remote school children a dedicated education portal to complete their studies.

Keywords Rural women · Social marketing · Policy · Advocacy · Agriculture · Technology

13.1 Introduction

Australia covers approximately 7.692 million km^2 (Geoscience Australia 2004). Around 81% of this is broadly defined as encompassing rural, regional, and remote areas of Australia, with much of the area made up of rangelands. Rangelands are characterised by "eucalypt savannah and native grasslands, small areas of cleared land and scattered settlements, and rivers and wetlands that sustain ecosystems" (Dept. of Environment and Energy 2005). The rangelands are the home of Australian beef cattle production, where the industry produces around 27 million head of cattle to the value of $17.87 billion (2014/15) (Meat and Livestock Australia 2016). Agricultural products from Australia are held in high regard, as such, the beef industry looks to information communication technology (ICT) to help boost production to meet projected global food demand goals (Linehan et al. 2012).

R. Hay (✉) · L. Eagle
James Cook University, Townsville, Australia
e-mail: rachel.hay@jcu.edu.au

R. Hay et al. (eds.), *Broadening Cultural Horizons in Social Marketing*,
https://doi.org/10.1007/978-981-15-8517-3_13

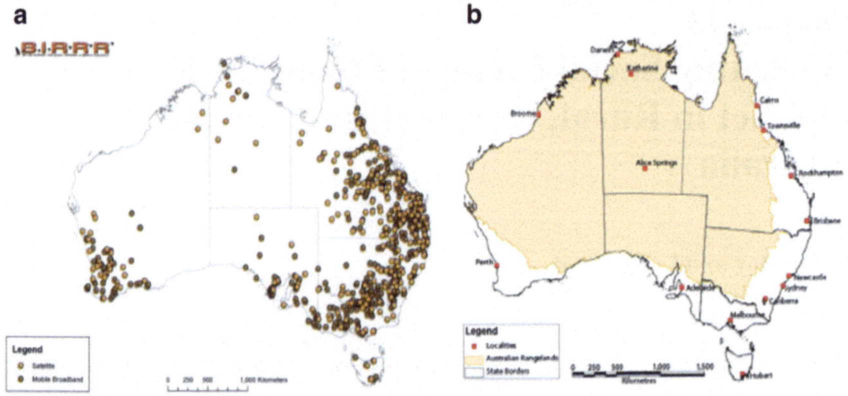

Fig. 13.1 **a** Responses mapped from the Better Internet for Rural Regional and Remote Australia (BIRRR) Regional Internet Access Survey showing access to internet in Australia (Hay 2016). *Note* Not all addresses could be mapped, and **b** location of Australian Rangelands (https://www.environment.gov.au/land/rangelands)

ICTs have the potential to transform how people live in rural, regional, and remote areas. New internet and mobile phone technology allows producers to keep in contact, not only with friends and relatives but also with markets, suppliers, telehealth services, weather, flood and fire services, and banking, as well as remote education. However, access to networks in Australian rangelands is challenging (Hay 2016; Curtin 2001). ICT connectivity across the rangelands is limited. For example, Fig. 13.1 demonstrates a lack of overlap between the Australian rangelands and internet responsiveness to a national survey of internet usage (Hay 2016). In many sparsely populated pastoral regions, download speeds can be as low as 0.7 Mbps (BIRRR Regional Internet Access Survey 2016). Expensive and unattainable access to either mobile or internet connectivity is adding to the digital divide currently experienced by those on the rangelands (Curtin 2001). However, a small group of cattle-producing women are using social marketing to advocate for change.

13.2 Case Study

Two remote-area cattle-producing women who were experiencing unexplained excessive usage on their mobile broadband data and were struggling to educate their children through distance education as well as run their businesses with very limited access to the internet, have taken action to change policy, surrounding Rural, Regional and Remote (RRR) internet connectivity. Together they founded the Better Internet for Rural Regional and Remote Australia (BIRRR) action group. The women set up a Facebook page in 2014 and a website and a Twitter (#DataDrought; #fixbushinternet) account in 2015 and started to respond to media interest about the #DataDrought.

BIRRR is a not-for-profit group, which has grown to include five administrators, several provider and technical volunteers, and more than 12,000 members. BIRRR provides information and support about telecommunications services and delivers it via their website and Facebook group to people in RRR areas of Australia. BIRRR uses survey data collected from their Facebook members, their website and from media coverage, to lobby for better internet for RRR Australia and regularly engages with community, government, and other stakeholders to highlight the deficits in internet access in RRR Australia (Hay 2016). The team employs social marketing practices and advocacy (Novelli 2011) and uses community connectedness through their social networks (Lefebvre 2013, p. 46) to focus on fixing bush internet (#fibush-internet) and putting an end to the data drought (#datadrought) experienced by RRR communities in Australia. BIRRRs target audience is members of RRR Australia. The sample, the BIRRR group members, is segmented into broadband and satellite users as each offering has both similar and unique issues. The programme goal was to get members in RRR areas to recognise common issues with internet access so that they would then join the campaign, elevating rural internet access as a priority. The behavioural objective was to get those affected by the deficits to make informed decisions based on their knowledge and to have BIRRR recognised as a not-for-profit organisation working for change on behalf of RRR communities. The primary objective was to get members to see the issues more clearly, so that as a community the members could influence government and telecommunications suppliers to table the issues. This chapter aimed to analyse how BIRRR used social marketing and advocacy to create change that benefits children, families, and businesses in rural, regional, and remote Australia.

13.3 Literature Review

Social Marketing: Social marketing is designed to benefit the target audience and is defined by Andreasen (1995) as "the application of commercial marketing technologies to the analysis, planning, execution, and evaluation of programs designed to influence the voluntary behaviour of target audiences in order to improve their personal welfare and that of their society". More recently social marketing has been defined as "seeking to develop and integrate marketing concepts with other approaches to influence behaviours that benefit individuals and communities for the greater social good" (iSMA, ESMA, & AASM 2013). "Social Marketing is about making the world a better place for everyone…" (Andreasen 2006). This chapter discusses "the integration of advocacy into behaviour change theory" (Lefebvre 2013, p. 90) to elicit change in internet access for RRR communities. Text from BIRRR's Facebook page will be measured using content analysis against five principles of social marketing (Lee and Kotler 2011). Firstly, that the programme must benefit both individual and society. Secondly, that it must influence behaviour. Thirdly, that the audience must play a role. Fourthly, the programme considers the four principles of marketing (product, price, place, and promotion—4Ps), and lastly, that there is evidence and evaluation.

Community Engagement in Change: The BIRRR team gathers information for its online community across the often-confusing landscape of "bush" broadband. The online community activates the information requests via the BIRRR Facebook page. Responses (or activation) are operationalised by volunteers, both self-selected and recruited (Roncarat, Lefebvre & Carleton 1989). Through social media, the advocacy group has built a "natural helper network" (Lefebvre 2013, p. 388) of internet providers and technical experts. Natural helper networks are a volunteer delivery system, whereby volunteers are recruited, trained, and supervised to provide a volunteer service (Israel 1985; Lefebvre 2013). BIRRR admin responds to 150–200 enquiries per week. Without the help of "natural helper networks" (Lefebvre 2013) the group could not keep up with demand and positive change would not occur. By assisting customers/members of their group in topics such as dropouts, lagging, outages, unexplained data use, high costs, and poor connectivity (Hay 2016, 2017), BIRRR has become the customers advocate. Customer advocacy is a powerful tool that involves the provision of information including advantages and disadvantages of a product to a customer with the view of representing them (Urban 2005). The role of the advocate is to "adopt a stance, advance a cause and attempt to produce a result on behalf of an interest of a person, group or cause" (Cohen 2004). The goal of BIRRR's advocacy is to provide RRR people with open, honest, transparent, and complete information (Urban 2005). BIRRR uses trust and advocacy within their network to engage public participation (Carpini et al. 2004) in research on RRR internet access (Regional Access Survey Results 2016) and Sky Muster™ services (Sky Muster Survey Results 2017). Public participation included peer support, surveys, providing supporting evidence of issues with "bush" internet to submissions, lobbying and attending meetings, public hearings, and conferences. These tools were used to shift from engagement to action (Lefebvre 2013), which was motivated both by the group members need for support for issues with the internet and it was motivated by the BIRRR administrators being their own customers (Grimm 2006). The type of motivation is important to action, for without motivation the audience will not have the will to act (Urban 2005). Therefore, BIRRR being its own customer created an internal motivation to act.

The data collected were used to create reports and government submissions that provided transparent and relevant information to inform policy makers about "bush" internet (Sparrow 2015; Sparrow and Gowen 2017). Technical information from experts, providers, and users was given in an open forum, which ensured that interested parties could discuss or challenge the information to provide solutions to issues raised by members creating transparent two-way communication and in turn inducing trust (Urban 2005). These actions were supported heavily by targeted media campaigns to highlight issues and to reinforce recommendations made by BIRRR. By embracing advocacy on behalf of RRR internet consumers, BIRRR has become a powerful and supported representative for RRR people, gaining strong public and political support and commitment (Urban 2005; Lefebvre 2013). The BIRRR team is successful because they have a customer-centred mindset; this is partly because they are their own customer. However, sometimes the group slips into the wrong mindset of being organisation-centred as they charge on to satisfy the BIRRR group goals. For example, seeing their mission as inherently good, where the primary goal of fixing bush internet and stopping the data drought is intertwined with both being customer-centred and organisation-centred (Andreasen 1995). The group is ethical in its conduct with a focus on both the good of the individual and RRR society as a whole (Eagle et al. 2013).

13.4 Theory Development

This chapter aims to increase the literature on advocacy used in social marketing. Urban (2005) states that customer advocacy "does not speak at people", rather that it is a "mutual discussion that assumes that if you advocate for your customer they will advocate for you". Figure 13.2 gives an example of customer advocacy using the BIRRR group.

BIRRR follows the principles of the Advocacy Pyramid (Urban 2005) (Fig. 13.3). Prior to BIRRR there was no one-stop shop for information about "bush" internet. The information that BIRRR provides is of high quality is backed up by technical experts and is bounded by total quality management (TQM) and customer satisfaction, which in turn increased the trustworthiness of the group's advocacy. BIRRR's relationship management saw their Facebook administrators manage member contributions with strict and sensible rules, which exclude political statements and abuse of members, administrators, or providers, but allows open discussion of issues encountered by members. Customer relationship management allows BIRRR to personalise their advocacy with each customer/member, forming relationships and creating trust. CRM extends to build relationships with relevant stakeholders such as technical experts, media, other advocacy groups and policy makers to maximise interest in the issues surrounding "bush" internet.

13.5 Methodology

The BIRRR Facebook page and website was chosen for content analysis for a case study of advocacy marketing (Novelli 2011). The data was gathered by accessing

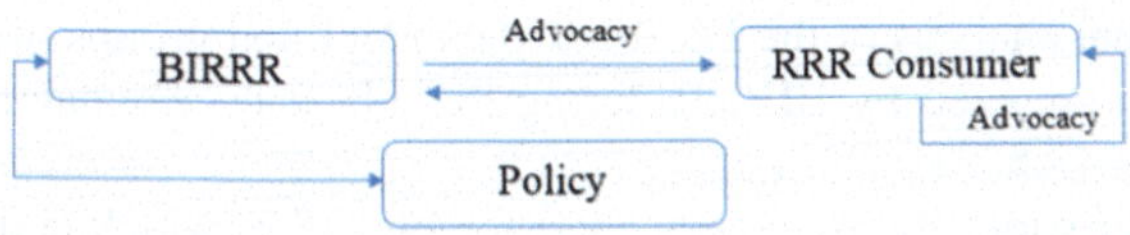

Fig. 13.2 BIRRR advocates for consumers, consumers advocate to consumers, consumers advocate for BIRRR, which then feeds into policy

Fig. 13.3 The advocacy pyramid

BIRRR Facebook posts using www.gritics.com. The data set included 445 posts and 6578 comments from 22/10/2014 to 2/02/2017. The researcher, with permission, notified BIRRR group members to alert them of the study prior to collecting data. Content analysis (Krippendorff 2013) allowed the researcher to uncover conceptual insights in detail by moving the level of analysis from the posts and their comments to the actual text used by the group members. This facilitates the systematic discovery of key themes and concepts surrounding internet issues in RRR Australia.

13.6 Data Analysis

Leximancer 4.5 (see https://info.leximancer.com/) was used to explore the conceptual layout of unstructured social media text generated from the BIRRR Facebook page. Text extracts were mined to show a connection between the advocate Kristy Sparrow (Founder of BIRRR) and the Principles of Social Marketing (PSM). In Fig. 13.4, the advocate is directly linked to nearly all of the concepts, which shows the relevance of the advocate to the social marketing campaign despite only being represented in 9% of the text extracts.

The conceptual analysis of the Facebook data identified groups of concepts that meet each PSM engaged by BIRRR during the advocacy campaign are visually portrayed in Fig. 13.5.

Figure 13.6 provides the frequency chart for Fig. 13.5, which demonstrates the strength of occurrence between each concept and "service" (the idea most central

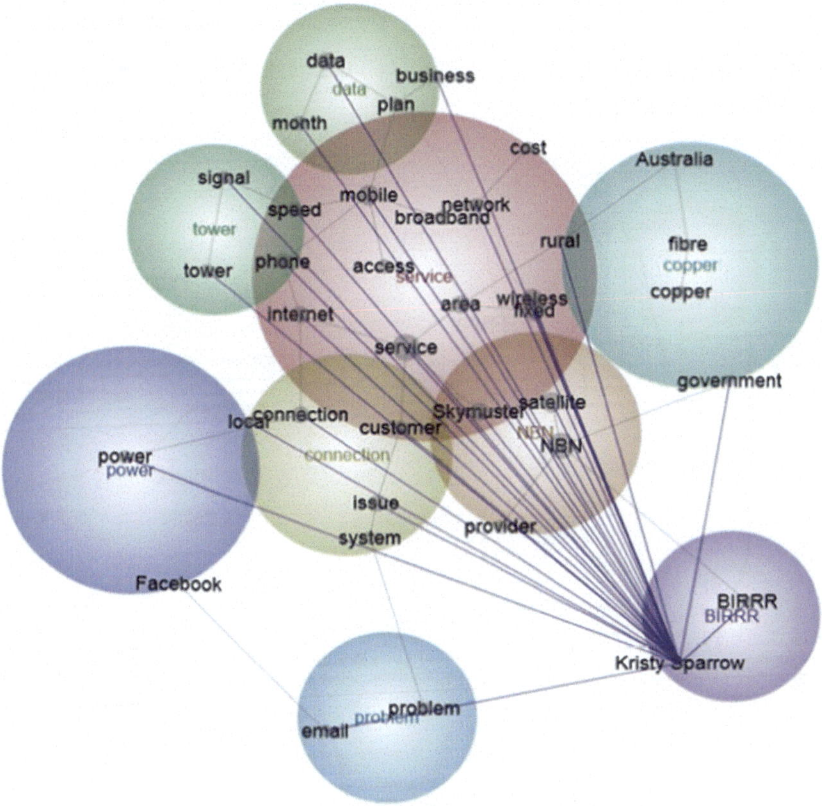

Concepts List from Leximancer showing name-like concept rankings

Name-Like	Count	Relevance
NBN	938	100%
Skymuster™	457	49%
BIRRR	179	19%
Australia	101	11%
Kristy Sparrow	88	09%
Facebook	25	03%

Fig. 13.4 Leximancer map showing the elements of connection between BIRRR and the principles of social marketing

to the conversation). Figure 13.5, which provides a visual of the Leximancer Gaussian map, shows that BIRRR had a stronger role in influencing behaviour (PSM2), addressing the four Ps (PSM4), and providing evidence (PSM5) compared to allowing the audience to play a role (PSM3) and benefitting both the individual and society (PSM1).

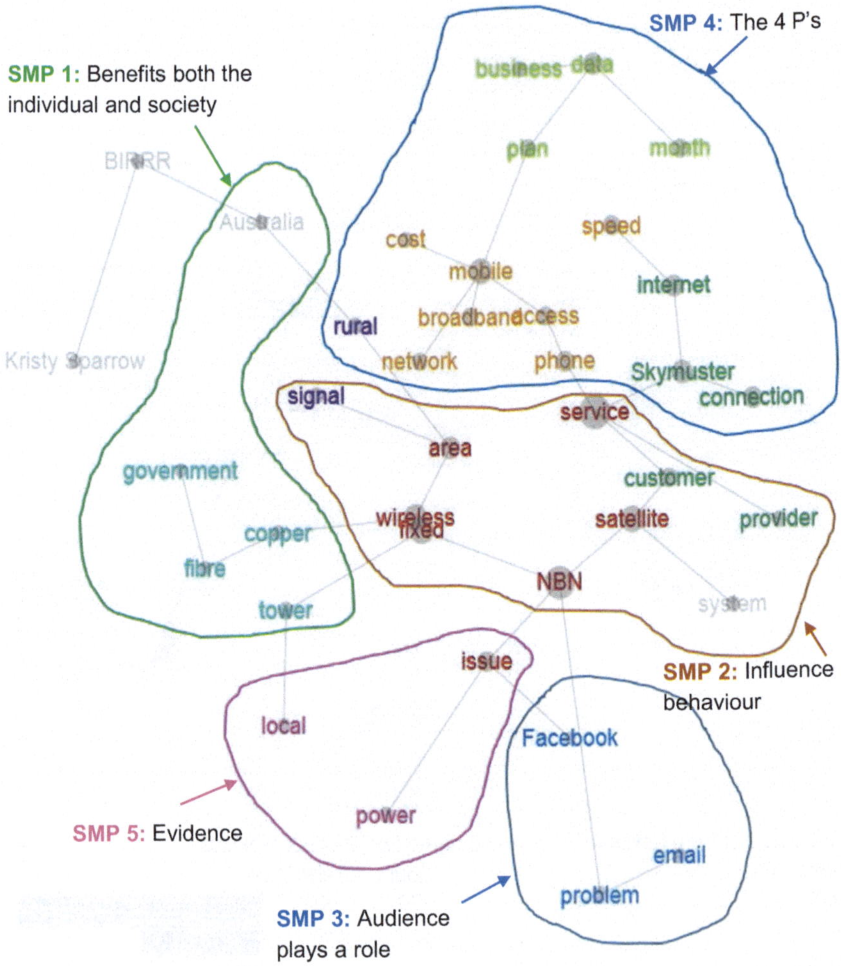

Fig. 13.5 Leximancer concept map showing concepts relevant the principles of social marketing

13.7 Discussion/Conclusion

BIRRR uses five of the social marketing principles in their social marketing advocacy campaign.

1. *PSM1*: *It benefits both the individuals and society in rural, regional, and remote areas*—BIRRR focusses on improving accessibility and in turn well-being of rural, regional, and remote people. The campaign uses accurate, relevant, and clear messages based on research to communicate in creative and impactful ways. Concepts revealed in the Leximancer analysis that relate to PSM 1 include that the group is Australian, and that the text discusses government, copper (runs

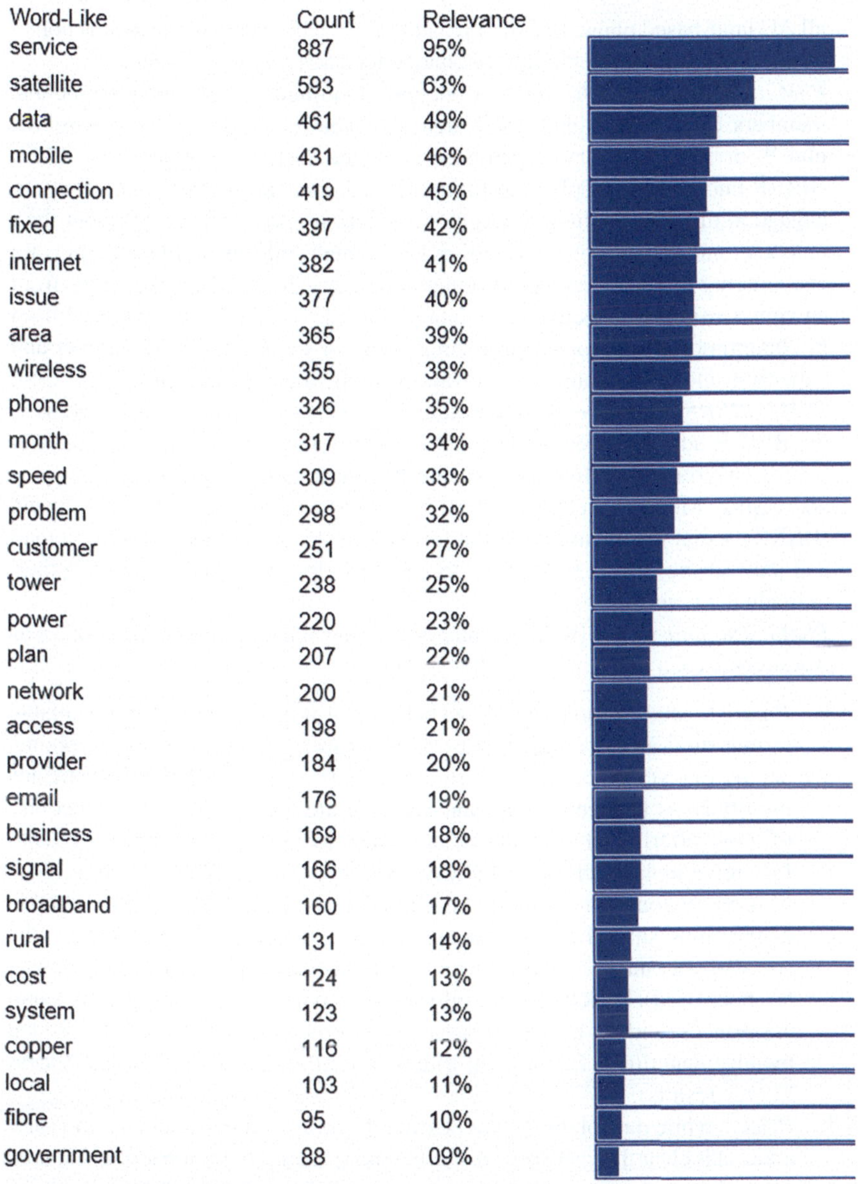

Fig. 13.6 Concepts list from Leximancer showing word-like concept rankings

all AU land-based phone lines), fibre, and tower, which both relate to new nbn™ products (Fig. 13.5). Lobbying for change focusses on these concepts.

2. *PSM2*: *Influencing behaviour*—Concepts identified in the analysis include members discussing signal, their area and what is available (fixed wireless, nbn™, or satellite), service providers, and the systems in place (Fig. 13.5). BIRRR encourages people with little knowledge about internet connectivity to engage in the conversation, to search for solutions, and to speak up about their unique connectivity problems. Rather than remain information takers, they are encouraging them to become information makers, to tell their own story from an emic (inside) perspective. The data produced then feeds into evidence used to communicate the broader problem of poor or no access to the internet and expensive and inadequate data allowances, to the intended audience.
3. *PSM3*: *BIRRRs audience plays a principle role in the social marketing process*—the BIRRR administrators understand their customer because as a customer of their own group they live and breathe the same issues that the campaign is addressing. Members relate to identified issues and play an active role using BIRRRs social media and other platforms (Fig. 13.5: Facebook, Twitter, email and the website) to tell their story, to ask for help, and to find support in their individual circumstances.
4. *PSM4*: *The four Ps*—BIRRR is immersed in the standard marketing four Ps as it addresses each element (Fig. 13.5: PSM4):

    a. *Product*—comparisons of product between providers and of what is available in non-rural areas versus rural areas in terms of plans, data, speed, phone, and mobile versus satellite and rural access (Fig. 13.5). BIRRR investigates the offerings between broadband and satellite and reports on the inequity of those offerings. For example, customers located in city's have access to fast, reliable, large cheap and always available internet plans, whereas those in rural, regional, and remote areas pay more for less data, have unreliable connections, shaped accounts and peak and off peak data (shaped is where the speed of your internet connection is lowered when the planned data allowance has been used). Peak data is the allocation used in peak times (typically small allowances) and off-peak data has a larger allowance that can only be used at inconvenient times (for more information, see the Hay 2016 Regional Access Survey results).
    b. *Price*—while data plans cannot be directly compared between city and rural areas, it is clear that city folk are getting more "bang for their buck" compared to rural regions. The price of internet connection not only affects the purse of rural, regional, and remote people in Australia, it also affects access to education, business tools, tele-health, emergency information, and communication. This has an increased cost on time, effort, and overall well-being of people in rural, regional, and remote areas.
    c. *Place*—government and communication providers often quote that a core product is available in RRR areas or they may quote what their policy says that the core product should be supplying. For example, the nbn™ Sky Muster™

Fair Use Policy limits end users to 75 GB of peak data (nbn™ Sky Muster™ is the service provider for broadband connectivity to regional and remote Australia). However, in reality the average peak data available to users across all providers is 55 GB (Hay 2016). BIRRR lobbies government, relevant industry bodies and telecommunications providers to highlight shortfalls in service provision. These shortfalls would have previously gone unnoticed. BIRRR's main aim is to ensure that people in RRR areas have equitable access to telecommunications.

d. *Promotion*—prior to the BIRRR group there was no single way to seek and find information about internet connectivity. Information came from a large number of diverse sources. Many people were experiencing problems and they knew that others were experiencing the same problems, but they did not have a contact point or a united voice. The BIRRR group calls on community participation to send messages through social media, their website, news media, public speaking, word of mouth and by lobbying about the deficits of internet in the bush.

5. *PSM5*: *Evidence and evaluation*—BIRRR conducts regular surveys and polls and participates in community engagement to gather evidence required to lobby government and communication providers to implement change for the broadband needs of RRR people in Australia. Government ranks last in the frequency of terms because, compared to troubleshooting members problems, lobbying for change takes up a small proportion of the advocates time. Typically, policy change is addressed in the beginning months of each year. The group celebrate their achievements through social media, awards nominations, and news media (https://birrraus.com/press-releases/). The group is recognised by telecommunications providers and government departments as an expert in their field and a force to be reckoned with. Evidence on their return on investment include:

   1. Un-metering of education specific sites for Telstra (Australia's leading telecommunications company (Telstra™ 2017)) mobile broadband distance education. Previously distance education would use the entire data allowance purchased prior to business or communication needs being met.
   2. Education ports for distance education and home schoolers using nbn™ Sky Muster™ (2017)—dedicated education ports that allocated 50 GB per student (up to 3 students) in addition to standard data allowances.
   3. Establishment of nbn contact team for RRR users—dedicated service providers trained in rural technology and are aware of the challenges felt by RRR people.
   4. Widespread survey of bush telecommunications—produced base line data on actual usage and issues affecting RRR people.

This analysis has demonstrated BIRRR's use of five social marketing principles to advocate for equity in internet provision for people in rural, regional, and remote Australia. The analysis has highlighted that there is more work to be done in terms of identifying the extent of social marketing practices in the BIRRR group. Research identifying the significance of overall conversations and exploring the role of SMP in BIRRR chats and further investigation of the role of the advocate are in progress.

References

Andreasen AR (1995) Marketing social change: changing behaviour to promote health, social development, and the environment, 1st edn. Jossey-Bass, San Francisco
Andreasen AR (2006) Social marketing in the 21st century. Thousand Oaks CA, Sage
BIRRR (2016) Better internet for rural, regional and remote Australia: regional internet access—survey results, 2016. In: Hay R (ed). https://birrraus.files.wordpress.com/2016/05/birrr-report-2016-survey-results-final.pdf
Carpini MXD, Cook FL, Jacobs LR (2004) Public deliberations, discursive participation and citizen engagement: a review of the empirical literature. Ann Rev Politi Sci 7:315–344
Cohen ES (2004) Advocacy and advocates: definitions and ethical dimensions. Generations 28(1):9–16
Curtin JD (2001) A digital divide in rural and regional Australia? Curr Iss Brief (2001–2). https://www.aph.gov.au/About_Parliament/Parliamentary_Departments/Parliamentary_Library/Publications_Archive/CIB/cib0102/02CIB01. Accessed from Parliament of Australia
Dept. of Environment and Energy (2005) Introduction to Australia's rangelands. Outback Australia—the rangelands. https://www.environment.gov.au/land/rangelands. Accessed 2 Jan 2017
Eagle L, Dahl S, Hill S, Bird S, Spotswood F, Tapp A (2013) Social marketing. Pearson, Harlow, England
Geoscience Australia (2004) Area of Australia—states and territories. National location information, 16 December 2016. https://www.ga.gov.au/scientific-topics/national-location-information/dimensions/area-of-australia-states-and-territories. Accessed 2 Jan 2017
Grimm K (2006) Discovering the activation point. Communication Leadership Institute. https://www.rwjf.org/files/publications/other/Activation-point1.pdf
Hay R (2016) Better internet for rural, regional and remote Australia: regional internet access survey results 2016, BIRRR. https://birrraus.files.wordpress.com/2017/04/birrr-report-2016-survey-results-final.pdf
Hay R (2017) Better internet for rural, regional and remote Australia: skymuster survey results 2017, BIRRR. https://birrraus.files.wordpress.com/2017/04/birrr-report-2016-survey-results-final.pdf.
iSMA, ESMA, AASM (2013) Consensus definition of social marketing. https://www.i-socialmarketing.org/social-marketing-definition. Retrieved from International Social Marketing Association
Israel BA (1985) Social networks and social support: implications for natural helper and community level interventions. Health Edu Quart 12(1):65—80. https://doi.org/10.1177/109019818501200106
Krippendorff K (2013) Content analysis: an introduction to its methodology. Sage Publications, Thousand Oaks
Lee NR, Kotler P (2011) Social marketing: influencing behaviors for good Sage
Lefebvre RC (2013) Social marketing and social change: strategies and tools for improving health, well-being, and the environment. Wiley
Linehan V, Thorpe S, Andrews N, Yeon K, Beaini F (2012) Food demand to 2050: opportunities for Australian agriculture. Paper presented at the ABARES Outlook Conference, Canberra, ACT
Meat and Livestock Australia (2016) Fast facts: Australia's beef industry. Meat and Livestock Australia. Online at https://www.mla.com.au/Prices-markets/Trends-analysis/Fast-Facts
nbn™ (2017) Sky muster satellite. Accessed 17 Jan 2017
Novelli WD (2011) The Sage handbook of social marketing. In: Workman B (ed) Sage Publications Ltd., London

Roncarati DD, Lefebvre RC, Carleton RA (1989) Voluntary involvement in community health promotion: the Pawtucket Heart Health Program. Health Promo (Oxford, England) 4(1):11

Sparrow K (2015) Regional telecommunications review submission: better internet for rural, regional & remote Australia (BIRR). Better Internet for Rural Regional and Remote Australia. https://www.google.com.au/url?sa=t&rct=j&q=&esrc=s&source=web&cd=1&ved=0ahUKEwjxh63xmfDRAhXFoZQKHWsDAM0QFggeMAA&url=https%3A%2F%2Fwww.communications.gov.au%2Fsites%2Fg%2Ffiles%2Fnet301%2Ff%2FSparrow%2C%2520Kristy%2520-%2520Public%2520Submission%2520RTIRC%25202015.docx&usg=AFQjCNE5P7yLwQMWA8TIP7_67_UjrqOzXw&cad=rja

Sparrow K, Gowen R (2017) Submission to the productivity commission (PC) telecommunications universal service obligation (TUSO) report. Better Internet for Rural Regional and Remote Australia. https://birrraus.com/

Telstra™ (2017) Telstra. Our company. Accessed 17 Jan 2017

Urban G (2005) Don't just relate, advocate: a blueprint for profit in the era of customer power. Pearson Education, New Jersey

Dr. Rachel Hay is a Lecturer in Marketing with the College of Business Law and Governance, James Cook University. Dr. Hay recently completed her Ph.D. 'The engagement of women and technology in agriculture'. Her research interests centre on technology adoption and agriculture, social marketing, and readability. Within this area, she is active in research relating to technology adoption by women in agriculture, and marketing communication effects and effectiveness, including readability in populations sectors that face literacy and numeracy challenges. Her recent projects focussed on trans-disciplinary approaches to sustained behaviour change in social marketing and environmental protection interventions.

Prof. Lynne Eagle is a Professor in Marketing at James Cook University, Townsville, Australia and an Adjunct Professor at Canterbury University, Christchurch, New Zealand. She holds a Ph.D. from the University of Auckland in her native New Zealand. Her research interests centre on marketing communication effects and effectiveness, including the impact of persuasive communication on children; the impact of new, emerging and hybrid media forms and preferences for the use of formal and informal communications channels, along with trans-disciplinary approaches to sustained behaviour change in social marketing/health promotion/environmental protection campaigns. She has published in a wide range of academic journals, including the Journal of Advertising and European Journal of Marketing, led the development of both Marketing Communications and Social Marketing texts and contributed several book chapters for other texts as well as writing commissioned expert papers and presenting numerous research papers at international conferences.

The manufacturer's authorised representative in the EU is Springer Nature Customer Service Centre GmbH, Europaplatz 3, 69115 Heidelberg, Germany. If you have any concerns regarding our products, please contact ProductSafety@springernature.com

Printed and bound by CPI Group (UK) Ltd, Croydon, CR0 4YY
30/09/2026
02225506-0003